Cognitive Work Analysis

Applications, Extensions and Future Directions

Cognitive Work Analysis

Applications, Extensions and Future Directions

Edited by

Neville A. Stanton, Paul M. Salmon, Guy H. Walker and Daniel P. Jenkins

CRC Press is an imprint of the
Taylor & Francis Group, an **informa** business

CRC Press
Taylor & Francis Group
6000 Broken Sound Parkway NW, Suite 300
Boca Raton, FL 33487-2742

Printed on acid-free paper

International Standard Book Number-13: 978-1-138-74942-9 (Paperback)
978-1-4724-4392-2 (Hardback)

Library of Congress Cataloging-in-Publication Data

Names: Stanton, Neville A. (Neville Anthony), 1960- editor. | Salmon, Paul M., editor. | Walker, Guy., editor. | Jenkins, Daniel P., editor.
Title: Cognitive work analysis : applications, extensions and future directions / [edited by] Neville A. Stanton, Paul M. Salmon, Guy H. Walker, and Daniel P. Jenkins.
Other titles: Cognitive work analysis (Boca Raton, Fla.)
Description: Boca Raton : Taylor & Francis, CRC Press, [2017] | Includes bibliographical references.
Identifiers: LCCN 2016058576| ISBN 9781472443922 (hardback : alk. paper) | ISBN 9781315572536 (ebook)
Subjects: LCSH: Human-machine systems. | Human-computer interaction. | Work environment. | Cognitive psychology.
Classification: LCC TA167 .C635 2017 | DDC 620.8/2--dc23
LC record available at https://lccn.loc.gov/2016058576

Visit the Taylor & Francis Web site at
http://www.taylorandfrancis.com

and the CRC Press Web site at
http://www.crcpress.com

Contents

SECTION V Risk and Resilience

Preface

This book has arisen from the desire to share our insights into the practical application of the cognitive work analysis (CWA) framework. Over the past decade, we have applied CWA in many domains and have been impressed by the insights that the products of the analysis afford. CWA offers a systemic and systematic approach to the analysis, design and evaluation of sociotechnical systems. As a formative method, we have used it to design new concepts of operations, new interfaces and new ways of working.

CWA was originally developed by Jens Rasumussen and colleagues at the Riso National Laboratory in Denmark (see the Recommended Further Reading list). This work began in the 1960s, when they were tasked with designing and evaluating the efficacy of nuclear power control systems for energy production. They quickly realized that there was no single method available for such a complex sociotechnical systems design. This gave rise to the development of a suite of approaches. Initially these methods were stand-alone, but they soon realized that they could be integrated to provide the CWA framework we recognize today. The importance of those original pioneers of CWA cannot be overstated. They have performed a great service to the disciplines of cognitive engineering, human factors and ergonomics.

The history of the CWA would not be complete without mentioning Kim Vicente (see the Recommended Further Reading list) who, as a second generation of CWA enthusiast, worked with Jens to provide a user-friendly introduction to the framework. Kim also led a team of researchers who have shown how CWA can form the basis of ecological interface design (EID). The bridge between system analysis and design is often a difficult one to cross, but this pioneering work has inspired the research community and contributions to this book.

Our own modest contributions to CWA came in the form of a practical how-to-do-it book and accompanying software (see the Recommended Further Reading list). The software has been disseminated widely as free-of-charge shareware. We have run tutorials in CWA all over the world and continue to teach it to our students and colleagues at our respective institutions. We have also used the framework in the many projects that we have been involved in over the past decade, some of which are contained in this book. It seemed timely and fitting to edit a text on the applications of CWA (as an update to what is already been published – see the Recommended Further Reading list), to which we have invited contributions from all over the world. We are pleased to have contributions from Australia, Canada, Sweden and the United Kingdom. It is our hope that readers of this book will be inspired by the practical nature of CWA, and feel empowered to apply the framework in their own work.

Neville A. Stanton
Professor of Human Factors Engineering
University of Southampton

Editors

Professor Neville A. Stanton, PhD, DSc, is a chartered psychologist, chartered ergonomist and a chartered engineer. He holds the chair in human factors engineering in the Faculty of Engineering and the Environment at the University of Southampton in the United Kingdom. He has degrees in psychology, applied psychology and human factors and has worked at the Universities of Aston, Brunel, Cornell and MIT (Massachusetts Institute of Technology). His research interests include modelling, predicting, analysing and evaluating human performance in systems as well as designing the interfaces and interaction between humans and technology. Professor Stanton has worked on design of automobiles, aircraft, ships and control rooms over the past 30 years, working on a variety of automation projects. He has published 35 books and over 270 journal papers on ergonomics and human factors. In 1998 he was awarded the Institution of Electrical Engineers Divisional Premium Award for research into System Safety. The Institute of Ergonomics and Human Factors awarded him The Otto Edholm Medal in 2001, The President's Medal in 2008 and The Sir Frederic Bartlett Medal in 2012 for his contributions to basic and applied ergonomics research. The Royal Aeronautical Society awarded him and his colleagues the Hodgson Prize and Bronze Medal in 2006 for research on design-induced, flight-deck error published in *The Aeronautical Journal*. The University of Southampton has awarded him a Doctor of Science in 2014 for his sustained contribution to the development and validation of human factors methods.

Professor Paul M. Salmon holds a chair in Human Factors and is creator and director of the Centre for Human Factors and Sociotechnical Systems at the University of the Sunshine Coast. He currently holds a prestigious Australian Research Council Future Fellowship and has almost 15 years of experience in applied human factors research in a number of areas, including defence, transportation safety, sports and outdoor recreation and disaster management. Paul currently leads major research programs in the areas of road and rail safety, identity theft and cybersecurity, and led outdoor recreation accidents. Paul has co-authored 11 books, over 140 peer-reviewed journal articles, and numerous conference articles and book chapters. He has received various accolades for his contributions to research and practice, including the Australian Human Factors and Ergonomics Societies 2016 Cumming Memorial medal, the UK Ergonomics Society's President's Medal, the Royal Aeronautical Society's Hodgson Prize for best research and paper, and the University of the Sunshine Coast's Vice Chancellor and President's Medal for Research Excellence. His current research interests relate to extending Human Factors and Sociotechnical Systems theory and methods to support the optimisation of systems in many areas. Specific areas of focus include accident prediction and analysis, systems thinking in transportation safety, the development of systemic accident countermeasures, human factors in elite sports and cybersecurity.

Guy H. Walker is an associate professor in the Institute for Infrastructure and Environment at Heriot-Watt University in Edinburgh. He lectures on human factors and is the author/co-author of over one hundred peer-reviewed journal articles and 13 books. He and his co-authors have been awarded the Institute for Ergonomics and Human Factors President's Medal for the practical application of Ergonomics theory and the Peter Vulcan prize for best research paper by the 2013 Australasian Road Safety Research Conference. In 2011, Guy also won Heriot-Watt University's Graduate's Prize for inspirational teaching. Dr Walker has a BSc honours in psychology from the University of Southampton and a PhD in human factors from Brunel University. His research interests are wide ranging, spanning driver behaviour and the role of feedback in vehicles, using human factors methods to analyse black-box data recordings, the application of sociotechnical systems theory to the design and evaluation of civil engineering systems through to safety, risk and reliability. His research has been featured in the popular media, from national newspapers, TV and radio through to an appearance on the Discovery Channel.

Daniel P. Jenkins leads the research team at DCA Design International, one of the world's leading product design and development consultancies. Within multidisciplinary teams, he supports the design of a diverse range of products and services, from trains to toothbrushes. Working with clients and key stakeholders, Dan specialises in the application of human factors and systems thinking tools to ensure system performance is maximised and that products stand the greatest chance of commercial success. Notable projects include patient administered drug delivery devices, next-generation radiotherapy equipment, personal protective equipment, airline interiors, train cabs, train passenger areas and a wide range of fast moving consumer goods. Dan has co-authored 10 books and over 50 journal papers, he also contributes to a wide range of blogs and magazines with the aim of disseminating the role of human factors in the design process to a wider audience. Dan has received numerous awards for his work in human factors including the HFES User-Centred Design Award and the CIEHF President's Medal.

Contributors

Chris Baber
School of Engineering
University of Birmingham
Birmingham, United Kingdom

Lindsay Beevers
School of Energy, Geoscience, Infrastructure and Society
Institute for Infrastructure and Environment
Heriot-Watt University
Edinburgh, United Kingdom

Kevin Bessell
BAE Systems
Defence Information, Training and Services
Yeovil, United Kingdom

Simon Blainey
Boldrewood Innovation Campus
University of Southampton
Southampton, United Kingdom

Catherine M. Burns
Systems Design Engineering
University of Waterloo
Waterloo, Ontario, Canada

Mhairi Cooper
Abbott Risk Consulting (ARC) Ltd
Glasgow, United Kingdom

Anandhi Dhukuram
School of Psychology
University of Cambridge
Cambridge, United Kingdom

Catherine Harvey
Faculty of Engineering
The University of Nottingham
Nottingham, United Kingdom

Adrian Hickford
Boldrewood Innovation Campus
University of Southampton
Southampton, United Kingdom

Antony Hilliard
Cognitive Engineering Laboratory
Mechanical and Industrial Engineering
University of Toronto
Toronto, Ontario, Canada

Robert Houghton
Faculty of Engineering
The University of Nottingham
Nottingham, United Kingdom

Greg A. Jamieson
Cognitive Engineering Laboratory
Mechanical and Industrial Engineering
University of Toronto
Toronto, Ontario, Canada

Sean W. Kortschot
Cognitive Engineering Laboratory
Mechanical and Industrial Engineering
University of Toronto
Toronto, Ontario, Canada

Stas Krupenia
Scania CV AB
Södertälje, Sweden

Michael G. Lenné
Monash University Accident Research Centre
Monash University
Victoria, Australia

Ida Löscher
Peace Research Institute Oslo (PRIO)
Oslo, Norway

Rich C. McIlroy
Boldrewood Innovation Campus
University of Southampton
Southampton, United Kingdom

Christine M. Mulvihill
Monash University Accident Research Centre
Monash University
Victoria, Australia

Katherine L. Plant
Civil, Maritime, Environmental Engineering and Science Unit
University of Southampton
Southampton, United Kingdom

John M. Preston
Boldrewood Innovation Campus
University of Southampton
Southampton, United Kingdom

Gemma J. M. Read
Centre for Human Factors and Sociotechnical Systems
Faculty of Arts, Business and Law
University of the Sunshine Coast
Queensland, Australia

and

Monash University Accident Research Centre
Monash University
Victoria, Australia

Aaron P. Roberts
Civil, Maritime, Environmental Engineering and Science Unit
University of Southampton
Southampton, United Kingdom

Brendan Ryan
Faculty of Engineering
The University of Nottingham
Nottingham, United Kingdom

Nicholas Stevens
Centre for Human Factors and Sociotechnical Systems
Faculty of Arts, Business and Law
University of the Sunshine Coast
Queensland, Australia

Ailsa Strathie
School of Energy, Geoscience, Infrastructure and Society
Institute for Infrastructure and Environment
Heriot-Watt University
Edinburgh, United Kingdom

Natalie Taylor
Centre for Human Factors and Sociotechnical Systems
Faculty of Arts, Business and Law
University of the Sunshine Coast
Queensland, Australia

T. Glyn Thomas
Civil, Maritime, Environmental Engineering and Science Unit
University of Southampton
Southampton, United Kingdom

Pauline Thompson
School of Energy, Geoscience, Infrastructure and Society
Institute for Infrastructure and Environment
Heriot-Watt University
Edinburgh, United Kingdom

Fiona F. Tran
Cognitive Engineering Laboratory
Mechanical and Industrial Engineering
University of Toronto
Toronto, Ontario, Canada

Cole Wheeler
Cognitive Engineering Laboratory
Mechanical and Industrial Engineering
University of Toronto
Toronto, Ontario, Canada

Kristie L. Young
Monash University Accident Research Centre
Monash University
Victoria, Australia

Aimzhan Zhunussova
Cognitive Engineering Laboratory
Mechanical and Industrial Engineering
University of Toronto
Toronto, Ontario, Canada

Recommended Further Reading on Cognitive Work Analysis

We can recommend the following books on CWA and EID as we have found them extremely helpful in our own work.

Bisantz, A. M. and Burns, C. M. 2004. *Applications of Cognitive Work Analysis.* CRC Press, Boca Raton, FL.

This was the first book to bring together a wide range of example applications of CWA. It shows how the different phases of CWA can be applied to a wide range of problems and domains. There is no system that will not benefit from CWA.

Burns, C. M. and Hajdukiewicz, J. R. 2004. *Ecological Interface Design.* CRC Press, Boca Raton, FL.

This was the first book to bring together examples of ecological interface design (EID) from a wide spectrum of research applications. The examples help to convey the ideas behind EID and the potential usefulness revealing work domain constraints to users of systems.

Jenkins, D. P., Stanton, N. A., Walker, G. H. and Salmon, P. M. 2009. *Cognitive Work Analysis: Coping with Complexity.* Aldershot, Ashgate.

This is our own how-to-do-it book on CWA. We have demonstrated how to work through all five phases in a practical manner with plenty of examples and guidance. We introduce the shareware that we have developed to support the analysis. A study on the reliability and validity of CWA is also presented.

Naikar, N. 2013. *Work Domain Analysis: Concepts, Guidelines and Cases.* CRC Press, Boca Raton, FL.

The majority of this book focuses mainly on the first phase of CWA, namely work domain analysis, although the other phases are represented. A rich source of everyday examples makes it easy to grasp the concepts being presented. The guidelines will help the reader to embark on their own analysis.

Rasmussen, J., Pejtersen, A. M. and Goodstein, L. P. 1994. *Cognitive Systems Engineering.* Wiley, New York.

The book that brought all of the approaches together into the framework that is now known as cognitive work analysis. This is the foundation for all the work that has followed since. This book really launched the discipline of cognitive systems engineering as we know it today.

Vicente, K. J. 1999. *Cognitive Work Analysis: Toward Safe, Productive, and Healthy Computer-Based Work*. Lawrence Erlbaum Associates, Mahwah, NJ.

The book that brought CWA to the people. Kim once described this book to me as 'CWA for dummies', but it is much more than that. It brings the whole framework together and explains how to use CWA for design of complex sociotechnical systems.

Section I

Overview of Cognitive Work Analysis

1 Application of Cognitive Work Analysis to System Analysis and Design

Neville A. Stanton and Daniel P. Jenkins

CONTENTS

1.1 INTRODUCTION

This aim of this chapter is to introduce cognitive work analysis (CWA) in the context of system analysis and design. CWA is a structured framework for considering the development and analysis of complex sociotechnical systems. The framework provides a systematic approach to analysing systems, by explicitly identifying the purposes and constraints. By focussing on constraints, rather than on particular ways of working, CWA aims to support workers in adapting their behaviour to maintain performance and safety in a variety of situations, including unanticipated events. The five main phases of CWA (i.e. work domain analysis [WDA], control task analysis [ConTA], strategies analysis [StrA], social organisation and cooperation analysis [SOCA] and worker competencies analysis [WCA]) can feed supportive information into all aspects of system design.

The benefits of CWA are realised when the framework is used to design new and first-of-a-kind systems. Rasmussen et al. (1990) introduced CWA as a methodology for design and specification of the functionality of integrated, complex, multimedia information systems, which enables designers to predict the kind of behaviour of individuals and organisations to be expected in response to changes in work conditions, such as those caused by the introduction of new information systems.

The CWA framework has been applied to a variety of application domains, as illustrated by the chapters in this book. These domains include the following:

- Nuclear power (Chapters 1, 10 and 18)
- Aviation (Chapters 2, 15 and 16)
- Domestic and consumer products (Chapters 3 and 5)
- Healthcare (Chapters 4 and 8)
- Energy grid distribution (Chapter 6)
- Rail transportation (Chapters 9, 11 and 12)
- Urban planning (Chapters 11 and 13)
- Automotive (Chapters 11 and 14)
- Head-up display design (Chapters 14 and 15)
- Criminal network analysis (Chapter 18)
- Environmental flood protection (Chapter 19)

This book is laid out in five sections. In Section I, this chapter provides an overview of the CWA framework and its application to sociotechnical systems design. Section II contains four chapters on using CWA to develop a specification of design requirements. There are five chapters in Section III that use CWA to analyse complex sociotechnical systems. In Section IV, five chapters use CWA to design and evaluate systems. Finally, Section V presents three chapters that show how CWA can be used to assess risk and resilience.

This chapter shows the workings of the CWA framework through all of the phases. This reveals the inputs and products of the process of applying CWA. As a test case, CWA was applied to the Trafalgar-class power plant. CWA has five main phases, which were explored throughout the project. In particular, the project defined where the products of these phases could be used to inform the design process, and what form this information might take. When reading other chapters in this book, the reader might find it helpful to refer back to this chapter in order to understand the details of the framework and products.

1.2 INTRODUCTION TO THE TRAFALGAR-CLASS SUBMARINE

Trafalgar-class submarines displace 5,208 tonnes submerged and measure 85.4 m in length and 9.8 m in across. They are powered by a single pressurised water-cooled reactor (PWR1), which is controlled in the propulsion plant by a complement of up to five people, can travel at a speed of 32 knots and dive to more than 985 feet. There are seven submarines in the class as indicated in Table 1.1.

This chapter on the activities within the propulsion plant, which is located towards the rear of the submarine from which the reactor, propulsion, power generation and associated systems are controlled. A general indication of the layout of a nuclear submarine is shown in Figure 1.1. The position of the human interface to the propulsion plant is indicated by the highlighted circle.

The controls within the control room are organised into a large horseshoe arrangement, as indicated in Figure 1.2 – from left to right, the auxiliary machine panel, the electrical panel, the propulsion panel, the reactor panel and the reactor plant auxiliary panel. The three control room operators sit in the centre of the horseshoe, with the throttle control panel operator (TCPO) in the centre, the electrical panel operator (EPO) on the left and the reactor panel operator (RPO) on the right. The supervisors (nuclear chief of the watch [NCOTW] and engineer officer of the watch) stand behind the operators and are able to view each of their consoles.

A more detailed description of the different roles within the control room environment can be found in Chapter 10.

TABLE 1.1
Trafalgar-Class Submarines

Submarine	Pennant Number	Commissioned
Trafalgar	S107	27 May 1983
Turbulent	S110	28 April 1984
Tireless	S117	5 October 1985
Torbay	S118	7 February 1987
Trenchant	S91	14 January 1989
Talent	S92	12 May 1990
Triumph	S93	12 October 1991

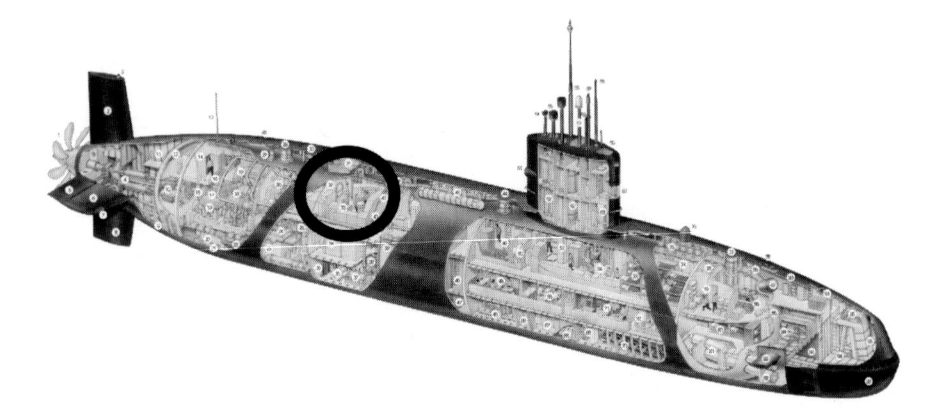

FIGURE 1.1 Layout of a Trafalgar-class nuclear submarine. (Adapted from Allaway, J. 2004. *Inside the Royal Navy: A Collection of Stunning Cutaway Drawings of Royal Navy Ships, Submarines and Aircraft – Past and Present*. Portsmouth, UK: Navy News [HMSO].)

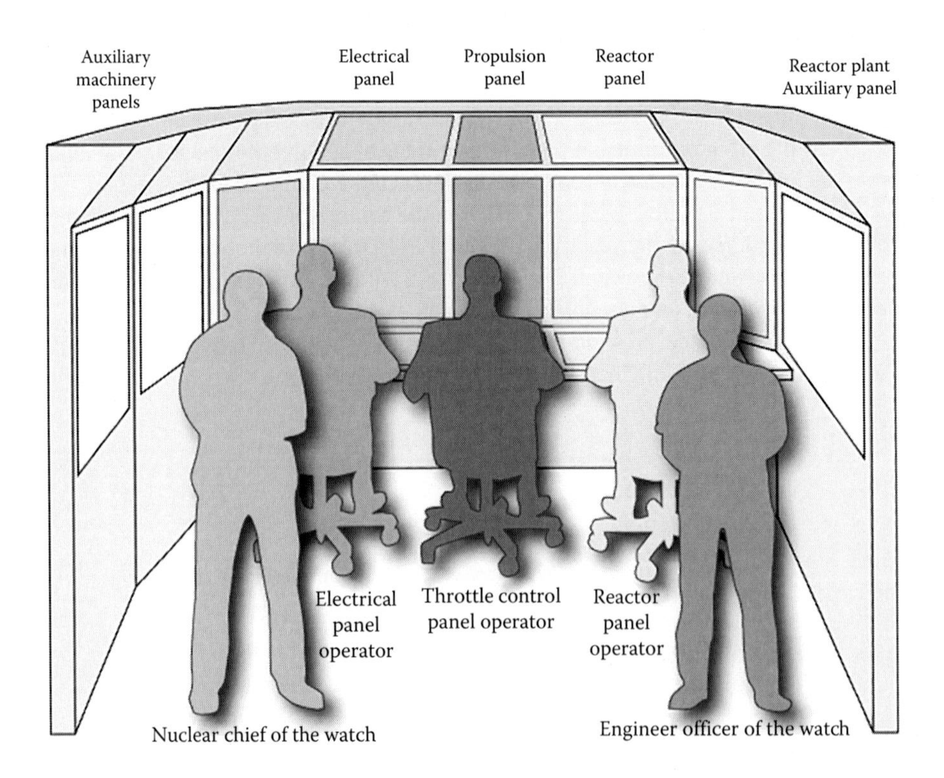

FIGURE 1.2 Layout of the control room.

1.3 COGNITIVE WORK ANALYSIS

1.3.1 Introduction to CWA

The propulsion plant of a nuclear submarine can be classified as a 'complex sociotechnical system'. Complex sociotechnical systems are systems made up of numerous interacting parts, both human and non-human, operating in dynamic, ambiguous and safety critical domains. The complexity embodied in these systems presents significant challenges for modelling and analysis. Traditional reductionist modelling techniques, which decompose activity into a set of task sequences, can rarely be extended beyond stable and repeatable systems. An analysis or modelling technique is, therefore, required that can handle this inherent complexity and adaptability. One means of achieving this is to concentrate on the constraints shaping the way work is conducted in a given domain.

CWA is a structured framework specifically developed for considering the development and analysis of these complex sociotechnical systems. The framework leads the analyst to consider the environment the task takes place within, and the effect of the imposed constraints on the way work can be conducted. The framework guides the analyst through the process of answering the question of why the system exists, what activities can be conducted within the domain as well as how these activities can achieved and who can perform them, identifying the competencies required.

CWA was originally developed at the Risø National Laboratory in Denmark (Rasmussen, 1986) for use within the nuclear power industry. The technique was developed as a result of the electronic departments' identified need for 'design for adaptation'. This need to design for new situations was determined from a study of industrial accidents; the Risø researchers found that most accidents began with non-routine operations. According to Fidel and Pejtersen (2005), CWA's theoretical roots are in general systems thinking, adaptive control systems and Gibson's ecological psychology. CWA is particularly appealing as it can be applied in both closed systems, in which operations are predictable and options for completing a task are normally limited, and open systems, in which task performance is subject to influences and disturbances that cannot always be foreseen. The approach is also described as being amenable for the design and development of systems that are not currently in existence (first-of-a-kind systems).

Vicente (1999) offers a description of the framework including abstraction hierarchies, decision ladders, information flow maps and the skills–rules–knowledge (SRK) framework. These tools can be seen listed on the right-hand side of Figure 1.3. The technique is divided into five phases, each focusing on different constraint sets; the names of phases are listed on the left-hand side in Figure 1.3 and also included, in the centre of the figure, is some guidance on the possible acquisition methods for these phases.

The CWA process is often criticised for being complex and time consuming, to address these concerns and in an attempt to provide some level of guidance and expedite the documentation process, the HFI-DTC (Human Factors Integration Defence Technology Centre; https://www.defencehumancapability.com/HFIDTCLegacy.aspx) has developed a CWA software tool. Built to run on the Microsoft .NET

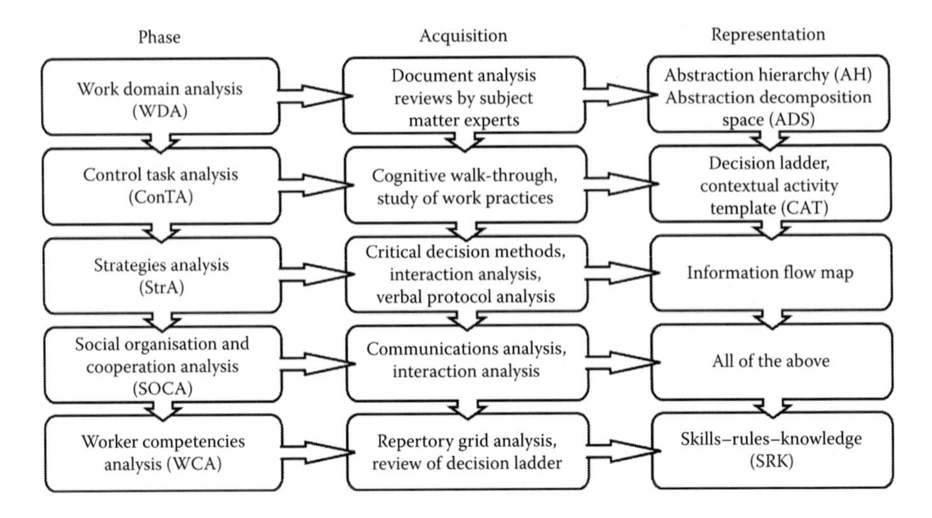

FIGURE 1.3 Five phases of CWA according to Vicente (1999). (Acquisition methods added from Lintern, G. et al. 2004. *Asymmetric Adversary Analysis for Intelligent Preparation of the Battlespace (A3-IPB).* U.S. Air Force Research Department Report.)

framework, the tool uses a familiar windows-based interface and interfaces directly with 'Microsoft Office' applications such as 'Word'. The tool presents templates to allow documents to be created that describe each of the five phases described in the CWA framework, providing a structure for those unfamiliar with the technique. The software tool allows data to be passed between these phases, expediting the documentation process and facilitating updates and changes. Thus, the tool provides structure to the analysis process and markedly expedites the documentation and presentation of the analysis results.

1.3.2 Work Domain Analysis

According to Naikar (2006a), WDA identifies the constraints on workers' behaviour that are imposed by the purposive and physical context, or problem space, in which workers operate. WDA is conducted at the functional, rather than behavioural level; it is used to define the environment within which the activity is conducted. WDA identifies a fundamental set of constraints on the actions of any system component, thus providing a solid foundation for subsequent phases.

WDA is the most commonly used component within CWA. Lintern (2005) describes WDA as a design artefact, commenting that it organises information in a systematic manner that will support design. The tool Vicente (1999) recommends for WDA is the abstraction–decomposition space (ADS). The ADS is a two-dimensional space representing an abstraction hierarchy (AH) and a decomposition hierarchy. The ADS represents the entire domain under analysis outlined by an analyst-defined boundary. Along the decomposition axis, the table shows the entire system under examination (total system), stepping down through levels of detail to a component level. The abstraction component of the diagram models the same system at a number of levels of abstraction; at the highest level, the overall functional purpose of the system is considered and at the lowest level, the individual components within the system are described. Generally, five levels of abstraction are used; these levels have been summarised in Table 1.2 by Naikar et al. (2005). The old labels refer to the original labels generated by Rasmussen, and used by Vicente (1999), these were subsequently updated. To avoid confusion for the remainder of this document, the 'new labels' have been adopted as a standard.

An ADS is a matrix that is an activity-independent description of a work domain. The ADS focuses on the purpose and constraints of the work domain, and for this reason, analyses are typically performed with data collected from discussions with engineering experts and with the review of design and engineering documents. This provides an understanding of how and why the structures work together to enact the desired purposes. Sharp and Helmicki (1998) point out that if the data sources are inadequate, the analysis will be correspondingly inadequate, but even partial and incomplete knowledge can be used to provide a helpful understanding of the work domain.

According to Burns and Hajdukiewicz (2004), the concept for the ADS was developed at Risø National Laboratory as a result of observations of problem solvers, and this led to the realisation that people reasoned a problem using a structure basically asking how (how does this work) and why (why is this here). This was brought together to form means–ends links within a hierarchy; this hierarchy forms a basis for storing the relationships within a system. Means–ends relationships form the basis of decision-making and are captured in the process for creating an ADS. These relationships capture the affordances of the system; the process involves examining what needs to be done within a system and understanding the available means for completing these ends. In complex work domains, there are often many-to-many relationships; these require the user to select from a number of physical arrangements to meet a specific purpose. In the same way, one specific physical arrangement may be the only method for two independent purposes requiring the user to make allocation of resources decisions. In routine or highly familiar situations, the means–ends relations that must be considered by workers are usually well

TABLE 1.2
Descriptions of Levels of Abstraction

'Old Labels'	'New Labels'	Description
Functional purposes	Functional purposes	The purposes of the work system and the external constraints on its operation
Abstract functions	Values and priority measures	The criteria that the work system uses for measuring its progress towards the functional purposes
Generalised functions	Purpose-related functions	The general functions of the work system that are necessary for achieving the functional purposes
Physical functions	Object-related processes	The functional capabilities and limitations of physical objects in the work system that enable the purpose-related functions
Physical form	Physical objects	The physical objects in the work system that afford the object-related processes

established and stable. Decision-making in these situations is relatively straightforward and reflects rule-based reasoning (Rasmussen et al., 1994). An unfamiliar decision, brought about by an unfamiliar situation, will require more thought and can be relatively challenging or demanding, requiring knowledge-based reasoning (Rasmussen et al., 1994).

Naikar et al. (2005) detail a nine-step methodology for completing an ADS; this nine-step list can be explained in detail by supplementing it with information from Burns and Hajdukiewicz's (2004) advice for completing a WDA:

1. *Establish the purpose of the analysis.* At this point, it is important to consider what is sought, and expected, from the analysis.
2. *Identify the project constraints.* Project constraints such as time and resources will all influence the fidelity of the analysis.
3. *Identify the boundaries of the analysis.* The boundary needs to be broad enough to capture the system in detail; however, the analysis needs to remain manageable. It is difficult to supply heuristics; the size of the analysis will in turn define where the boundary is set.
4. *Identify the nature of the constraints in the work domain.*
5. *Identify the sources of information for the analysis.* Information sources are likely to include, among others, engineering documents (if the system exists), the output of structured interviews with subject matter experts (SMEs) and the outputs of interviews with stakeholders. The lower levels of the hierarchy are most likely to be informed from engineering documents and manuals capturing the physical aspects of the system. The more abstract functional elements of the system are more likely to be elicited from stakeholders or literature discussing the aims and objectives of the system.
6. *Construct the ADS with readily available sources of information.* Often, the easiest and most practical way of approaching the construction of the AH is to start at the top and the bottom and meet in the middle.
    a. In most cases, the functional purpose for the system should be clear; it is the reason that the system exists. This functional purpose should be independent of any specific situation or time.
    b. By considering ways of determining how to measure the success of the functional purpose(s), the values and priority measures can be set.
    c. In many cases, at least some of the physical objects are known at the start of the analysis. By creating a list of each of the physical objects and their affordances, the bottom two levels can be partially created.
    d. Often, the most challenging part of an AH is creating the link in the middle between the physical description of the systems components and functional description of what the system should do. The purpose-related functions level involves considering how the physical functions can be used to have an effect on the identified values and priority measures. The purpose-related functions should link upward to the values and priority measures to explain why their requirement; they should also link down to the physical functions, explaining how they can be achieved.

e. The next stage is to complete the means–ends links, checking for unconnected nodes and validating the links. The why–what–how triad should be checked at this stage.

7. *Construct the ADS by conducting special data collection exercises.* After the first stage of constructing the AH, there are likely to be a number of nodes that are poorly linked, indicating an improper understanding of the system. At this stage, it is often necessary to seek further information on the domain from literature or from SMEs.
8. *Review the ADS with domain experts.* The validation of the AH/ADS is a very important stage. Although this has been listed as Stage 8, it should be considered as an iterative process, throughout the creation of the document. Often, the most systematic process for validating the AH is to step through node by node checking the language. Each of the present links should be validated along with each of the correct rejections.
9. *Validate the ADS.* Often, the best way to validate the ADS is to consider known recurring activity checking to see that the AH contains the required physical objects and that the values and priority measures captured cover the functional aims of the modelled activity.

Naikar et al. (2005) have generated some very useful prompts to aid the analyst in completing the ADS. These prompts assist in the classification of nodes into the specific levels of abstraction and decomposition (defined in Table 1.2). The prompts can be seen in Tables 1.3 and 1.4.

Using the process discussed previously, an AH was produced for the propulsion plant (see Figure 1.4). This was then decomposed to produce the ADS (see Figure 1.6). This, along with the other representations, was created using a CWA software tool (Jenkins et al. 2007a,b).

1.3.2.1 The Abstraction Hierarchy

A number of information resources were used to develop the AH. These included the following:

1. Nuclear Reactor Plant Introductory Course
2. Discussions with SMEs
3. Observations of trainees in propulsion plant simulators
4. A background understanding of land-based power stations

The AH was conducted by two analysts in collaboration with three SMEs (all ex-submariners on Trafalgar-class submarines). It was then independently validated by another analyst and SME. The completed AH is presented in Figure 1.4.

1.3.2.1.1 The Functional Purposes

The AH was created in a top-down and a bottom-up fashion. First, the overall functional purposes of the propulsion plant were captured. These capture the reasons why the system exists. These purposes are independent of time; they exist for as long as the system exists. In this case, they are to 'provide ship with adequate power'

TABLE 1.3
List of Prompts for Abstraction by Naikar et al. (2005)

	Prompts	Keywords
Functional purposes	Purposes • For what reasons does the work system exist? • What are the highest-level objectives or ultimate purposes of the work system? • What services does the work system provide to the environment? • What needs of the environment does the work system satisfy? • What role does the work system play in the environment? • What has the work system been designed to achieve? • What are the values of the people in the work system?	Purposes: Reasons, goals, objectives, aims, intentions, mission, ambitions, plans, services, products, roles, targets, aspirations, desires, motives, values, beliefs, views, rationale, philosophy, policies, norms, conventions, attitudes, customs, ethics, morals, principles
	External constraints • What kinds of constraints does the environment impose on the work system? • What values does the environment impose on the work system? • What laws and regulations does the environment impose on the work system? • What societal laws and conventions does the environment impose on the work system?	External constraints: Laws, regulations, guidance, standards, directives, requirements, rules, limits, public opinion, policies, values, beliefs, views, rationale, philosophy, norms, conventions, attitudes, customs, ethics, morals, principles
Values and priority measures	• What criteria can be used to judge whether the work system is achieving its purposes? • What criteria can be used to judge whether the work system is satisfying its external constraints? • What criteria can be used to compare the results or effects of the purpose-related functions on the functional purposes? What are the performance requirements of various functions in the work system? How is the performance of various functions in the work system measured or evaluated and compared? • What criteria can be used to assign priorities to the purpose-related functions? What are the priorities of the work system? How are priorities assigned to the various functions in the work system? • What criteria can be used to allocate resources (e.g. material, energy, information, people, money) to the purpose-related functions? What resources are allocated to the various functions of the work system? How are resources allocated to the various functions of the work system?	Criteria, measures, benchmarks, tests, assessments, appraisals, calculations, evaluations, estimations, judgements, scales, yardsticks, budgets, schedules, outcomes, results, targets, figures, limits Measures of: Effectiveness, efficiency, reliability, risk, resources, time, quality, quantity, probability, economy, consistency, frequency, success Values: Laws, regulations, guidance, standards, directives, requirements, rules, limits, public opinion, policies, values, beliefs, views, rationale, philosophy, norms, conventions, attitudes, customs, ethics, morals, principles

(*Continued*)

TABLE 1.3 *(Continued)*
List of Prompts for Abstraction by Naikar et al. (2005)

	Prompts	Keywords
Purpose-related functions	• What functions are required to achieve the purposes of the work system? • What functions are required to satisfy the external constraints on the work system? • What functions are performed in the work system? • What are the functions of individuals, teams and departments in the work system? • What functions are performed with the physical resources in the work system? • What functions coordinate the use of the physical resources in the work system?	Functions, roles, responsibilities, purposes, tasks, jobs, duties, occupations, positions, activities, operations
Object-related processes	• What can the physical objects in the work system do or afford? • What processes are the physical objects in the work system used for? • What are the functional capabilities and limitations of physical objects in the work system? • What physical, mechanical, electrical or chemical processes are afforded by the physical objects in the work system? • What functionality is required in the work system to enable the purpose-related functions?	Processes, functions, purposes, utility, role, uses, applications, functionality, characteristics, capabilities, limitations, capacity, physical processes, mechanical processes, electrical processes, chemical processes
Physical objects	• What are the physical objects or physical resources in the work system – both man-made and natural? • What physical objects or physical resources are necessary to enable the processes and functions of the work system? • What is the inventory (e.g. names, number, types) of physical objects or physical resources in the work system? • What are the material characteristics (e.g. external form including shape, dimensions, colour; internal configuration; material composition) of physical objects or physical resources in the work system? • What is the topography or organisation (e.g. layout or location of physical objects in relation to each other) of physical objects or physical resources in the work system?	Man-made and natural objects: Tools, equipment, devices, apparatus, machinery, items, instruments, accessories, appliances, implements, technology, supplies, kit, gear, buildings, facilities, premises, infrastructure, fixtures, fittings, assets, resources, staff, people, personnel, terrain, land, meteorological features Inventory: Names of physical objects, number, quantities, brands, models, types Material characteristics: Appearance, shape, dimensions, colour, attributes, configuration, arrangement, layout, structure, construction, make-up, design Topography: Organisation, location, layout, spacing, placing, positions, orientations, ordering, arrangement

TABLE 1.4
List of Prompts for Decomposition by Naikar et al. (2005)

Prompts	Keywords
Levels of decomposition: • What is viewed as the whole system in the work domain? • What is the coarsest level at which workers view the work system? • What is the whole system around which work is organised in the work domain? • What do workers view as the parts of the work system? • What is the most detailed level at which workers view the work system? • What are the different levels of detail at which workers view the work system? • What are the parts around which work is organised in the work system? Part–whole relations: • Are the entities at higher levels of decomposition composed of the entities at lower levels of decomposition? • Are the entities at lower levels of decomposition parts of the entities at higher levels of decomposition?	Names of wholes or parts of: Organisations, physical structures, physical spaces, conceptual structures, groups, teams, functions, positions, arrangements, aggregations, formations, assemblies, segments, pieces, units, components, systems, subsystems, divisions, branches, sectors, departments

and to 'maintain system effectiveness'. The purposes are represented as two nodes rather than single statement as they can often be in conflict. At times, there may be a demand to deliver a power that has the potential to damage the system. Both of these purposes need to be considered in the running of the system.

1.3.2.1.1.1 The Values and Priority Measures The next stage of the construction was to consider the level below the functional purposes, the values and priority measures. Here, further constraints on the functional purposes are more explicitly listed; these are measures for determining how well the system is achieving its functional purposes. Primarily related to providing the ship's power, there is a requirement to 'meet desired speed and torque at propulsor' and to 'meet electric power demand'. Again, the requirement for electrical as well as propulsive power may be, at times, in conflict. The way that system is configured to meet these needs is likely to be heavily contextually dependent.

A proportion of electrical power is required in order to maintain the effectiveness of the system. The system also needs to be operated with limits of power, reactivity, temperature, pressure, chemistry and radiation dose.

1.3.2.1.2 The Physical Objects

As previously stated, a mixture of a top-down and bottom-up approach has been employed in this analysis. Moving down to the lowest level of the hierarchy, the

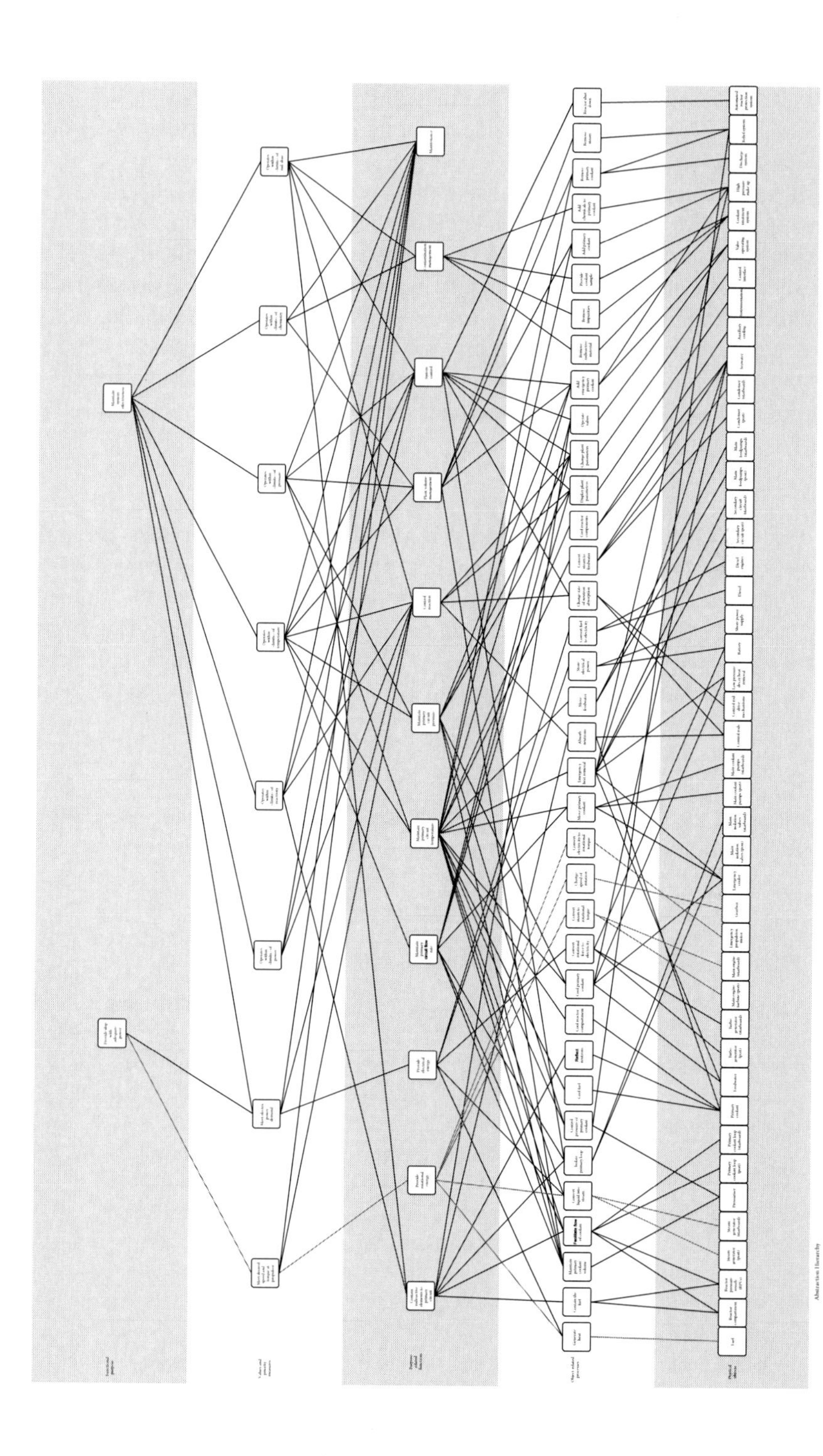

FIGURE 1.4 AH for propulsion plant.

physical objects within the system are listed. The boundaries of this analysis have limited this list to the elements under direct control of the propulsion plant. The exception to this is the inclusions of the diesel engines and fuel into the analysis. These have been included as they provide the propulsion plant with significant functionality in charging the batteries.

The boundaries of the analysis indicate the levels of fidelity applied here. Only the core objects have been included, and objects such as chairs, flooring and lighting all provide important affordances to the propulsion plant; however, in an attempt to keep the analysis manageable, the boundary has been set to omit them. It would also be possible to decompose many of the listed objects into their component parts and describe their affordances more concisely; however, this has also been classified as outside the remit of this initial evaluation. A complete list of the physical objects can be found at the base of Figure 1.4; to aid readability, this list has been reproduced in Table 1.5.

1.3.2.1.3 The Object-Related Processes

The second level from the bottom of the AH, the object-related processes, captures the processes that are conducted by the physical objects in order to perform purpose-related functions. Most importantly, they capture the affordances of the physical objects independently of their purpose. A complete list of the object-related processes can be found at the base of Figure 1.4 along with the links that indicate to which object they relate (Table 1.6). To aid readability, this list has been reproduced in Figure 1.7.

1.3.2.1.4 The Purpose-Related Functions

In the middle of the hierarchy, the purpose-related functions are listed. These functions have the ability to influence one or more of the values and priority measures.

TABLE 1.5
List of Physical Objects Considered in the Analysis Remit

Fuel	Gearbox	Main feedpumps (post)
Reactor compartment	Emergency cooler	Main feedpumps (starboard)
Reactor pressure vessels	Main isolation valves (post)	Condenser (port)
Steam generator (port)	Main isolation valves (stb)	Condenser (starboard)
Steam generator (starboard)	Main coolant pumps (port)	Seawater
Pressuriser	Main coolant pumps (stb)	Auxiliary cooling
Primary coolant loop (port)	Control rods	Instrumentation
Primary coolant loop (stb)	Control rod drive mechanisms	Control interface
Primary coolant	Low pres decay heat removal	Valve operating system
Feedwater	Battery	Coolant treatment system
Turbo generator (port)	Shore power supply	High-pressure make-up
Turbo generator (starboard)	Diesel	Discharge system
Main engine turbine (port)	Diesel engines	Relief system
Main engine (starboard)	Secondary circuit (port)	Automated reactor protection
Emergency propulsion motor	Secondary circuit (starboard)	

TABLE 1.6
List of Object-Related Processes

Generate heat	Convert steam to rotational torque	Display plant parameters
Contain the fuel	Change speed of rotation	Change plant parameters
Maintain primary coolant volume	Convert electricity to rotational torque	Operate valves
Facilitate flow of coolant	Move primary coolant	Add emergency primary coolant
Convert liquid into steam	Emergency heat removal	Remove radioactive material
Isolate primary loop	Absorb neutrons	Remove impurities
Control pressure of primary coolant	Move feedwater	Provide coolant sample
Cool fuel	Store electrical power	Add primary coolant
Reflect neutrons	Convert fuel to electricity	Add chemicals to primary coolant
Cool reactor compartment	Change rate of neutron absorption	Remove primary coolant
Cool primary coolant	Convert steam to feedwater	Remove steam
Convert rotational force to electricity	Cool reactor components	Reactor shutdown

They link the purpose-independent processes with the object-independent functions. They are listed as follows:

1. Contain radioactive elements to primary circuit
2. Provide rotational energy
3. Provide electrical energy
4. Maintain primary circuit flow rate
5. Maintain primary circuit temperature
6. Maintain primary circuit pressure
7. Control reaction
8. Plant volume management
9. System control
10. Contamination management
11. Maintenance

The function of maintenance has been included as a placeholder as it is considered a significantly important consideration in terms of maintaining system effectiveness. Although included at this level, it has not been extended to lower levels of abstraction as it lies outside the control of the propulsion plant, and therefore outside the remit of this analysis.

The use of the means–ends links and the utility of the AH can be described with the example shown in Figure 1.5. Figure 1.5 shows the node 'provide rotational energy' highlighted in the purpose-related functions level. Following the links out of the top of this node answers the question 'why is this needed' – in this case, to 'meet desired speed and torque at propulsor' in turn to 'provide ship with adequate power'. Following the links down from the 'provide rotational energy', it is possible

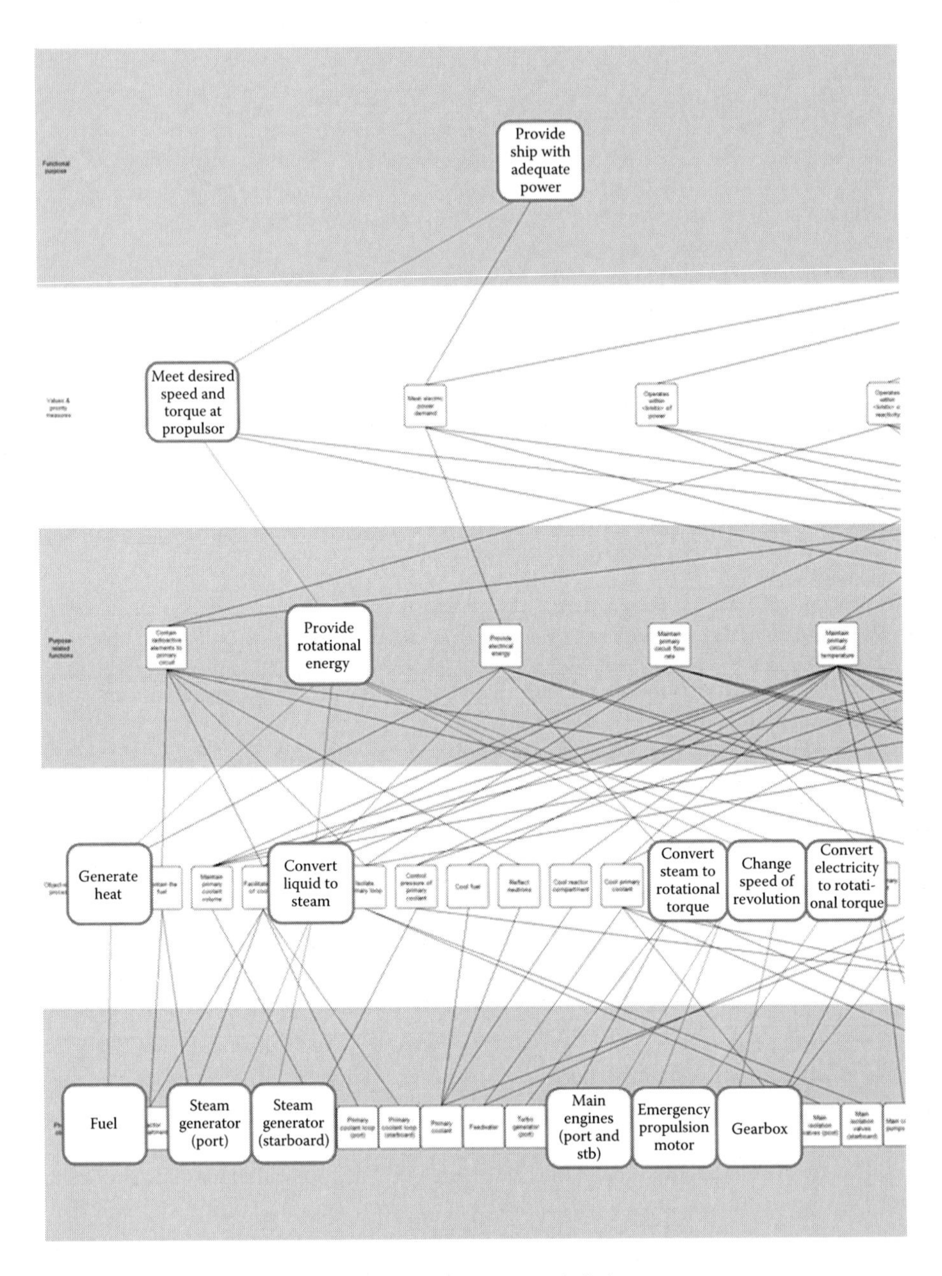

FIGURE 1.5 Enlargement of AH showing means–ends links.

to answer the question 'how can this be achieved' – in this case, by 'generating heat', 'converting heat to steam', 'converting steam to rotational torque', by 'changing the speed of revolution' and by 'converting electricity to rotational torque'.

The diagram does not prescribe a particular arrangement for providing this functionality; rather, it lists all of the components that can affect it. In this case, there is redundancy in the system. The system can be configured so that heat is generated by the fuel, converted to steam in one or both of the steam generators, then converted to rotational torque in one or both of the engines and then stepped down by changing the speed of revolution in the gearbox. Alternately, electricity from the batteries (not shown in Figure 1.5) can be converted to rotational torque in the emergency population motor and then stepped down in the gearbox.

1.3.2.2 The Abstraction–Decomposition Space

The ADS decomposes the nodes of the AH presented in Figure 1.4 into total system, subsystem or component (see Figure 1.6). This provides an additional dimension and presents the information in a more compact format.

1.3.3 Control Task Analysis

The second phase of the CWA framework, ConTA, allows the requirements associated with known, recurring classes of situations to be identified. Naikar (2006b) comments that, by identifying the activity that is necessary to achieve the objectives

Abstraction \ Decomposition	Total system	Subsystem	Component
Functional purpose	Provide ship with adequate power Maintain system effectiveness		
Values and priority measures	Operates within <limits> of rad. dose Operates within <limits> of chemistry Operates within <limits> of reactivity Operates within <limits> of temperature Operates within <limits> of power Meet desired speed and torque at propulsor Operates within <limits> of pressure Meet electric power demand		
Purpose-related functions		Maintenance Contain radioactive elements to primary circuit Maintain primary circuit temperature Contamination management System control Plant volume management Control reaction Maintain primary circuit pressure Maintain primary circuit flow rate Provide electrical energy Provide rotational energy	
Object-related processes		Change plant parameters Display plant parameters Cool reactor components Convert steam to feedwater Move feedwater Absorb neutrons Move primary coolant Isolate primary loop Emergency heat removal Convert electricity to rotational torque Cool primary coolant Cool fuel Control pressure of primary coolant Contain the fuel Generate heat Reactor shut down Remove steam Remove primary coolant Add chemicals to primary coolant Add primary coolant Provide coolant sample Remove impurities Remove radioactive material Add emergency primary coolant Operate valves Cool reactor compartment Facilitate flow of coolant Maintain primary coolant volume	Convert fuel to electricity Store electrical power Change speed of rotation Convert steam to rotational torque Convert rotational force to electricity Change rate of neutron absorption Reflect neutrons Convert liquid into steam
Physical objects		Control interface Auxiliary cooling Secondary circuit (starboard) Secondary circuit (port) Control rod drive mechanisms Primary coolant loop (starboard) Primary coolant loop (port) Automated reactor protection system Valve operating system Relief system Discharge system High pressure makeup Coolant treatment system	Shore power supply Diesel engines Diesel Instrumentation Seawater Condenser (starboard) Condenser (port) Battery Main feed-pumps (starboard) Main feed-pumps (port) Control rods Main coolant pumps (starboard) Main coolant pumps (port) Main isolation valves (starboard) Main isolation valves (port) Emergency cooler Gearbox Main engine (starboard) Main engine turbine (port) Turbo generator (starboard) Turbo generator (port) Feedwater Primary coolant Pressuriser Steam generator (starboard) Steam generator (port) Reactor pressure vessels (RPVs) Fuel Reactor compartment Low pressure decay heat removal Emergency propulsion motor

Abstraction decomposition space

FIGURE 1.6 ADS for propulsion plant.

of a system with a given set of physical resources, ConTA complements WDA. The phase identifies what needs to be done independently of how it can be conducted or who can conduct it. According to Sanderson (2003), control tasks emerge from work situations and they transform inputs (e.g. current state, targets) into outputs (decisions, control actions, etc.). Returning to the definition by Rasmussen et al. (1994), the ConTA phase can be considered as both 'activity analysis in work domain terms' and 'activity analysis in decision-making terms'.

1.3.3.1 Activity Analysis in Work Domain Terms

On the basis of Rasmussen et al.'s (1994) definition of activity analysis in work domain terms, Naikar et al. (2006) have developed the contextual activity template (CAT) (see Figure 1.7) for use in this phase of the CWA. This template is one way of representing activity in work systems that are characterised by both work situations and work functions. Work situations are situations that can be decomposed based on recurring schedules or specific locations. Rasmussen et al. (1994) describe work functions as activity characterised by its content independent of its temporal or spatial characteristics (Rasmussen et al. 1994); these functions can often be informed by the AH. Rasmussen et al. (1994) recommend that the analyst decomposes on either work functions or work situations; however, Naikar et al. (2006) plot these on two axes so that their relationship can be investigated, allowing the representation of activity in work systems that are characterised by both work situations and work functions. Typically, the work situations are shown along the horizontal axis and the work functions are shown along the vertical axis of the CAT. The circles indicate the work functions with the bars showing the extent of the table in which the activity typically occurs. The dotted boxes around each circle indicate all of the work situations in which a work function can occur (as opposed to must occur), thus capturing the constraints of the system.

Figure 1.7 shows the CAT for the purpose-related functions taken from the AH (see Figure 1.4), whereas Figure 1.8 shows the CAT for the object-related processes taken from the AH. The situations were derived with the assistance of an SME; they follow the submarine through the key stages of start-up to shutdown. These are listed as follows:

1. Alongside – the submarine in dock with the reactor shutdown
2. Start-up – starting up of the reactor
3. Warming through (plant state B) – the process of heating the system
4. Criticality – the point in which the system is ready to start driving the engines
5. TG running (plant state A) – main engines running
6. Half power – submarine running engines at half power
7. Full power – submarine running at full power
8. Hot and pressurised – the reactor shutdown, the system being still hot and containing steam
9. Depressurised – the system being cooled

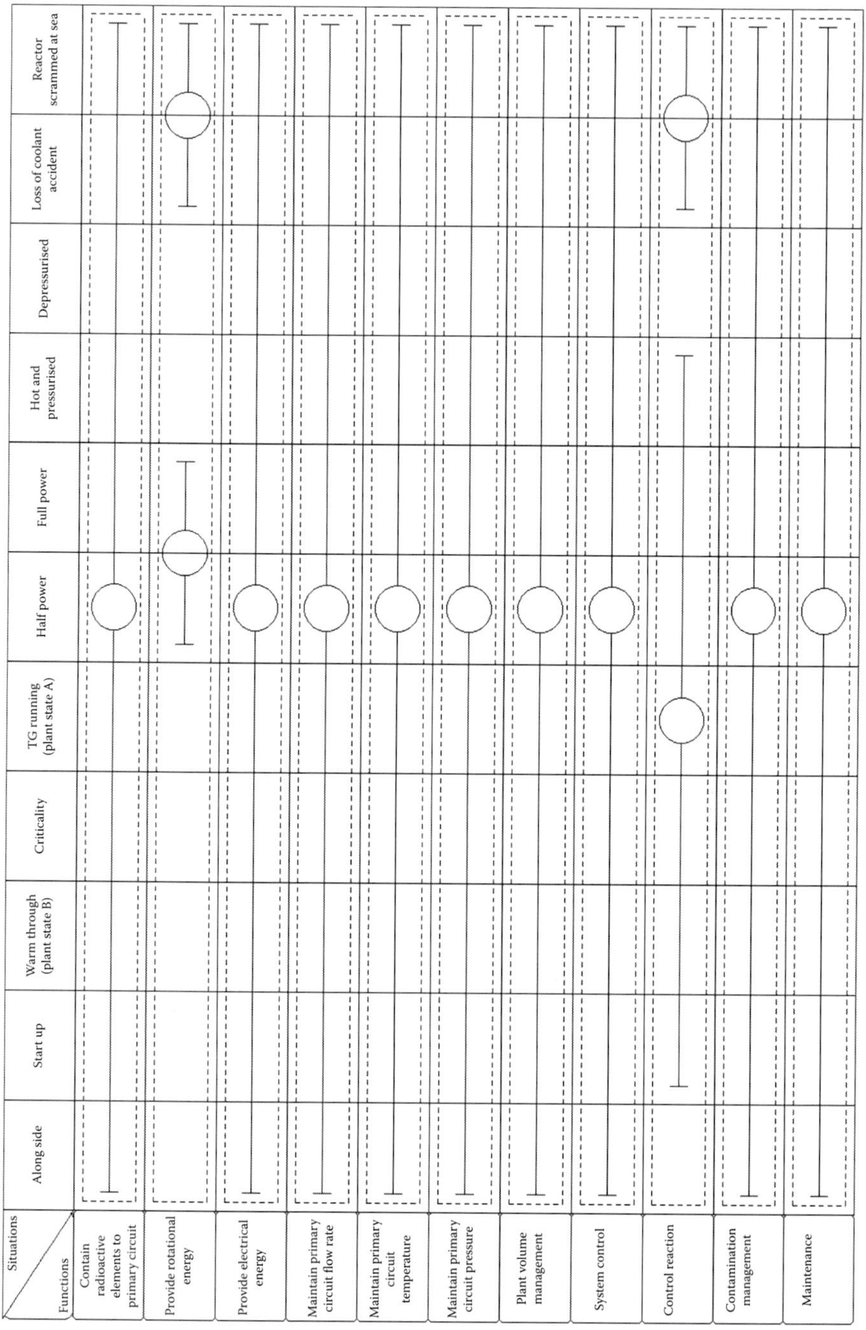

Situations

Functions

Along side

Start up

Warm through (plant state B)

Criticality

TG running (plant state A)

Half power

Full power

Hot and pressurised

Depressurised

Loss of coolant accident

Reactor scrammed at sea

Contain radioactive elements to primary circuit

Provide rotational energy

Provide electrical energy

Maintain primary circuit flow rate

Maintain primary circuit temperature

Maintain primary circuit pressure

Plant volume management

System control

Control reaction

Contamination management

Maintenance

FIGURE 1.7 CAT for purpose-related functions.

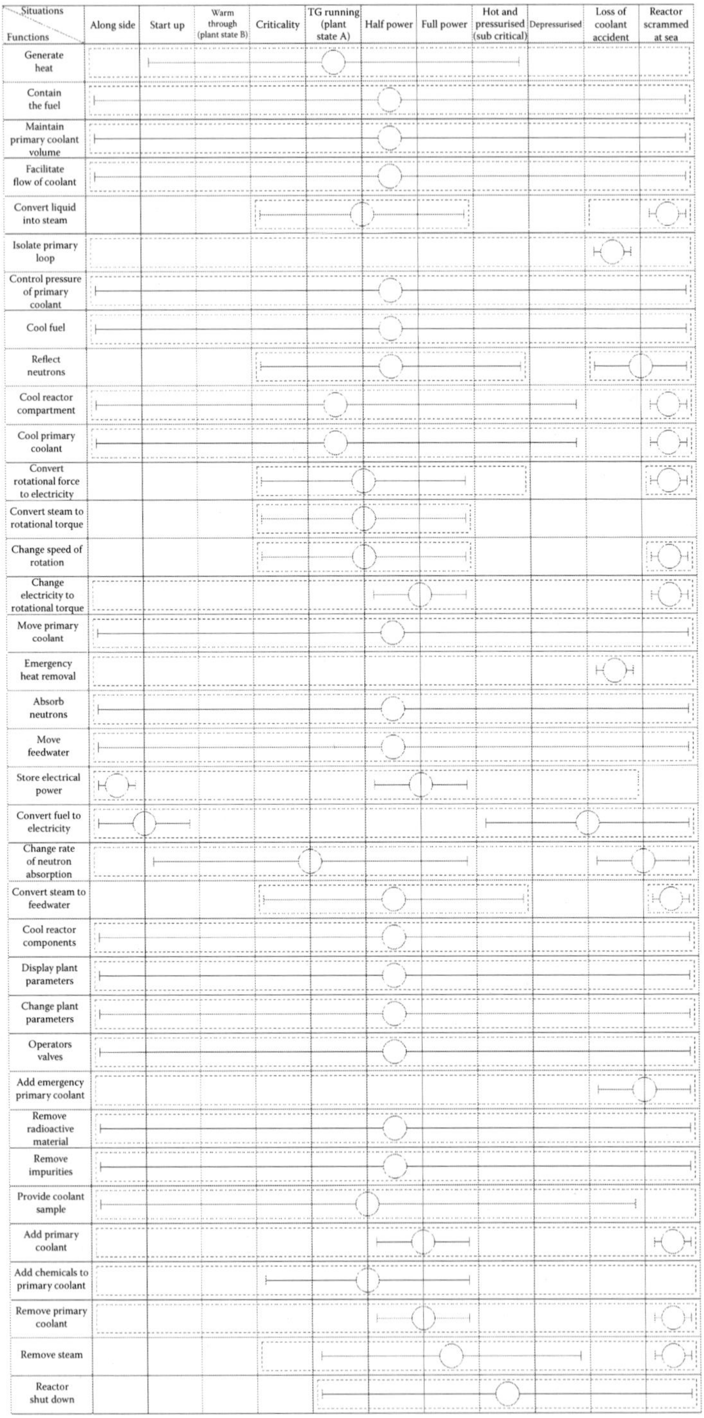

Situations / Functions

Along side | Start up | Warm through (plant state B) | Criticality | TG running (plant state A) | Half power | Full power | Hot and pressurised (sub critical) | Depressurised | Loss of coolant accident | Reactor scrammed at sea

Generate heat
Contain the fuel
Maintain primary coolant volume
Facilitate flow of coolant
Convert liquid into steam
Isolate primary loop
Control pressure of primary coolant
Cool fuel
Reflect neutrons
Cool reactor compartment
Cool primary coolant
Convert rotational force to electricity
Convert steam to rotational torque
Change speed of rotation
Change electricity to rotational torque
Move primary coolant
Emergency heat removal
Absorb neutrons
Move feedwater
Store electrical power
Convert fuel to electricity
Change rate of neutron absorption
Convert steam to feedwater
Cool reactor components
Display plant parameters
Change plant parameters
Operators valves
Add emergency primary coolant
Remove radioactive material
Remove impurities
Provide coolant sample
Add primary coolant
Add chemicals to primary coolant
Remove primary coolant
Remove steam
Reactor shut down

FIGURE 1.8 CAT for object-related processes.

Two additional emergencies were added to the analysis to demonstrate the approach further. In a more detailed analysis, additional situations should also be considered. The considered emergencies are as follows:

1. Loss of coolant accident
2. Reactor scrammed at sea

Looking first at Figure 1.7, perhaps the most salient feature is that, in this system, function constraints are not notably contextually influenced – they do not change significantly based upon context; in fact, the dotted boxes show that all functions can happen in all situations indicating that context does not heavily constrain the operation of the propulsion plant. The second function from the top 'provide rotational energy' is only likely to take place while at half power, full power or during a loss of coolant accident or a reactor scram at sea. In situations such as 'alongside' or 'start-up', rotational force is unlikely to be required; however, in the event that it would be required batteries could be used to provide this power. The findings apply only to the examples presented.

Figure 1.8 shows the CAT for a lower level of abstraction, the object-related processes. In this representation, constraints that are more physical are present. Taking the example of 'convert liquid into steam', the reactor has to be producing sufficient heat to generate steam; in a number of situations, this function is physically impossible. Other functions, such as 'convert fuel to electricity' (i.e. diesel), can happen in all situations (providing the submarine is on the surface); however, it is unlikely to take place while the system is producing steam.

Other functions such as 'add emergency coolant' can take place in all situations; however, they typically would occur in the event of a system fault such as a loss of coolant.

1.3.3.2 Activity Analysis in Decision-Making Terms

The second part of the ConTA phase is used to consider activity analysis in decision-making terms. Vicente (1999) recommends the use of the decision ladder (see Figure 1.9) for this phase of CWA. The decision ladder was developed by Jens Rasmussen who observed that expert users were relying on rule-based behaviour (RBB) to conduct tasks. Rasmussen (1974) states that the sequence of steps between the initiating cue and the final manipulation of the system can be identified as the steps a novice must necessarily take to carry out the sub-task.

The decision ladder, shown in Figure 1.9, can be seen to contain two different types of node: the rectangular boxes represent data-processing activities and the circles represent states of knowledge that result from data processing. According to Vicente (1999), the decision ladder represents a linear sequence of information-processing steps, which is 'bent in half'. Novice users are expected to follow the decision ladder in a linear fashion, whereas expert users are expected to link the two halves by shortcuts. According to Naikar and Pearce (2003), the left side of the decision ladder represents the observation of the current system state, whereas the right side of the decision ladder represents the planning and execution of tasks and procedures to achieve a target system state. Sometimes, observing information and

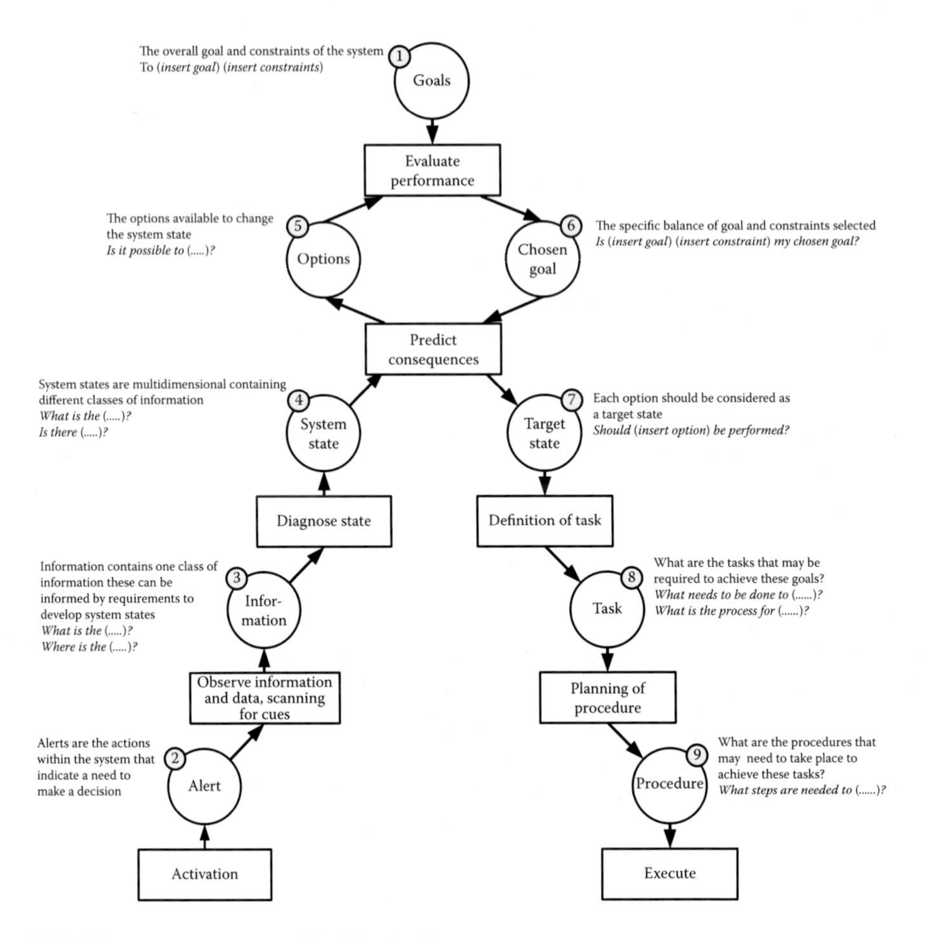

FIGURE 1.9 Description of decision ladder.

diagnosing the current system state immediately signal a procedure to execute. This means that rule-based shortcuts can be shown in the centre of the ladder. On the other hand, effortful, knowledge-based goal evaluation may be required to determine the procedure to execute; this is represented in the top of the ladder. There are two types of shortcut that can be applied to the ladder; 'shunts' connect an information-processing activity to a state of knowledge (box to circle) and 'leaps' connect two states of knowledge (circle to circle); this is where one state of knowledge can be directly related to another without any further information processing. It is not possible to link straight from a box to a box as this misses out the resultant knowledge state. Cummings and Guerlain (2005) point out that when a shortcut is taken, various information-processing actions are bypassed but the desired results are still achieved. The decision ladder not only displays these shortcut relationships in information-processing activities, but also highlights those states of knowledge that are bypassed if a shortcut is taken. According to Cummings and Guerlain (2005), the decision ladder maps rather than models the structure of a decision-making process. In the case of systems with computer-based decision support tools, the decision ladder represents the decision process and states of knowledge that must be addressed by the system, regardless of whether a computer or a human makes the decision.

In an attempt to better understand decision-making, the decision ladder can be used to develop prototypical models of activity. As Rasmussen et al. (1994) is keen to communicate, it is important to draw the distinction between typical and prototypical work situations. People tend to describe what they find to be normal, usual ways of doing things, representing an intuitive averaging across cases – typical situations. Conversely, prototypical work situations are developed from actual data from context-specific cases. This then forms a set of prototypical activity elements, defined by either problem to solve or situation to solve within, which in varying combinations can serve to characterise the activity within a work system.

Figure 1.10 shows the decision ladder for an upstream steam leak in the reactor compartment. The model was generated by eliciting information from an experienced operator and trainer using the following approach; the discussion took place in a simulator of the real working environment. The expert was encouraged to walk around the simulator, moving to the information elements he or she was discussing. The information was elicited in a two-phase process; the first stage of the process was to complete the decision ladder based upon a typical example supplied by the SME. The second phase of the process was to supplement this with additional information representing additional elements not captured in the typical model, thus converting the typical model in a prototypical one. The numbers in the description relate to the steps of the decision ladder shown in Figure 1.9.

1. The expert was introduced to the decision ladder model and asked to describe his overall goal in operating the system.
2. The expert was asked to talk to the analysts through the process of making a decision about where the leak was and how to deal with it. The expert was guided to start the description by indicating what first drew his attention to the problem (the alert).
3. The expert was then asked to list the artefacts that he used to gather information.
4. The expert was asked to explain how he used these information elements to diagnose the current system state, firstly that there was a steam leak and to locate it.
5. The expert then described the options available to him.
6. He then explained how he would balance the competing constraints on his goals.
7. On the basis of the goal selected, the expert then listed the target states available (his options) and selected the target state he would take.
8. This state was then broken down into a series of tasks.
9. The tasks were then broken down into procedures.

Once recorded, the notes were read back to the expert; at each stage of the decision ladder, the expert was asked to capture all other elements that would be available, for example, list all possible ways in which the operators could be alerted to a

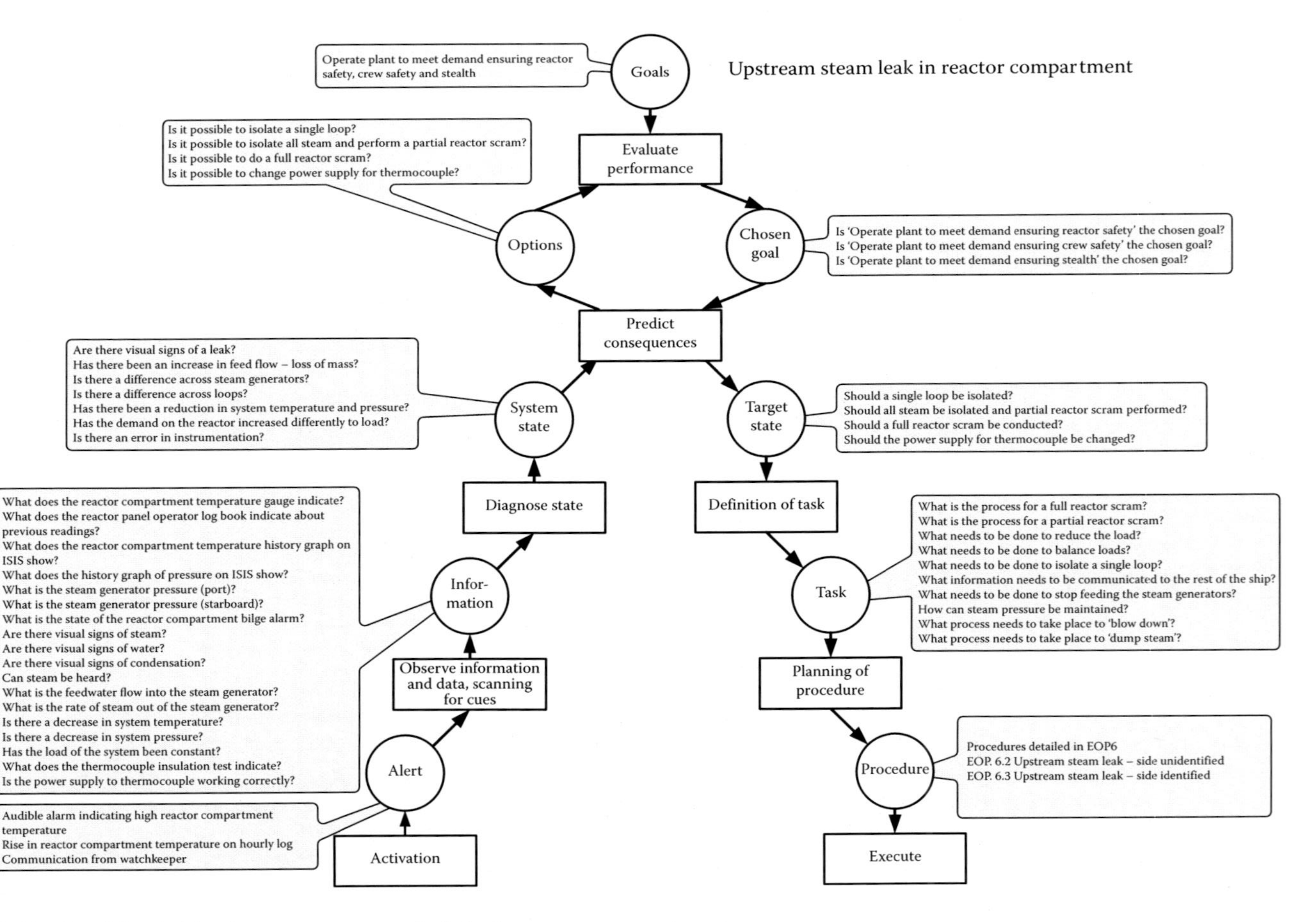

FIGURE 1.10 Decision ladder for upstream steam leak in the reactor compartment.

steam leak and list all information elements that may be of use in developing a system state. This information was used to generate the prototypical model presented in Figure 1.10. The annotation was structured using the guidance shown in the italic comments shown in Figure 1.9.

1.3.3.2.1 Stage 1: Determining the Goal

The first stage of the process was to structure the goal of the system, framing the remainder of the data collection exercise. The SME was asked to provide a high-order goal along with a number of constraints affecting it; the expert was reassured that the constraints could possibly be in conflict. On the basis of Elix and Naikar's (2008) guidance (shown in Figure 1.9), the information was placed in the format 'To (insert goal) (insert constraints)'. The goal decided upon was 'Operate plant to meet demand'; the constraints included 'ensuring reactor safety, crew safety and stealth'. These constraints could be in conflict in certain situations, where an element of risk is placed on one part of the goal to ensure the other. Elix and Naikar (2008) found that activity was usually best characterised by having only one goal per work function. In situations where more than one goal is required, they recommended representing the analysis on multiple decision ladders.

1.3.3.2.2 Stage 2: Alert

The SME was asked to begin the walk-through by starting at the chronological start of the process; the SME was asked to base the discussion on a typical encounter. In the first pass through the model, the SME identified the alert as an audible alarm indicating high reactor compartment temperature. Other possible alerts included a rise in reactor compartment temperature on hourly log and communication from the watchkeeper.

1.3.3.2.3 Stage 3: Information

In the first pass of the process, the expert was asked to list the information elements they would use to determine the cause of the alert, such as direct observation of the reactor compartment, observation of additional instrumentation and reliability testing of the gauges. As with the alert stage, the second pass was used to validate this and to add in additional information elements that could be used in the role of diagnosis. In situations where the expert stated talking about system states, a note was made and the expert was allowed to continue.

1.3.3.2.4 Stage 4: System State

The system states represent a perceived understanding of the system based upon the interpretation of a number of information elements. For example, the loss of mass could be informed by fusing information on the feedwater flow into the steam generator and the rate of steam out of the steam generator, also comparing across the port and starboard. The key distinction between an information element and a system state is that information states are formed of more than one quantifiably different elements of information.

1.3.3.2.5 Stage 5: Options

The options within the ladder can be described as the opportunities for changing the system state in an attempt to satisfy the overall goal. As the italics in Figure 1.9 show, the points are structured as questions in the form: 'is it possible to (...)?'

1.3.3.2.6 Stage 6: Chosen Goal

The chosen goal at any one time is determined by selecting which of the constraints to place the highest priority upon. In this case, a decision is required to ensure reactor safety, crew safety or stealth. As Elix and Naikar (2008) make clear, this does not have to be an absolute choice per se, rather than one takes a higher priority than the other does in the given situation.

1.3.3.2.7 Stage 7: Target State

The target states mirror the option available; once a particular option is selected, it becomes a target state.

1.3.3.2.8 Stage 8: Task

The listed task questions relate to the tasks required for achieving the target state while maintaining the overall goal.

1.3.3.2.9 Stage 9: Procedure

The procedure lists questions related to the steps required to achieving each of the listed tasks. In this example, the procedures are laid down in EOP 6.

1.3.3.2.10 Validating the Model

One way of validating the model is to view the previous and subsequent knowledge states, checking for a linkage between elements used at each level.

Once the model was completed based upon the expert experience, the model was validated against the EOPs. From the EOPs it was also possible to create a decision ladder for the more generic 'steam leak' situation; this is presented in Figure 1.11. A number of shortcuts have also been included for situation where steam is seen or heard; in such instances, it is possible to jump straight to the procedure. In other familiar situations once the system state has been diagnosed, it is not necessary to balance the goals and defined the target state; in these familiar situations, the operator can jump straight to the prescribed procedure.

Decision analysis of the upstream steam leak was undertaken to illustrate the decision ladder can be used to explore the activities undertaken in fault diagnosis. This example shows four decision ladders, braking the activity down into different parts of the decision cycle: from presentation of Air Temp High Alarm to viewing reactor window (see Figure 1.12); from reporting on condensation to checking mismatch between steam flow and feed flow (see Figure 1.13); from identifying reduced steam flow to requesting EPO to balance load (see Figure 1.14) and from detecting leaking circuit to sounding general alarm (see Figure 1.15).

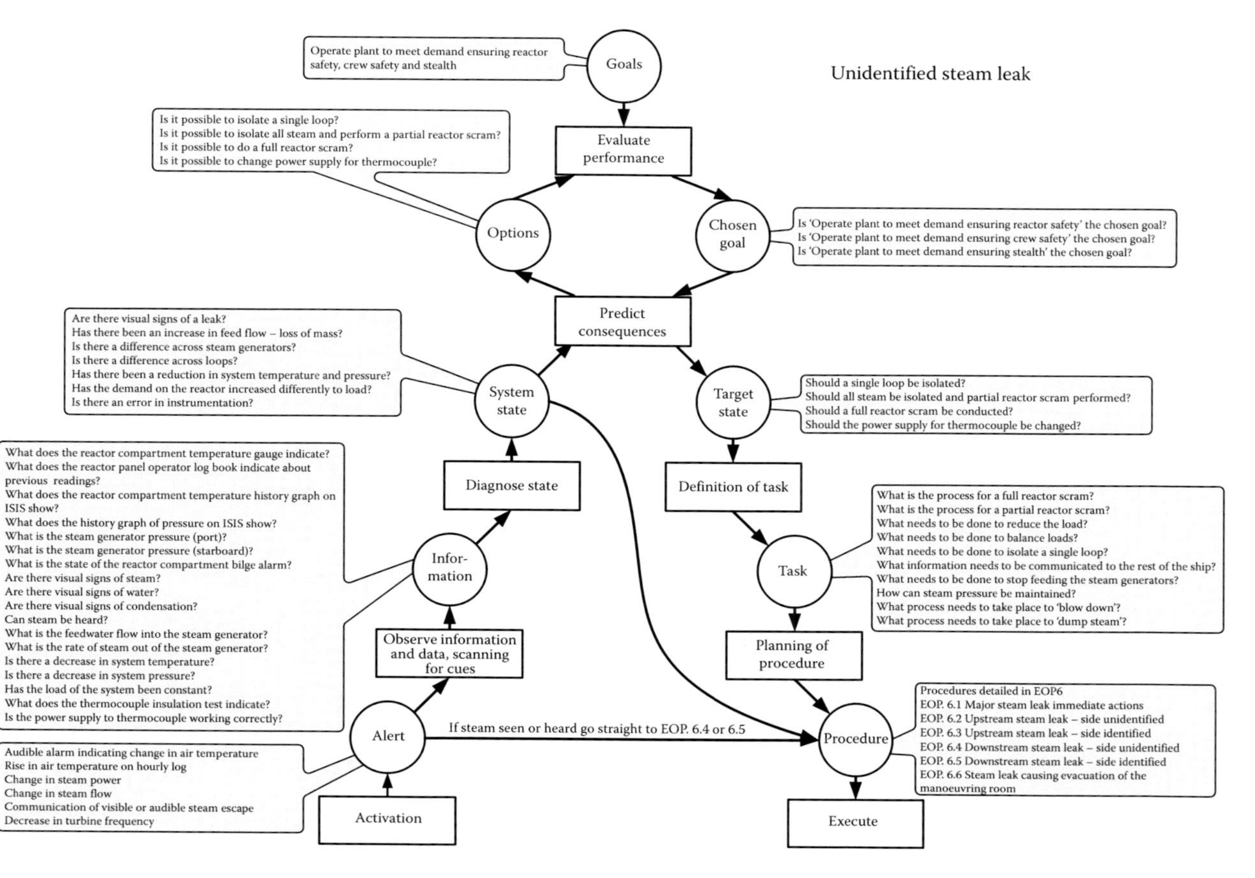

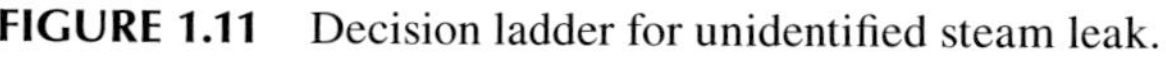
FIGURE 1.11 Decision ladder for unidentified steam leak.

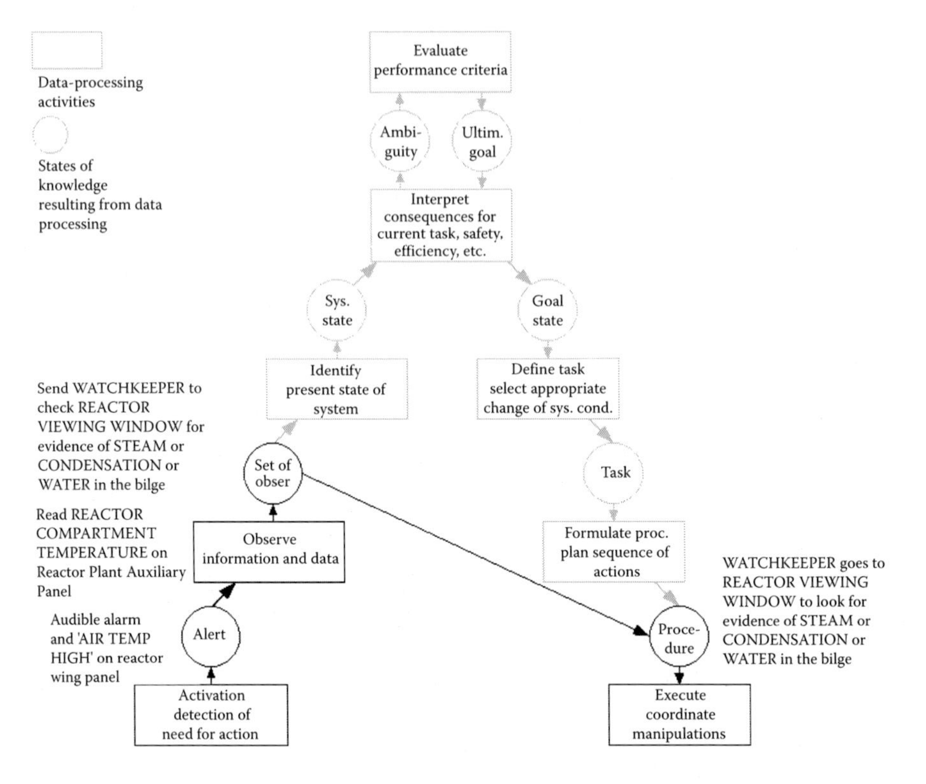

FIGURE 1.12 Decision ladder from presentation of Air Temp High Alarm to viewing reactor window.

1.3.4 Strategies Analysis

The StrA phase addresses the constraints influencing the way in which activity can be conducted. StrA looks at filling the 'black box' left in ConTA; it looks at different ways of carrying out the same task. Wherever the previous phase dealt with the question of what needs to be done, this phase addresses how it can be done.

The strategy adopted under a particular situation may vary significantly. The term 'agent' is used to signify that the activity could be performed by either a human or an automated technology. Different agents may perform tasks in different ways, and the same agent (either human or non-human) might perform the same task in a variety of different ways. Naikar (2008) comments that when an agent's work demands are low, agents may adopt a strategy that is cognitively more intensive, whereas when their work demands are high, agents may adopt a strategy that is cognitively less intensive. The strategy that the agent chooses to select will be dependent on a huge number of variables, including, among others, their experience, training, workload and familiarity with the current situation. To complicate things even further, the same agent may select different strategies on different occasions.

Ahlstrom (2005) offers a flow map for use in this phase. An example of this is shown in Figure 1.16; here, the situation is broken down to a 'start state' and an 'end state', and connecting the two, in the middle, are a number of strategies for the transformation.

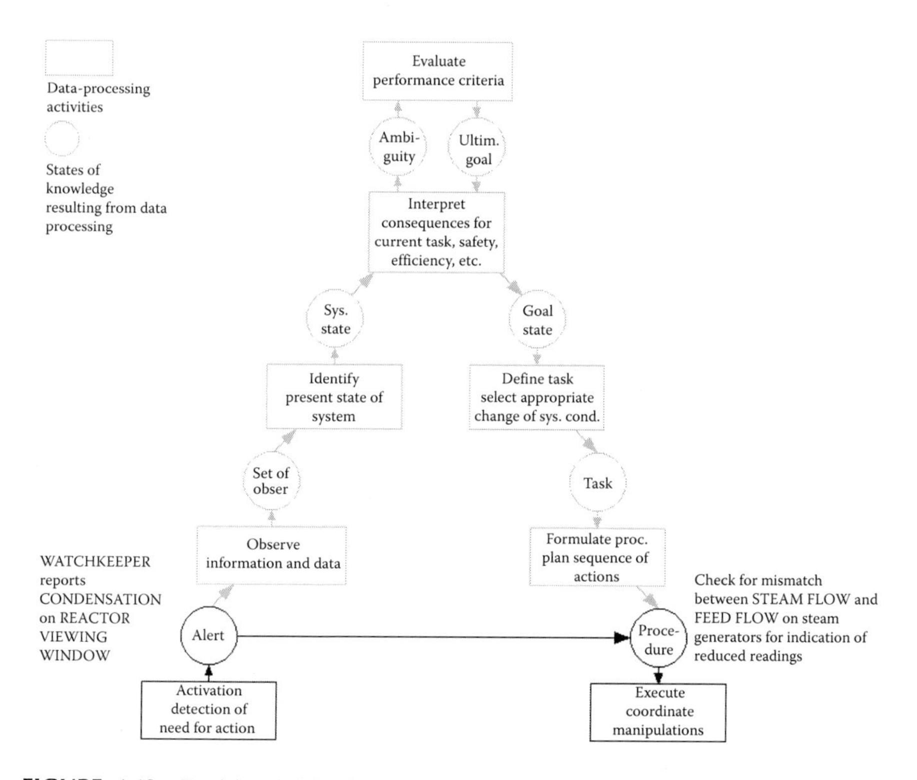

FIGURE 1.13 Decision ladder from reporting on condensation to checking mismatch between steam flow and feed flow.

Naikar (2006b) supports the adoption of structured flow maps commenting that StrA is not concerned with defining detailed sequences of actions or mental processes. Instead, StrA is concerned with identifying general 'categories of cognitive procedures', which are idealised, abstract descriptions of particular sequences of operations. Naikar (2006b) further discusses StrA pointing out that it recognises that workers will often switch between multiple strategies while performing a single activity in order to deal with task demands.

The CAT can be used as basis for eliciting the StrA required. The matrix of functions and situations illustrates the possible activity within the domain (see Figure 1.17) to perform a complete analysis. A StrA should be created for each of the cells where activities can take place (marked by the dotted line); the differences to the strategies available, effected by the situational constraints, can then be examined.

Within some specialist environments, the way activities are conducted is often mandated by Standard Operating Procedures and Emergency Operating Procedures. For this reason, systems are often configured to limit the available strategies open to an agent. While a lack of flexibility can be perceived as a negative, there are also clear advantages. Procedures make activity predictable; in situations with distributed team working and shift working, this is clearly an advantage. Therefore, the options available for an agent at a given time are limited. Nevertheless, there are two options for removing steam from the steam generator, as indicated in Figure 1.18.

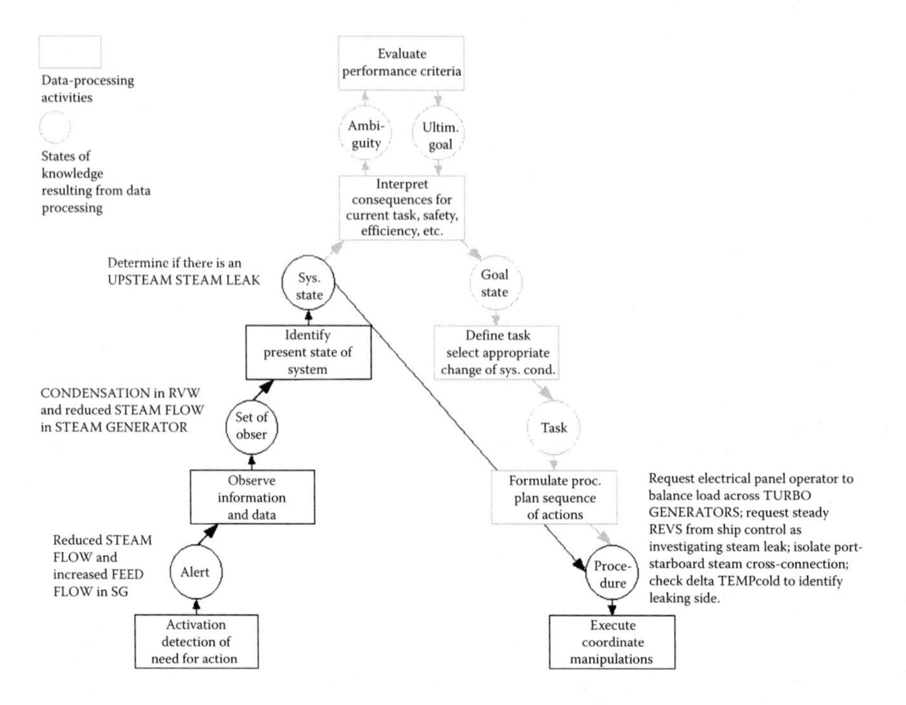

FIGURE 1.14 Decision ladder from identifying reduced steam flow to requesting EPO to balance load.

As Figure 1.18 shows, steam can be removed from the steam generator by either blowing down to the sea (the top path) or dumping steam into the condenser (the bottom path). The trigger to choose between these two alternatives is whether or not the captain wishes to remain stealth (in which case the bottom path is chosen) or whether it is more important to get rid of the steam quickly (in which case the top path is chosen). Even after blowing down the steam to sea, any remaining steam in the steam generator will still have to be removed by dumping it into the condenser.

1.3.5 Social Organisation and Cooperation Analysis

SOCA addresses the constraints imposed by organisational structures or specific agent roles and definitions. SOCA investigates the division of task between the resources and looks at how the team communicates and cooperates. The objective is to determine how the social and technical factors in a sociotechnical system can work together in a way that enhances the performance of the system as a whole.

SOCA recognises that organisational structures in many systems are generated online and in real time by multiple, cooperating agents responding to the local context (e.g. Beuscart, 2005). In the words of sociotechnical theory, this would be a demonstration of the autonomy granted to groups, and the freedom members of a group have to regulate their own internal states, relating themselves to the wider system. It is not necessarily concerned with planning upfront the nature of organisational structures that should be adopted in different situations. It is instead concerned

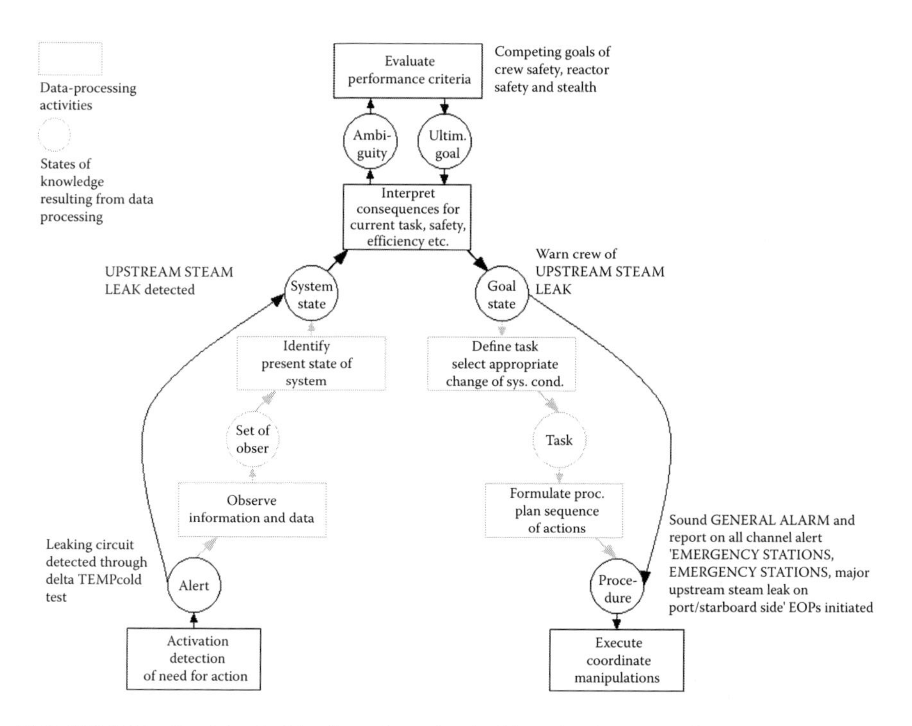

FIGURE 1.15 Decision ladder from detecting leaking circuit to sounding general alarm.

with identifying the set of possibilities for work allocation, distribution and social organisation. SOCA explicitly aims to support flexibility and adaptation in organisations (the sociotechnical principle of 'equifinality'; Bertalanffy, 1950) by developing designs that are tailored to the requirements of the various possibilities (the sociotechnical principle of 'multifunctionality'; Cherns, 1987). Ironically, SOCA is one of the more neglected phases of CWA (most emphasis being given to WDA).

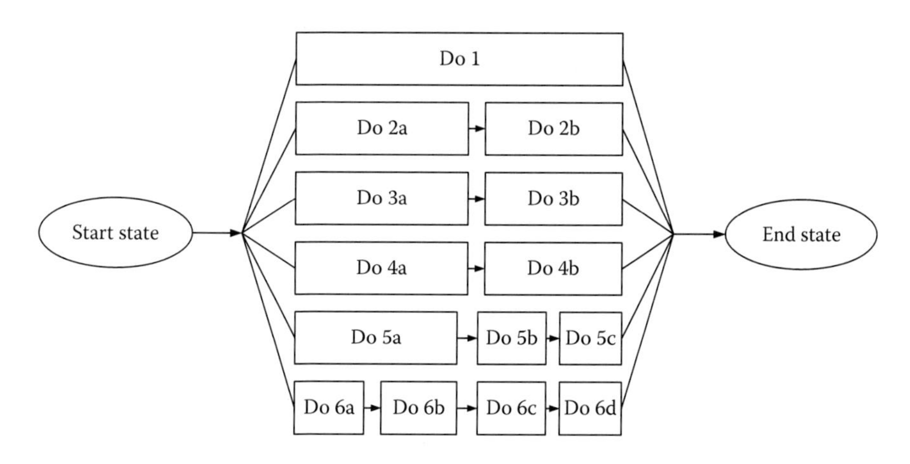

FIGURE 1.16 Strategies analysis.

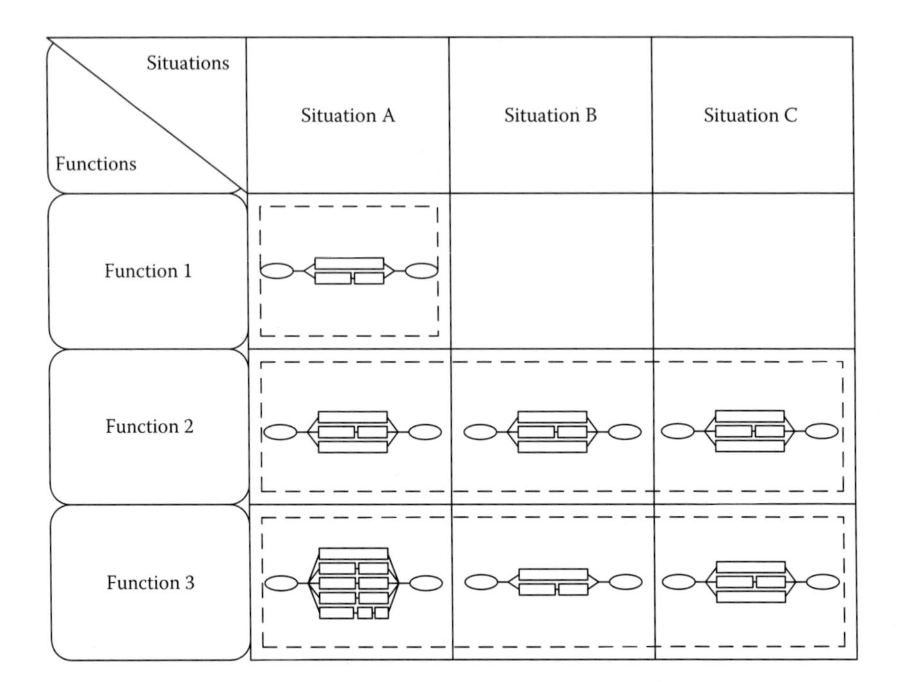

FIGURE 1.17 StrA plotted on the CAT.

Naikar (2006b) discusses the application of SOCA beyond the description of Vicente (1999) and Rasmussen et al. (1994), applying it to the design of new organisational structures. Naikar (2006b) argues that SOCA recognises that in many systems, flexible organisational structures that can be adapted to local contingencies are essential for dealing with unanticipated events. Rather than defining a single or best organisational structure, SOCA is concerned with identifying the criteria that may shape or govern how work might be allocated across agents. Naikar (2006b) goes on to list the criteria that may shape how work is allocated across a system.

- Agent competencies
- Access to information or means for action
- Level of coordination
- Workload
- Safety and reliability

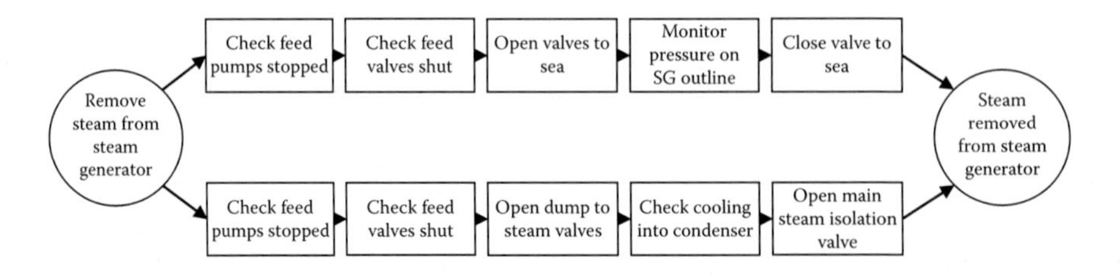

FIGURE 1.18 Course of action for removing steam.

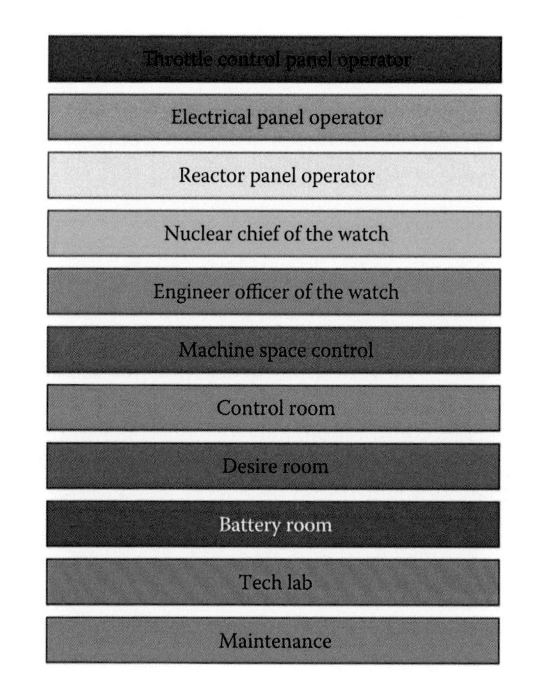

FIGURE 1.19 List of agent roles.

- Regulatory compliance
- Enjoyments
- Availability

The first stage of the process is to define the key agent roles in the system. At this stage, it is important to expound what is meant by agent role; the role reflects the job at any given time but not the person or machine. In environments, such as the submarine, where individuals perform different roles, it is important to avoid confusion. A particular individual may be able to perform a multitude of tasks; however, they may be required to change role in the process.

A list of the key roles and their related coding can be seen in Figure 1.19; the roles are as follows:

- *Throttle control panel operator (TCPO)* – This is the most junior role in the propulsion plant; the main responsibility of the TCPO is controlling the amount of steam entering the main engines.
- *Electrical panel operator (EPO)* – This is a role adopted by an experienced TCPO; the EPO is responsible for balancing the electrical power needs of the submarine, switching between generators, batteries and shore power supplies.
- *Reactor panel operator (RPO)* – The RPO is the most senior of the three operators already discussed; the RPO will have served time as a TPO and an EPO. The RPO is responsible for the running of the reactor as well as the primary and secondary loops.

- *Nuclear chief of the watch (NCOTW)* – The NCOTW works in a supervisory capacity; they oversee and coordinate the work of the TCPO, EPO and RPO. They are capable of assuming any of these roles if required. They provide decision rights to the TCPO, EPO and RPO.
- *Engineering officer of the watch* – Senior to the NCOTW, the engineering officer also works in a supervisory capacity.
- *Machine space controller* – The machine space controller works externally to the propulsion plant; they are able to make direct observations of the system and to make system manipulations that are not possible from the propulsion plant.
- *The control room* – The control room contains the captain of the submarine; they make demands on the propulsion plant for required speeds on the propulsor shaft. The control room may also place constraints on the propulsion plant in terms of steam blowdown.
- *The diesel room* – While surfaced, the diesel room is able to generate electricity in the event of a loss of steam.
- *The battery room* – The battery room is able to provide electricity into the system.
- *The tech lab* – The tech lab contains chemists who can provide information on the composition of the primary coolant; they provide a useful source of information in diagnosing system faults.
- *Maintenance* – Maintenance staff are able to test and repair different parts of the system.

It is possible to map each of the role types onto the existing products in order to show who has the capability of doing what, thus providing a mechanism for informing decisions related to the allocation of function. Using shading, it is possible to colour the existing products developed in the first three phases to show where each of the roles can conduct tasks.

1.3.5.1 The Abstraction–Decomposition Space

Figure 1.20 shows the ADS, from the WDA phase, coded to indicate which of the roles can have an effect on which of the elements. The first thing that is evident from the examination of Figure 1.20 is the high level of redundancy that has been designed into the system. The multicoloured cells indicate that more than one role can have an effect on the particular part of the system. The NCOTW and the engineering officer of the watch have the ability to influence all aspects of the system within the propulsion plant. A number of nodes eternal to the propulsion plant, related to the diesel room or the activities of the tech lab, are outside of their direct control.

The second salient observation is that a number of cells (display plant parameters, contain the fuel, automated reactor protection and the gearbox) are not coded. This is because they are not under direct human control.

1.3.5.2 The Contextual Activity Templates

The allocation of roles in the ADS can be explored in detail by considering specific situations in the CAT. As the ConTA showed, the activities of the propulsion plant

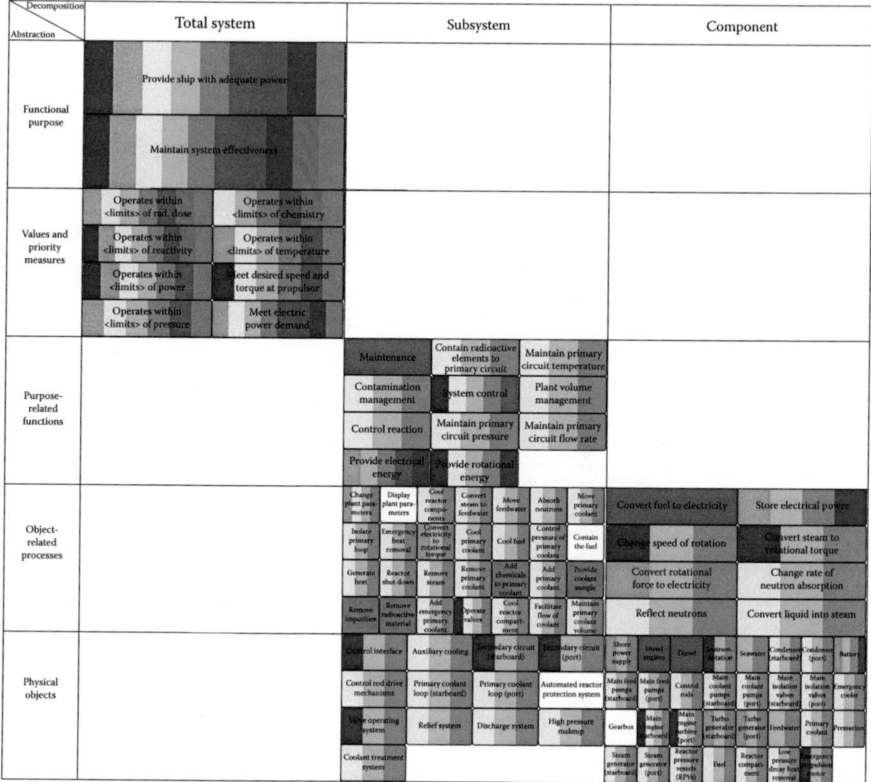

FIGURE 1.20 ADS coded to indicate which roles can have an effect on different elements of the system.

are not heavily influenced by context; therefore, unsurprisingly, the way the roles can perform in different situation is unlikely to change. This highlights the high level of flexibility and redundancy currently embodied in the system. The only notable difference in Figure 1.21 is in 'provide rotational energy' (Figure 1.22).

1.3.5.2.1 The Decision Ladders

The decision ladder for an unidentified steam leak, introduced previously, can also be coded to indicate which of the roles can affect the decision-making process. The diagram highlights a number of salient features: first, the high level of redundancy in the system – no single step of the decision-making process is limited to a single role; the second salient point is that the decision rights remain with the supervisors in the room. The role of the TCPO, EPO and RPO is to provide information and assist in diagnosing the system state; the role of determining the action to be taken lies with the supervisors, and the operators then execute the ordered tasks.

Each of the elements is also associated to individual agent roles in the 'call-out bubbles' using small coded circles. Un-coded information elements relate to

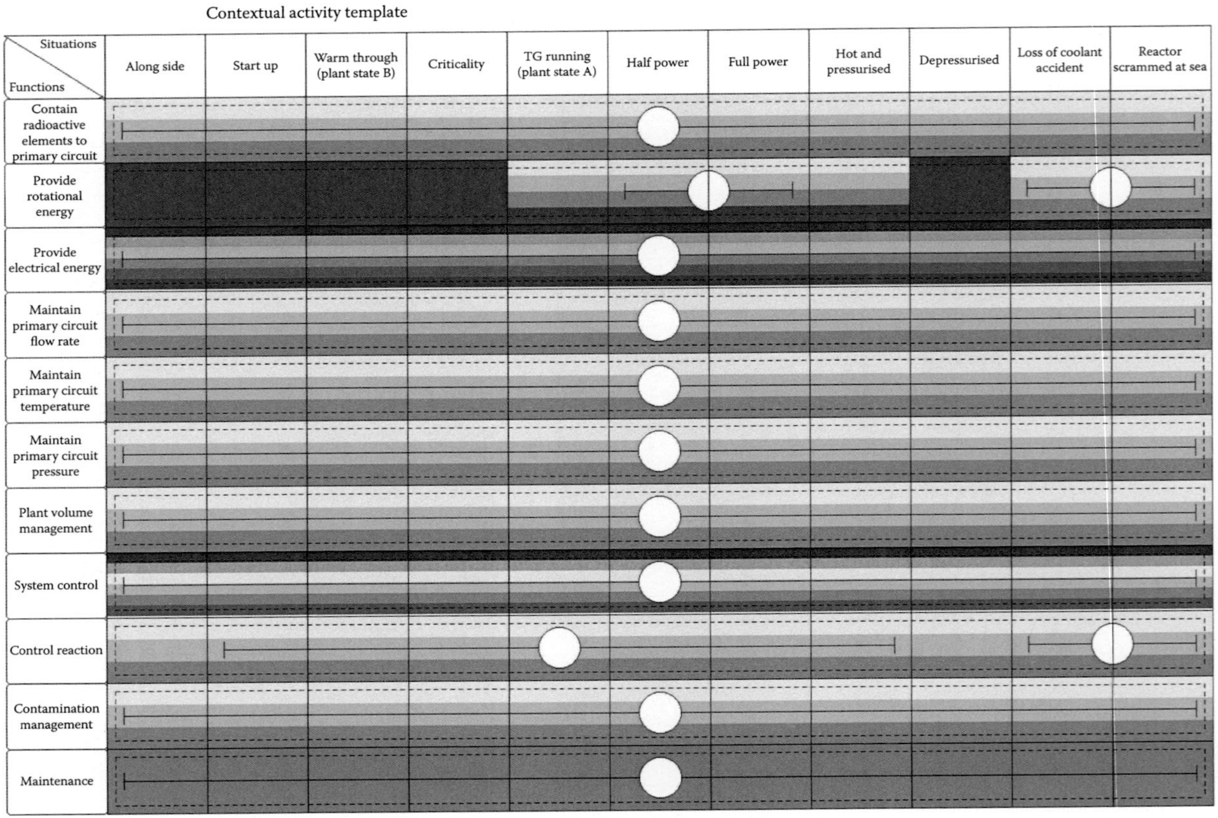

Contextual activity template

Functions \ Situations	Along side	Start up	Warm through (plant state B)	Criticality	TG running (plant state A)	Half power	Full power	Hot and pressurised	Depressurised	Loss of coolant accident	Reactor scrammed at sea
Contain radioactive elements to primary circuit											
Provide rotational energy											
Provide electrical energy											
Maintain primary circuit flow rate											
Maintain primary circuit temperature											
Maintain primary circuit pressure											
Plant volume management											
System control											
Control reaction											
Contamination management											
Maintenance											

FIGURE 1.21 Contextual activity temple coded to indicate which roles can perform activities in which situations (purpose-related functions).

Functions \ Situations	Along side	Start up	Warm up (plant state B)	Criticality	TG run up (plant state A)	Half power	Full power	Hot and pressurised (sub critical)	Depressurised	Loss of coolant accident	Reactor scrammed at sea
Generate heat											
Contain the fuel											
Maintain primary coolant volume											
Facilitate flow of coolant											
Convert liquid into steam											
Isolate primary loop											
Control pressure of primary coolant											
Cool fuel											
Reflect neutrons											
Cool reactor compartment											
Cool primary coolant											
Convert rotational force to electricity											
Convert steam to rotational torque											
Change speed of rotation											
Change electricity to rotational torque											
Move primary coolant											
Emergency heat removal											
Absorb neutrons											
Move feedwater											
Store electrical power											
Convert fuel to electricity											
Change rate of neutron absorption											
Convert steam to feedwater											
Cool reactor components											
Display plant parameters											
Change plant parameters											
Operators valves											
Add emergency primary coolant											
Remove radioactive material											
Remove impurities											
Provide coolant sample											
Add primary coolant											
Add chemicals to primary coolant											
Remove primary coolant											
Remove steam											
Reactor shut down											

FIGURE 1.22 Contextual activity temple coded to indicate which roles can perform activities in which situations (object-related processes).

information elements outside of the propulsion plant. The procedures have not been coded as allocation of function is captured in the relevant EOP documents (Figure 1.23).

1.3.6 Worker Competencies Analysis

As the final phase of the CWA framework, WCA addresses the constraints dictating the possible agent behaviour within different situations. WCA investigates the behaviour required by the humans within the system required to complete tasks. Typically, this behaviour is modelled using Rasmussen's (1983) SRK taxonomy.

According to Kilgore and St-Cyr (2006), the goal of WCA is to identify psychological constraints applicable to systems design. As the final phase of the CWA framework, WCA inherits all requirements identified through the four previous phases. It is important to note that the output of WCA is not a finished design. Instead, the entire CWA process feeds constraints for developing information requirements used in subsequent interface design activities. Kilgore and St-Cyr (2006) go on to say that several theories and models are relevant to identifying the implications of human characteristics in system design (e.g. manual control models, sampling theory, signal detection theory); however, each of these models focuses on specific, narrow psychological traits. Instead, an integrated model is needed to aid designers in deriving practical implications for system interfaces. Vicente (1999) proposes the use of the SRK taxonomy to address this need; Figure 1.24 shows each of the three types of behaviour along with their cues.

According to Vicente (1999), skill-based behaviour (SBB) is performed without conscious attention. SBB typically consists of anticipated actions and involves direct coupling with the environment. SBB is mainly dependent on automatic responses and neuromuscular control. RBB is based on a set of stored rules (if–then) that can be learned from experience or from protocol. Individual goals are not considered; the user is merely reacting to an anticipated event using familiar perceptual cues. Unlike SBB, users can verbalise their thoughts, as the process is cognitive. When decisions are made that consider explicitly the purpose or goal of the system, the behaviour can be considered knowledge-based behaviour (KBB). KBB is slow, serial and effortful because it requires conscious, focal attention. According to Vicente (1999), interfaces for complex sociotechnical systems should encourage SBB and RBB, while allowing for operators' seamless transition to KBB during problem solving.

Tasks rarely fall into one class of behaviour; the three levels interact. Novice users may find that they start to reason at a knowledge-based level considering each task they carry out. Through experience, this KBB can turn to RBBs as familiar perceptual cues are recognised. According to Vicente (1999), the outcome of a WCA should be a number of detailed implications for interface design and training, and using knowledge of the signs, signals and symbol distinctions, information can be presented in a form that will make it more likely that particular levels of cognitive control will be activated. As with the strategies and SOCA components, the WCA component has received only minimal attention to date.

To support this phase, Kilgore and St-Cyr (2006) developed an SRK inventory (see Figure 1.25); the conventional table used to capture SRK information (shown in white) is accompanied by additional boxes to capture information about the specific

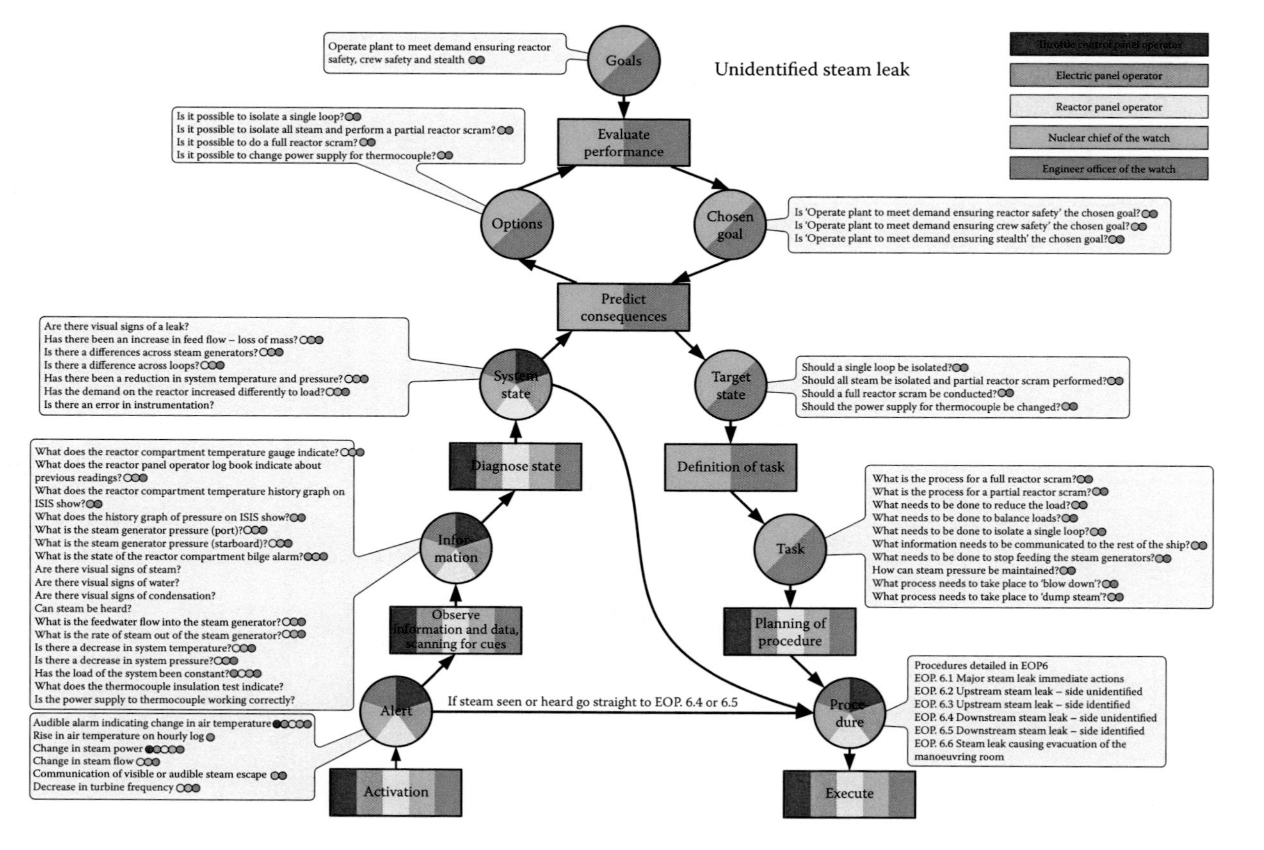

FIGURE 1.23 Decision ladder coded to indicate which roles can perform different elements of the decision process.

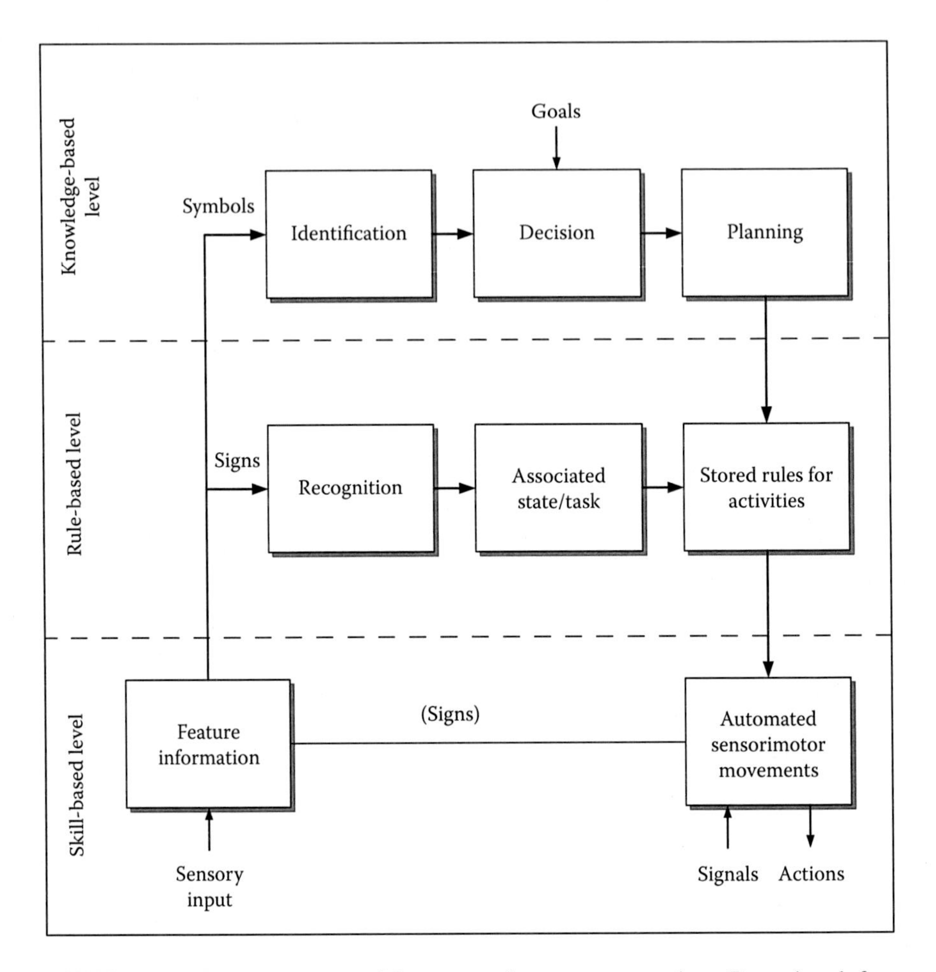

FIGURE 1.24 SRK taxonomy of human performance categories. (Reproduced from Vicente, K. J. 1999. *Cognitive Work Analysis: Toward Safe, Productive, and Healthy Computer-Based Work*. Mahwah, NJ: Lawrence Erlbaum Associates.)

activity taking place (shown in grey). The cells within the first two columns describe the information-processing activities and resultant knowledge states. The population and organisation of the rows within these columns can be informed directly by decision ladders generated during ConTA activities, thus creating a direct link between the outputs of the WCA and previous CWA phases. Kilgore and St-Cyr (2006) argue that if there are sections of the table that are incomplete, then the interface is not performing as well as it could be. An empty cell in the knowledge column indicates that the interface may not provide an operator with sufficient resources to engage in effective troubleshooting, or problem solving, for a particular information-processing step. Kilgore and St-Cyr (2006) go on to point out that completed cells in the SRK inventory can also be used to generate profiles of competencies that operators must possess to perform control tasks adequately. In this manner, the SRK inventory can be used to inform worker selection and training. The diagram proposed by

	Information processing step	Resultant state of knowledge	Skill-based behaviour	Rule-based behaviour	Knowledge-based behaviour
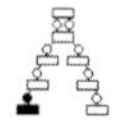	Audible alarm heard	Audible alarm indicating change in air temperature	The sound of the alarm is automatically associated with the air temperature gauge	The sound of the alarm is associated with an unknown gauge, the gauges are then scanned to find the alarming gauge	The sound is unknown but associated with a need for action, through systematic diagnosis the cause of the alarm is identified
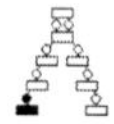	Observation of the air temperature gauge	Rise in air temperature on hourly log	Upon inspection the current value is instantly recognised as high	When compared to the previous readings the current value is recognised as high	The reading is compared with prevous reading and put in context of other system operations
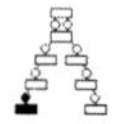	Observation of gauges relating to steam power	Change in steam power	Without direct relation to a single dial the operator is aware of a reduction in steam power	Whilst monitoring a single measure an operator is aware of a loss in steam power	Combining a series of readings, and through a systems understanding a loss of steam power is diagnosed
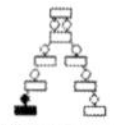	Observation of gauges relating to steam flow	Change in steam flow	Without direct relation to a single dial the operator is aware of a reduction in steam flow	Whilst monitoring a single measure an operator is aware of a loss in steam flow	Combining a series of readings, and through a systems understanding a loss of steam flow is diagnosed
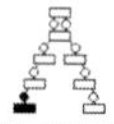	Communication recieved	Communication of visible or audible steam escape	A communication is delivered specifically indicating an observation of steam	A communication delivered indicating secondary effects of steam (condensation) along with location that corresponds to operator's model	A communication delivered indicating secondary effects of steam (condensation) is combined with location and understanding of the system
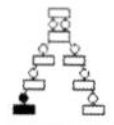	Observation of gauges relating to turbine frequency	Decrease in turbine frequency	Without direct relation to a single dial the operator is aware of a reduction in turbine frequency	Whilst monitoring a single measure an operator is aware of a loss in turbine frequency	Combining a series of readings, and through a systems understanding a loss of turbine frequency is diagnosed
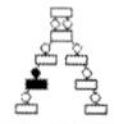	Observation of the reactor compartment temperature gauge	What does the reactor compartment temperature gauge indicate?	The operator glances at the gauge and understands if it shows an abnormal value	The operator glances at the gauge and the specific value it points to by applying heuristics he understands if the gauge is showing an abnormal value	The operator studies the gauge, value and limits considering them in terms of wider system parameters to diagnose if the value is abnormal
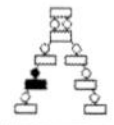	Study of the RPO log book	What does the reactor panel operator log book indicate about previous readings?	The operator scans the log, identifying patterns and relationships between past values	The operator scans the log, applying heuristics to establish patterns and relationships between past values	The operator considers the log readings in relationship to their model of the system, past system states and steps through ruling out hypotheses for differences
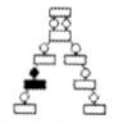	Study of the reactor compartment temperature history graph on ISIS	What does the reactor compartment temperature history graph on ISIS show?	The operator scans the graph identifying trends and relating them to the system performance	The operator scans the graph and applies learnt heuristics to identify trends related to performance	The operator studies the graph and diagnoses changes in relation to an understanding of the system
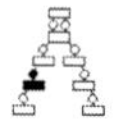	Study of the pressure history graph on ISIS	What does the history graph of pressure on ISIS show?	The operator scans the graph identifying trends and relating them to the system performance	The operator scans the graph and applies learnt heuristics to identify trends related to performance	The operator studies the graph and diagnoses changes in relation to an understanding of the system
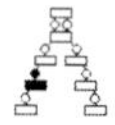	Observation of gauges relating to steam generator pressure	What is the steam generator pressure (port)? What is the steam generator pressure (starboard)?	The operator glances at the gauge and understands if it shows an abnormal value	The operator glances at the gauge and the specific value it points to by applying heuristics he understands if the gauge is showing an abnormal value	The operator studies the gauge, value and limits considering them in terms of wider system parameters to diagnose if the value is abnormal
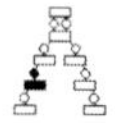	Study of the bilge alarm panel	What is the state of the reactor compartment bilge alarm?	The operator glances at the gauge and understands if it shows an abnormal value	The operator glances at the gauge and the specific value it points to by applying heuristics he understands if the gauge is showing an abnormal value	The operator studies the gauge, value and limits considering them in terms of wider system parameters to diagnose if the value is abnormal
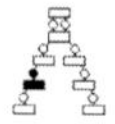	Send crew to directly observe suspected sites for signs of steam	Are there visual signs of steam, water, condensation? Can steam be heard?	Upon inspection cues are automatically recognised as a steam leak	Heuristics are applied to diagnose cues as a steam leak	A structured process of considering all other causes of the cues is put in place to determine if there is a steam leak
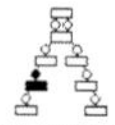	Observation of gauges relating to feedwater flow into the steam generator	What is the feedwater flow into the steam generator?	The operator glances at the gauge and understands if it shows an abnormal value	The operator glances at the gauge and the specific value it points to by applying heuristics he understands if the gauge is showing an abnormal value	The operator studies the gauge, value and limits considering them in terms of wider system parameters to diagnose if the value is abnormal
	Observation of gauges relating to steam flow out of the steam generator	What is the rate of steam out of the steam generator?	The operator glances at the gauge and understands if it shows an abnormal value	The operator glances at the gauge and the specific value it points to by applying heuristics he understands if the gauge is showing an abnormal value	The operator studies the gauge, value and limits considering them in terms of wider system parameters to diagnose if the value is abnormal
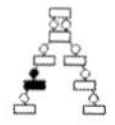	Observation of current value of gauges, compared with historical values stored in memory or on logs	Is there a decrease in system temperature/pressure?	The operator glances at the gauge and understands if it shows an abnormal value	The operator glances at the gauge and the specific value it points to by applying heuristics he understands if the gauge is showing an abnormal value	The operator studies the gauge, value and limits considering them in terms of wider system parameters to diagnose if the value is abnormal
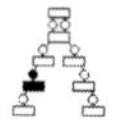	Observation of current load, and power within the system	Has the load of the system been constant?	The operator glances at the gauge and understands if it shows an abnormal value	The operator glances at the gauge and the specific value it points to by applying heuristics he understands if the gauge is showing an abnormal value	The maintainance crew studies the gauge, value and limits considering them in terms of wider system parameters to diagnose if the value is abnormal
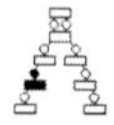	Test of the thermocouple insulation by maintanace crew	What does the thermocouple insulation test indicate?	The maintainance crew glances at the gauge and understands if it shows an abnormal value	The maintainance crew glances at the gauge and the specific value it points to by applying heuristics he understands if the gauge is showing an abnormal value	The sound of the alarm is automatically associated with the air temperature gauge
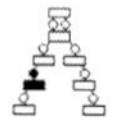	Test of the thermocouple power supply by maintanace crew	Is the power supply to thermocouple working correctly?	The operator inspects the thermocouple, quickly identifying if it is working correctly	The operator inspects the thermocouple, systematically stepping through a known procedure of tests to diagnose if it is working	The operator uses a fundamental understanding of the component along with testing equipment to diagnose if it is working

FIGURE 1.25 SRK inventory for steam leak (alerts and information).

Kilgore and St-Cyr (2006) has been extended to show graphically the processing steps on mini decision ladders.

Each of the process-related functions can also be considered in relation to the required skill-based, rule-based and knowledge-based understanding (see Figure 1.8). As stated previously, RBB can be taught and learnt by rote and does not require a comprehensive understanding of the system. KBB relies on a much more detailed understanding of the system. In summary, it is normally necessary to educate for KBB and train for RBB. The individual usually then compiles this understanding into SBB based upon practice and experience (Table 1.7).

1.3.7 Section Summary

This section has identified the need for a framework capable of modelling complex sociotechnical systems; CWA has particular applicability to product designers, and to this project, as it allows the modelling of 'first of a kind' systems. The WDA phase, in particular, leads the designer to focus, in a structured way, on the reason for developing the system. The framework encourages the designer to work based on the constraints of the system, while focusing on the key criteria by which the system will be evaluated. The process of considering the functions in abstract terms allows for creative thinking and problem solving. This encourages the designer to consider the need they are addressing, rather than jumping straight into solving the problem. The different phases and tools within the CWA framework can be applied throughout the design life cycle. CWA applications have previously been used for purposes ranging from the design of novel systems to the analysis of operational systems. This section has highlighted the differences between formative approaches, such as CWA and normative approaches; it is contended that CWA is better equipped to cope with the levels of flexibility within complex sociotechnical systems. According to Naikar (2008), the highly dynamic nature of network-centric warfare (NCW) means that the goals and work requirements of participating systems, and the humans operating those systems, are frequently changing – often in ways that were not, or even cannot be, anticipated by engineering designers or by professional, trained workers. Furthermore, rapid developments in the information and communication technologies that enable NCW means that the nature of human work is continually evolving. Standard techniques for work or task analysis are not well suited for modelling activity in network-centric operations. Instead, a formative approach, such as CWA, is necessary. Lintern et al. (2004) state that the constraint-based approach of CWA is one that can cope comfortably with the scale-up problem. Indeed, the potential benefits from a CWA grow as systems become larger, more technologically sophisticated and more complex.

CWA is not without its limitations; its formative nature, the great strength that separates it from other methods, can also be perceived as its weakness. CWA is notoriously non-prescriptive, which has implications for its ability to be taught and understood. According to Fidel and Peijtersen (2005), because CWA investigates information behaviour in context, individual studies create results that are only valid for the design of information systems in the context investigated, rather than for the design of general information systems. Results from a variety of studies, however, can be combined together and generalised to inform the design of other information

TABLE 1.7
SRK Table for Process-Related Functions

	Skill	Rule	Knowledge
Contain radioactive elements to primary circuit	Isolate leaking area	A heuristic-based understanding to isolate the cause of any leak by shutting down isolation valves and secondary circuits connected to it	A detailed understanding of the route the primary circuit takes, along with the effect on the system of a leak. An ability to diagnose leaks and identify their location based upon system parameters
Provide rotational energy	Open and close throttle in response to control room demand. Alternatively, select electric motor, as conditions require	Drive main engines, if steam not available for both engines, isolate one and run one, if steam still insufficient, use electric motor	A fundamental understanding of thermodynamics to determine the amount of steam the system can produce in its current configuration, an understanding of how the system can be manipulated to increase steam. An understanding of the most efficient way to use the available steam
Provide electrical energy	Select batteries, shore line or turbo generator as conditions require	An understanding of the methods of generating electrical power. A heuristic-based model of what source to use in what context	A detailed understanding of the present and future state of the system. Balancing current requirements with anticipated future power requirements
Maintain primary circuit flow rate	Ensure pumps are running and valves are open	An understanding that the primary circuit flows around two loops and that each loop can be independently isolated to ensure that coolant flows. An understanding that if flow rate drops it may be necessary to partially scram the reactor	An understanding of fluid dynamics and thermodynamics. An appreciation of the flow rates required to cool the reactor in different situations. An understanding of the implications of running the system with limited cooling in specific situations
Maintain primary circuit temperature	Controlling core temperature; running steam generators	An understanding that the primary circuit temperature is influenced by the core temperature and the amount of steam being generated. An understanding that the primary circuit flows around two loops and that each loop can be independently isolated to ensure that coolant flows. An understanding that if flow rate drops it may be necessary to partially scram the reactor	An understanding of fluid dynamics and thermodynamics. An appreciation of the flow rates required to cool the reactor in different situations. An understanding of the implications of running the system with limited cooling in specific situations. A full understanding of the relationship between temperature, pressure and volume. An understanding of the required changes in volume and pressure to maintain a desired temperature

(*Continued*)

The higher levels of abstraction in the AH (see Figure 1.4) provide some description for the CONOPS. The WDA phase can help answer questions about why a new system should exist, what functions it should implement and what physical devices are necessary. As the upper levels of the AH consider the purpose of the system, independently of any technology, the system can be considered in a revolutionary way. Thinking bottom-up, by considering the available technology and its affordances, existing technology can be considered independently of its current purpose allowing systems to be diversified to perform new objectives.

Figure 1.26 illustrates the top-down and bottom-up approaches to giving consideration to new concepts of operations. For example, inserting a new technology at the bottom level of the AH will give rise to new physical functions, which will contribute to generalised functions, priorities and values, and ultimately the functional purpose of the system. Alternatively, developing new functional purposes for the system will lead to new values and priorities and new generalised functions. These, in turn, will call for new physical functions with will need to be met by the (new) technologies.

The functional purposes ('provide ship with adequate power' and to 'maintain system effectiveness') provide a concise description of why the system exists. The means–ends links map the relationships between the physical objects at the base of the hierarchy and the purpose at the top (see Figure 1.5).

1.3.10 Function Analysis

According to MAP-01-011 (2006), the function analysis and the project-specific target audience description (PSTAD) provide information to allow functions to be allocated between personnel and equipment. However, this definition has considerable overlap with function allocation. For the purposes of this chapter, function analysis will be considered independent of any agent role, be it human or non-human.

As it is independent of any specific agent or scenario, the initial phase, WDA, is the most concise in informing function analysis; it relates the physical affordances of objects in the system with high-order functional purposes. These relationships are shown graphically within the AH means–ends links (see Figure 1.5). The data within this representation can be used to consider the domain within a clearly defined boundary. As previously stated, the representation is independent of agent and situation, meaning that it remains applicable for varying system design, expected events, crew allocation and training design. Figure 1.27 shows the functional analysis of the total system, subsystems and components.

1.3.11 Target Audience Description

According to MAP-01-011 (2006), user characteristics are described in the TAD. Within the context of procuring royal naval systems, there are two primary forms of TAD: the royal navy generic target audience description (RNGTAD) and the PSTAD.

To avoid unnecessary duplication of effort and cost associated with repeated development of TADs, Security Services Group (SSG) has commissioned the production

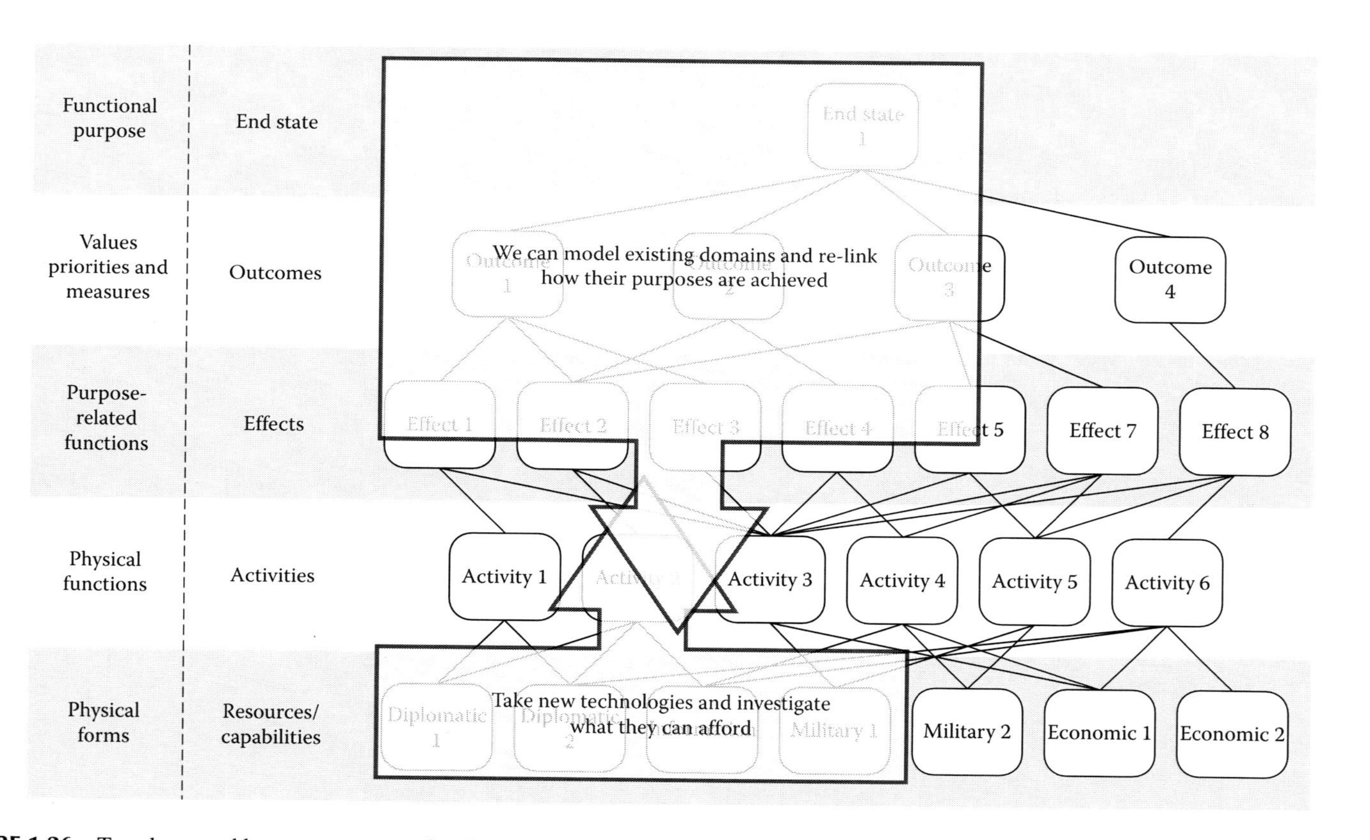

FIGURE 1.26 Top-down and bottom-up approaches for considering new concepts of operations.

Abstraction \ Decomposition	Total system	Subsystem	Component
Functional purpose	Provide ship with adequate power Maintain system effectiveness		
Abstract function/values and priority measures	Operates within <limits> of rad. dose Operates within <limits> of chemistry Operates within <limits> of reactivity Operates within <limits> of temperature Operates within <limits> of power Meet desired speed and torque at propulsor Operates within <limits> of pressure Meet electric power demand		
Generalised function/purpose-related functions		Maintenance Contain radioactive elements to primary circuit Maintain primary circuit temperature Contamination management System control Plant volume management Control reaction Maintain primary circuit pressure Maintain primary circuit flow rate Provide electrical energy Provide rotational energy	
Physical function/object-related processes		Change plant parameters Display plant parameters Cool reactor components Convert steam to feedwater Move feedwater Absorb neutrons Move primary coolant Isolate primary loop Emergency heat removal Convert electricity to rotational torque Cool primary coolant Cool fuel Control pressure of primary coolant Contain the fuel Generate heat Reactor shut down Remove steam Remove primary coolant Add chemicals to primary coolant Add primary coolant Provide coolant sample Remove impurities Remove radioactive material Add emergency primary coolant Operate valves Cool reactor compartment Facilitate flow of coolant Maintain primary coolant volume	Convert fuel to electricity Store electrical power Change speed of rotation Convert steam to rotational torque Convert rotational force to electricity Change rate of neutron absorption Reflect neutrons Convert liquid into steam
Physical form/physical objects		Control interface Auxiliary cooling Secondary circuit (starboard) Secondary circuit (port) Control rod drive mechanisms Primary coolant loop (starboard) Primary coolant loop (port) Automated reactor protection system Valve operating system Relief system Discharge system High pressure makeup Coolant treatment system	Shore power supply Diesel engines Diesel Instrumentation Seawater Condenser (starboard) Condenser (port) Battery Main feed pumps (starboard) Main feed pumps (port) Control rods Main coolant pumps (starboard) Main coolant pumps (port) Main isolation valves (starboard) Main isolation valves (port) Emergency cooler Gearbox Main engine (starboard) Main engine turbine (port) Turbo generator (starboard) Turbo generator (port) Feedwater Primary coolant Pressuriser Steam generator (starboard) Steam generator (port) Reactor pressure vessels (RPVs) Fuel Reactor compartment Low pressure decay heat removal Emergency propulsion motor

FIGURE 1.27 Functional analysis of the system, subsystems and components.

of a generic reference document. This is called the RNGTAD and is intended for use by those involved with the design and procurement of naval capability, including integrated project teams (IPTs), naval staff, prime contractors and so on (p. 6-3).

According to MAP-01-011 (2006, p. 2-13), the PSTAD typically comprises the following:

- Rank, rate and branch
- Body size and strength (anthropometric) characteristics
- Physical skills
- Knowledge and mental skills
- Educational background
- Experience
- Summary of training objectives (operational performance statement [OPS] and training performance statement [TPS]) and career progression

An understanding of the training level can be gained from the WCA phase of the CWA framework (see Table 1.8).

As a rough heuristic, the RBBs can be taught training, whereas the KBBs need to be taught through education.

TABLE 1.8
Extract from Table 1.7

	Skill	Rule	Knowledge
Contain radioactive elements to primary circuit	Isolate leaking area	A heuristic-based understanding to isolate the cause of any leak by shutting down isolation valves and secondary circuits connected to it	A detailed understanding of the route the primary circuit takes, along with the effect on the system of a leak. An ability to diagnose leaks and identify their location based upon system parameters

1.3.12 Normal and Emergency Operations

The initial phase of CWA models the system independent of any particular situation. It is not until the second phase, ConTA, that the constraints that govern how the system can perform in different situations are presented. Within the ConTA phase, the CAT can be used to explore normal and emergency operations in detail. In Figures 1.6 and 1.7, a number of normal situations are considered along with two emergency situations; loss of coolant accident and reactor scrammed at sea.

From the example shown in Figure 1.28, the normal and emergency operations are shown in terms of the functions that are performed and the situations that they are performed in. The analysis shows the situations in which the functions could be performed (as indicated by the dotted box) and the situations in which the functions are normally performed (the ball and whiskers). This representation prompts the analyst to challenge the normative activities, that is, why functions are performed in those situations? This contrast between the dotted boxes and the ball-and-whisker plots is useful to question the way in which activities are normally done. It would be easy to compare different models of the activity using this form of representation.

The 'activity analysis in decision-making terms' can be used to explore either normal or emergency operations in more detail, as indicated in Figure 1.29. This example looks at the decisions involved in exploring an upstream steam leak. This fine-grained analysis would be used only for the most important activities, where a thorough understanding was required. This understanding could be used to generate procedures and consider error potential.

1.3.13 Allocation of Function

The social organisation and cooperation phase is the best-suited phase to addressing the allocation of function, rather than describing who does or should have an effect on the system; the representations used within this phase model which of the agents can have an effect on the system, thus describing the social constraints.

The SOCA phase of CWA can be used to describe agent roles, including the allocation of function and coordination structures. This phase is built upon the results from the ConTA and StrA phases of CWA. These are grounded with the WDA, which contains context for the functions.

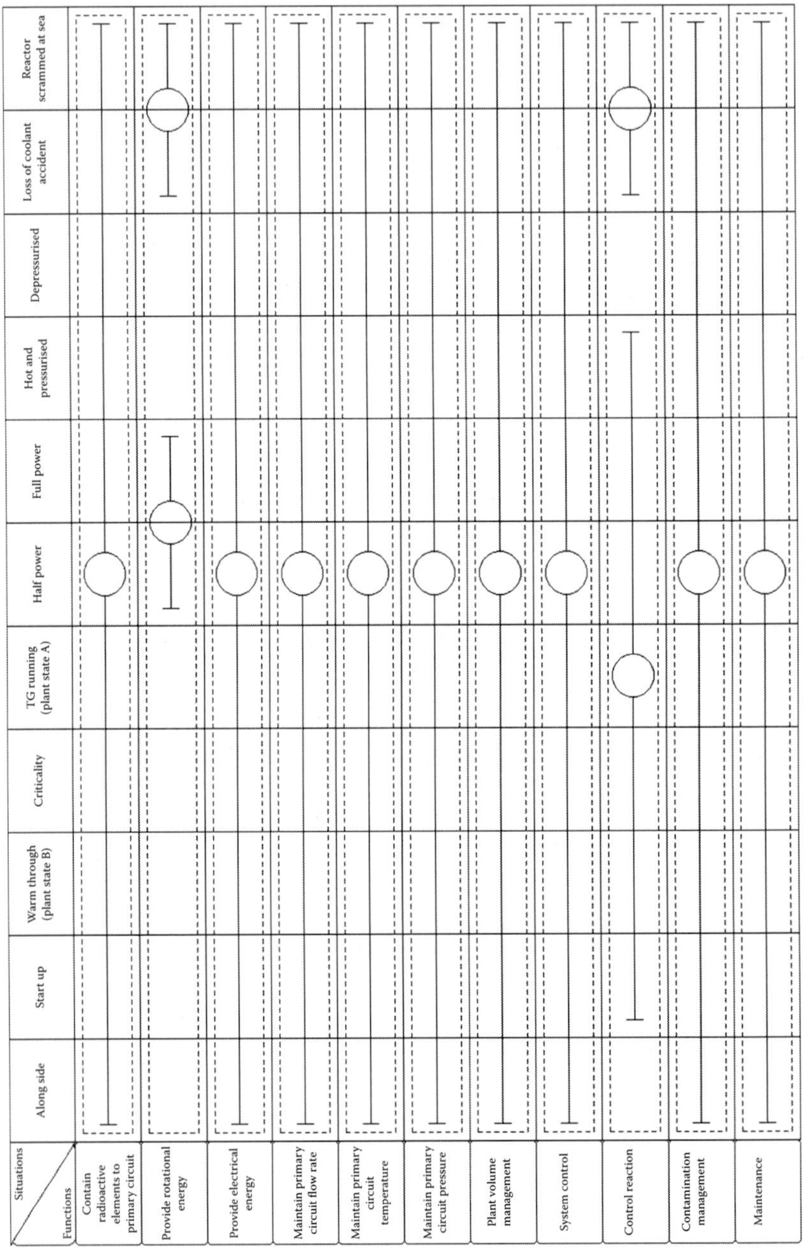

Situations / Functions	Along side	Start up	Warm through (plant state B)	Criticality	TG running (plant state A)	Half power	Full power	Hot and pressurised	Depressurised	Loss of coolant accident	Reactor scrammed at sea
Contain radioactive elements to primary circuit											
Provide rotational energy											
Provide electrical energy											
Maintain primary circuit flow rate											
Maintain primary circuit temperature											
Maintain primary circuit pressure											
Plant volume management											
System control											
Control reaction											
Contamination management											
Maintenance											

FIGURE 1.28 CAT analysis of normal and emergency operations.

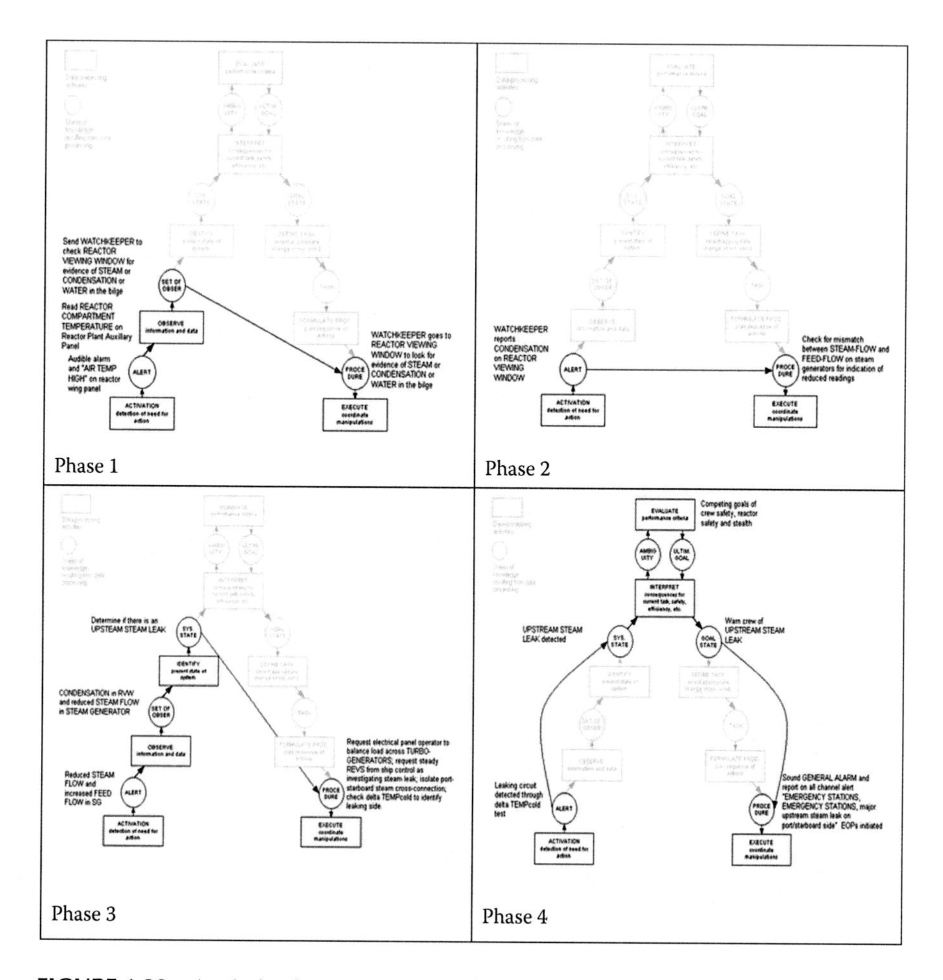

FIGURE 1.29 Analysis of emergency operations.

The models presented in Figures 1.20 to 1.23 (shown in Figure 1.30) have been coded based upon the constraints described by current operational descriptions and mandated staffing requirements. If these constraints were considered changeable, the representation could be used to postulate new manning function allocations and directly compare them with the status quo. For example, removal of one of the agents will highlight effects on delivering the system functions.

Throughout these representations, it is apparent that the roles of the NCOTW (shown in green) and the engineering officer of the watch (shown in blue) have identical capability. Both roles work in a supervisory capacity and have the potential to affect the same functions. The observation may warrant deeper study to establish the need for both of these roles.

The final phase of the framework, WCA, can be used to explore the competencies required for different functions. By breaking down the activities into skill-, rule- and knowledge-based reasoning, further insights can be gained. For example, if, in a

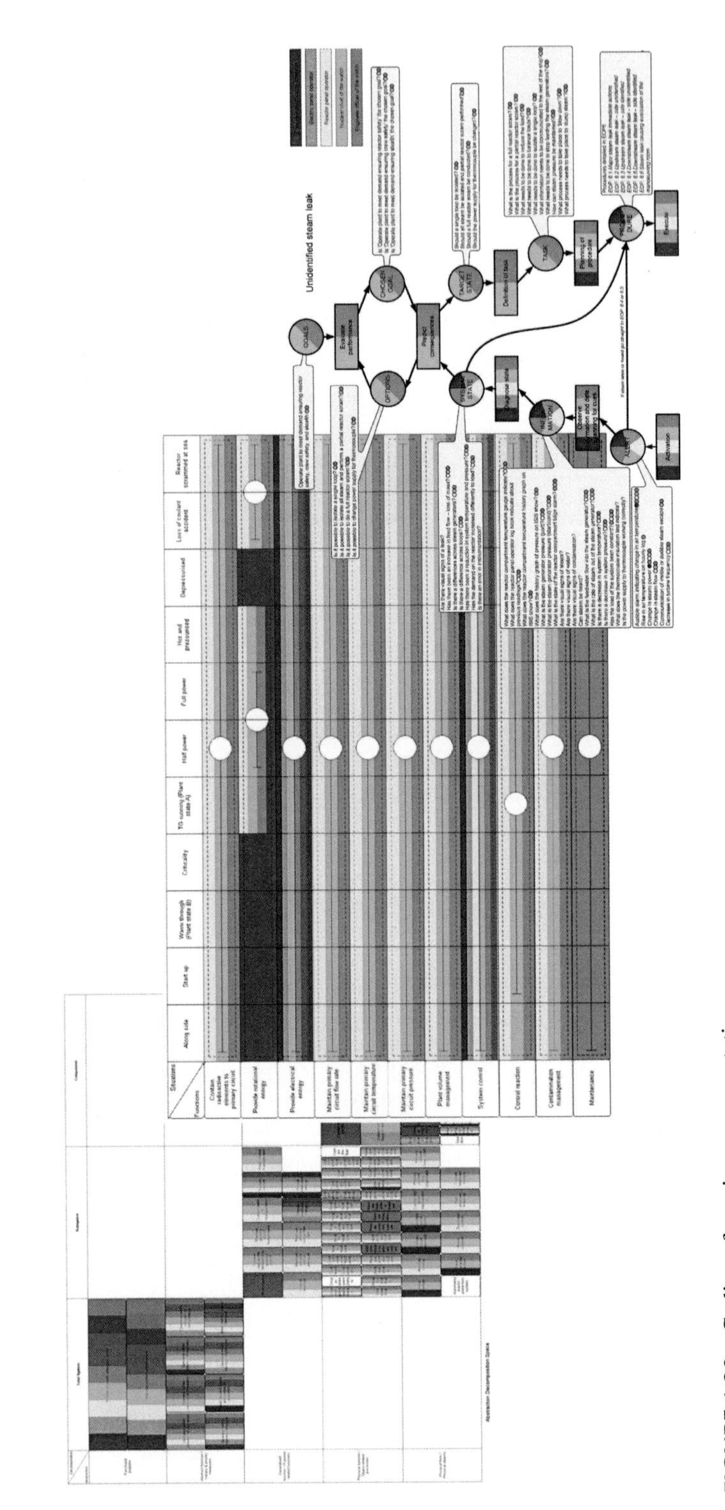

FIGURE 1.30 Coding of previous representations.

given situation, a function requires KBB, based upon a comprehensive understanding of the system and thermodynamics it may require an operator to have a high level of formal education. Alternatively, it may be sufficient to have an operator perform the role under supervision.

Due to the focus on constraints, the approach realises that allocation of function needs not be static. Dynamic allocation of function between people and automation can be considered based upon these representations by assigning agents (either people or automation) to the SOCA-CAT (the function–situation matrix coded by the SOCA) to indicate which agent (or set of agents) is responsible for that function in any given situation. Multiple assignments of agents may be used to indicate that any of the agents may perform the function. It may be necessary to go into a richer level of detail as afforded by the SOCA-DL (the decision ladder coded by the SOCA) to show the dynamics of the allocation in decision-making terms. Alternative allocations of function assignments for SOCA-CAT and SOCA-DL may be compared in an explicit manner, to consider the implications on workload and taskings.

1.3.14 Workload Assessment

The SOCA phase is able to provide an insight into workload assessment. As the approach is formative, it does not describe the exact workload of a given agent at a given time – this is believed to be unpredictable. What the approach can do is to provide a description of all of the activities an agent may be called to perform in a given situation along with which other agents can assist.

Table 1.9 shows a tally of the activities each role can be conducting in a given situation. From the examination of Table 1.9, it is clear that the TCPO has a limited

TABLE 1.9
Number of Object-Related Processes Each Agent Is Associated within Given Situations (Summed from Figure 1.23)

	Normal Situations									Emergency	
	Alongside	Start-Up	Warm-Up	Criticality	TG Run-Up	Half Power	Full Power	Hot and Pressurised	Depressurised	Loss of Coolant Accident	Reactor Scrammed at Sea
TCPO	1	1	1	3	3	3	3	1	1	1	2
EPO	3	3	3	4	4	4	4	4	3	3	3
RPO	19	19	19	23	24	24	24	23	21	23	24
NCOTW	20	20	20	27	28	28	28	22	22	24	27
Engineer	20	20	20	27	28	28	28	22	22	24	27
Diesel room	1	1	1	2	2	2	2	2	1	1	2
Tech lab	4	4	4	4	4	4	4	4	4	4	4
	68	68	68	90	93	93	93	78	74	80	89

remit in terms of the function he or she is called to contribute to; likewise, the EPO also has a limited remit. Comparably, the RPO, NCOTW and engineer all have the potential to contribute to more functions in a given situation. As previously stated, there is no definite relationship between workload and the number of functions a role can influence; one single function may demand a greater workload than a dozen others. However, Table 1.9 does provide a summary of the diversity required in a role that may prove to be a useful indicator to workload.

1.3.15 Training Needs Analysis

As the framework describes the work within the domain, it has great potential to inform training.

Training design can be informed by the SOCA phase. This phase can be used to provide lists of agents that can be involved with a function in a given situation. By considering which agents can influence a function, working practice can be considered in terms of data/information sharing, decision rights and type of shared data required. In turn, this understanding can inform decisions relating to collaborative and cooperative training, distributed training and part training. Each agent's training can be configured based upon an appropriate mix of training alone, in teams, in real world and in synthetic environments. For example, as expected, the RPO will be required to conduct a proportion of his or her training alone, learning the location and function of the available controls along with developing an understanding of the plant under his or her direct control. Examination of the functions the RPO conducts within the coded CAT (Figures 1.21 and 1.22) shows that every function the RPO is involved in, the NCOTW and the engineering officer of the watch are also involved with. A small number of functions are also shared with the EPO and the TCPO. As a result, it is highly likely that some kind of collaborative training would be required.

Collaborative training requirements can be further explored by examining the means–ends links in the AH (see Figure 1.4). For clarity, the relationships between the 'object-related processes' and the 'purpose-related functions' have been captured in matrix form in Figure 1.31. The understanding of the processes influencing a function has the potential to inform the way it is trained.

1.3.16 Job Design

Job design is an extension of allocation of function. Once functions have been assigned to roles, it is a case of concisely defining the requirements placed upon individuals. For this reason, the allocation of functions described in Section 1.3.13 is also applicable here. A more detailed description can also be gained from Section 1.3.6. The WCA phase addresses the level of skill required by the human operator dependent on the system constraints and configuration.

Hackman and Oldham (1980) identified five critical aspects to the design of jobs, namely, skill variety (the range of skills the work requires), task identity (the visibility of the outcome from the work), task significance (the perceivable impact of the work on others), autonomy (the degree of discretion in scheduling work and how it is undertaken) and feedback (information on the effectiveness of the worker's efforts).

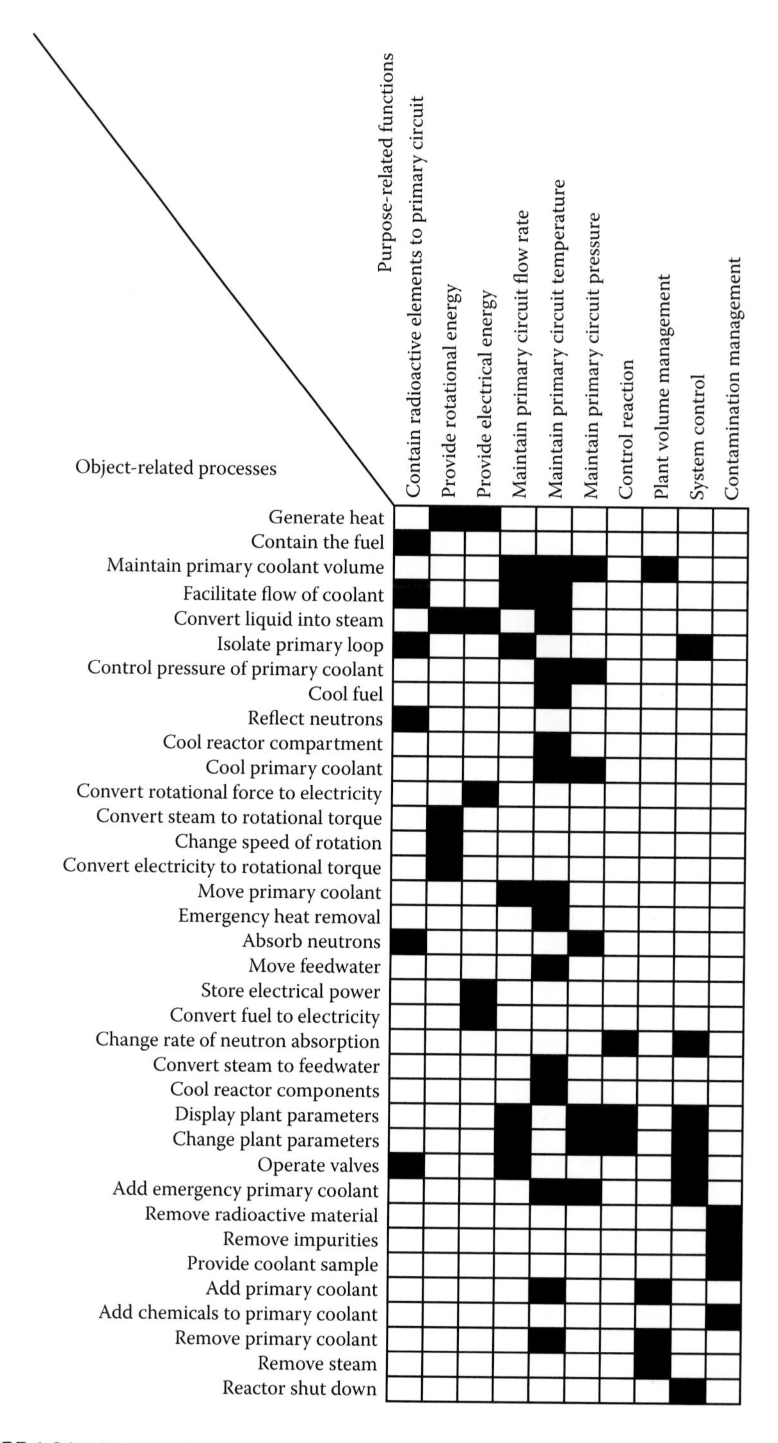

FIGURE 1.31 Relationship between object-related processes and purpose-related functions (black cell indicates a relationship).

According to the theory of the core job characteristics model, more meaningful work will rate high on the five dimensions. Some of these dimensions may be assessed using the products of the CWA. For example, skill variety may be assessed by looking at the number of functions and situations an agent's role is involved in. The SOCA-CAT in Figure 1.32 shows that the throttle panel operator (coloured in red) has very few functions assigned, indicating a low level of skill variety in that particular role.

Task identity for the throttle panel operator should be high, because there is a very visible outcome on the panel in front of the operator and in the change to the speed of the submarine – but CWA has no means to assess this.

Task significance may be assessed by looking at the relationship between the functions and the values and priorities of the system from the AH. More task significance would be indicated by being related to a greater number of system values and priorities. By way of an example of task significance, Figure 1.33 traces the links between individual processes conducted by the TCPO and the functional purpose. As the figure shows, the throttle panel operator's functions link to three of the eight values and priorities, suggesting a lower level of task significance.

Autonomy may be assessed by considering the social relationships in the SOCA-CAT (see earlier text). Again the throttle panel operator is supervised by others affording little opportunity for autonomy. Finally, feedback could be considered by looking again at SOCA-CAT (for feedback from others), which would appear to be high in the case of the throttle panel operator. As with task identity, there is no means of assessing feedback from the instrumentation to the throttle panel operator.

In summary, the analysis of the core job characteristics model used the SOCA-CAT and AH as the principle sources of data. These analyses were not complete, and other data sources would be required to complete the assessment of job design.

1.3.17 Operability Evaluation

Other than raising domain understanding, the CWA framework does not directly inform operability evaluation.

1.3.18 Maintainability Evaluation

Other than raising domain understanding, the CWA framework does not directly inform maintainability evaluation.

1.3.19 Operating Procedures Specification

In specification of operating procedures, the most important considerations are task complexity and user familiarity. Marsden (1996) suggests that low-complexity tasks with high operator familiarity require checklist-type procedures whereas high-complexity tasks with low operator familiarity require step-by-step procedures. Marsden (1996) presents a matrix for determining appropriate procedure format, as illustrated

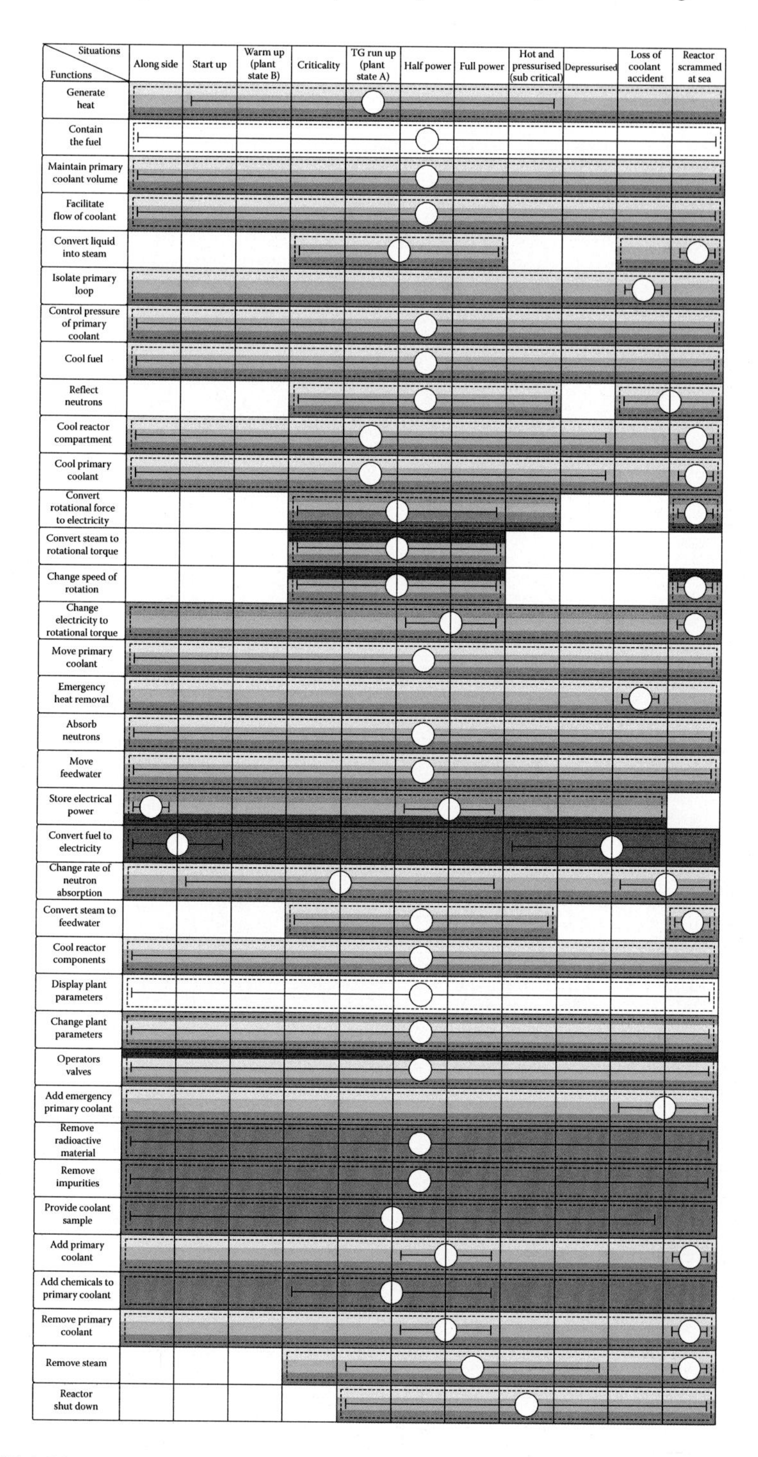

FIGURE 1.32 Using SOCA-CAT to identify skill variety.

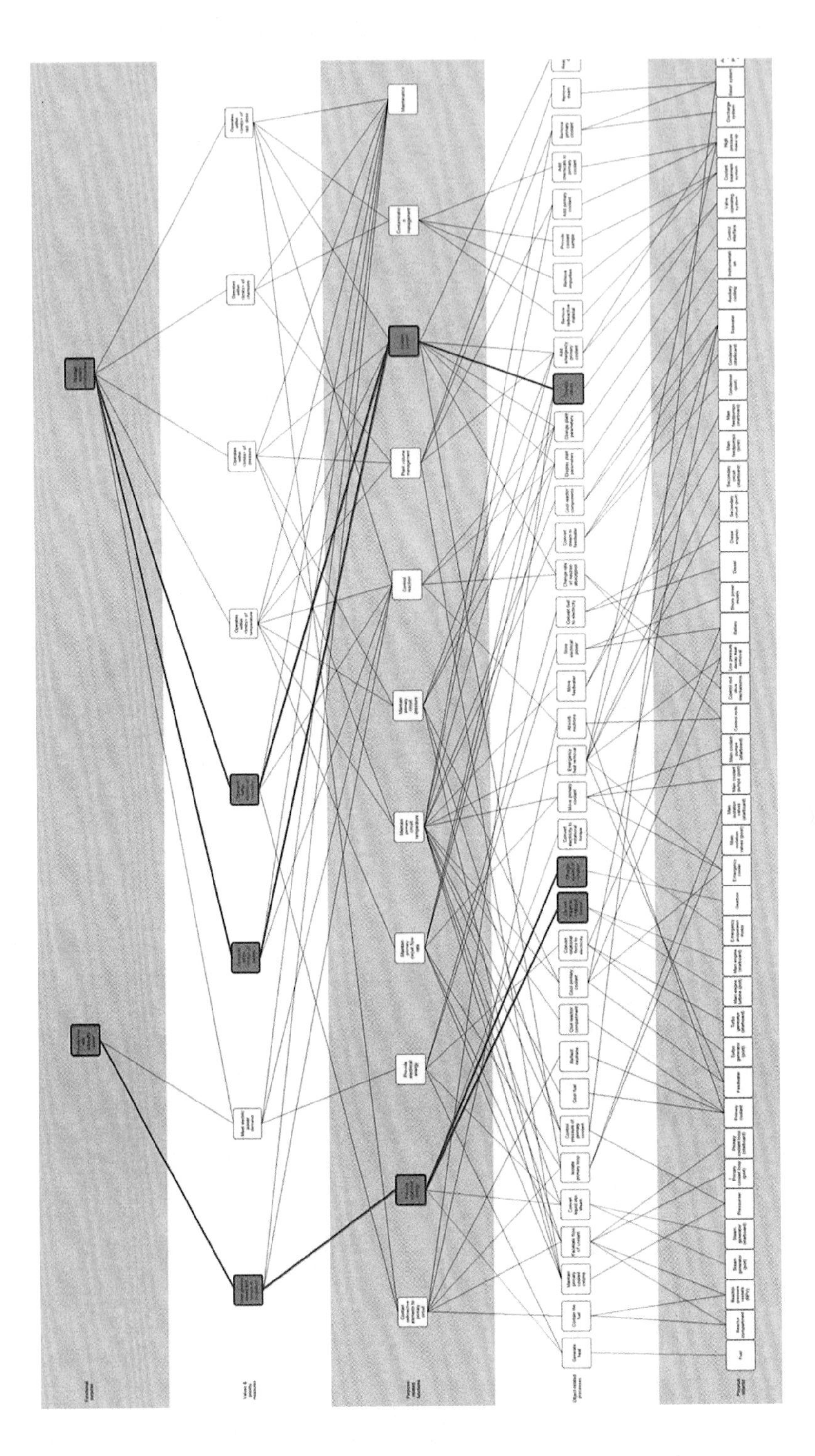

FIGURE 1.33 AH showing activities conducted by TCPO.

TABLE 1.10
Types of Decision Support Based upon Task Familiarity and Difficulty

	High Task Difficulty	Low Task Difficulty
High operator familiarity	Decision aid (e.g. based on decision ladder)	Checklist (e.g. based on SBB from WCA)
Low operator familiarity	Step-by-step procedure (e.g. based on StrA)	Narrative/fundamental system understanding (e.g. based upon AH and KBBs in the WCA)

in Table 1.10. The decision ladder model can be used to develop an understanding of the relationship between information elements and resultant states of knowledge.

As Table 1.10 shows, the source of data for development of the different forms the procedures may take can be found in the different CWA outputs.

1.3.20 Hazard Assessment

Other than raising domain understanding, the CWA framework does not directly inform hazard assessment.

1.3.21 Risk Assessment

Two of the mainstays in coping with risk are to build redundancy and diversity into systems; through its constraint-based approach, CWA provides a structure for modelling the level of redundancy and diversity in a system. Most notably, the AH captures the 'one-to-one', 'one-to-many' and 'many-to-many' relationships within the system. Figure 1.34 shows an example of a many-to-one relationship – electrical power can be provided by using stored power, converting fuel (diesel) to electricity or by combining the three processes of generating heat, converting liquid to steam and converting rotational force to electricity. Therefore, if heat steam or the turbines are lost, redundancy exists in the batteries or in the diesel engines.

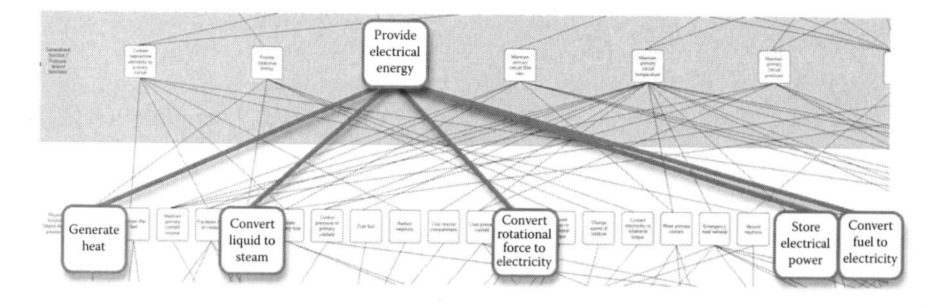

FIGURE 1.34 Example of many-to-one relationship.

1.3.22 Design Advice

The entire CWA framework is geared up to support design advice. The framework can support layout design (see Section 1.3.16), interface design (see Section 1.3.16), team design (e.g. Section 1.3.5 on SOCA-CAT) and training design (see Section 1.3.8).

1.3.23 Interface Design

CWA has long been associated with interface design. The framework offers a number of benefits in intelligent clustering of information. Furthermore, the approach has the potential to inform dynamic situation and agent-specific representations. Ecological interface design (EID) (Burns and Hajdukiewicz, 2004) is an approach, based upon CWA.

Much of the literature on EID (Burns and Hajdukiewicz, 2004) focuses on the initial phase (WDA). This approach captures the constraints governing the functions and purposes of the system, but does nothing to explicitly consider the allocation of function, nor does it consider the design for known recurring situations. By considering the proceeding phases, a more complete analysis of the system can be conducted, and in turn, a more considered interface. Table 1.11 captures the relationship between some of the key design questions posed and the phases of the framework (a shade box indicates a relationship).

Perhaps the most important and primary question in the design of an interface should be 'why is the information needed?' Through its description of the system purposes and functions, the WDA is the primary source for answering this question. As the WDA describes all of the functions performed by the system, it is also the main resource for answering the question of 'what information is required?' As the WDA addresses these questions independently of situation or agent, it is possible to consider these questions in a way that accounts for unexpected and unanticipated events. For most systems, the type of information displayed, at any one time, will be influenced by the current situation and the functions being performed by the system. The ConTA phase of the analysis describes each of the possible situation–function combinations within the CAT. An understanding of who could require different information sets can be informed from the SOCA phase. Once it is understood why the information is needed, what type of information required, when it should be

TABLE 1.11
Mapping of Design Questions to CWA Phases

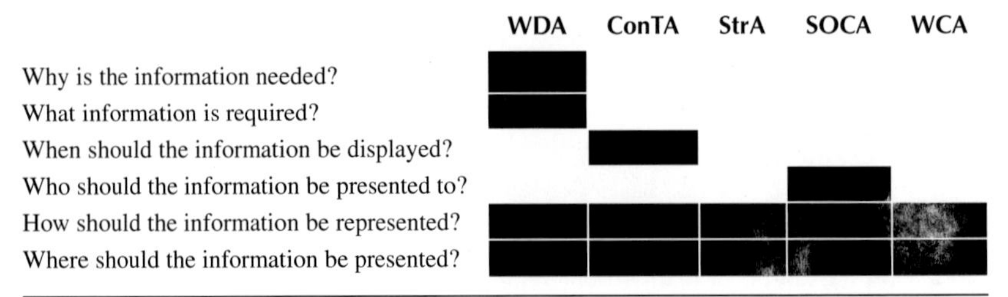

	WDA	ConTA	StrA	SOCA	WCA
Why is the information needed?	■				
What information is required?	■				
When should the information be displayed?		■			
Who should the information be presented to?				■	
How should the information be represented?	■	■	■	■	■
Where should the information be presented?	■	■	■	■	■

displayed and who it should be displayed to, the remaining tasks relate how and where the information should be arranged on the display. This clustering of the data is of fundamental importance and influenced by a wide number of constraints, hence the inclusion of all of the phases. The clustering of these data will be dependent on the type of system modelled. The optimum clustering technique may be based on spatial, functional, hierarchical or critical importance, or possibly a combination.

Why is the information needed? The AH informs the question 'why is the information needed?' at a number of levels of abstraction; specifically, the highest level of abstraction, the functional purposes, describes the overall purposes of the system – in this case, to provide the ship with adequate power and to maintain system effectiveness. Moving down the hierarchy in Figure 1.35, more specific questions can be answered; for example, information is needed to operate within limits of temperature, maintain primary circuit flow rate and operate valves.

What information is required? The information required is also informed by the AH; working from the bottom-up and exploring the means–ends links, it is possible to ascertain the kind of information that would be required within the system. Taking the

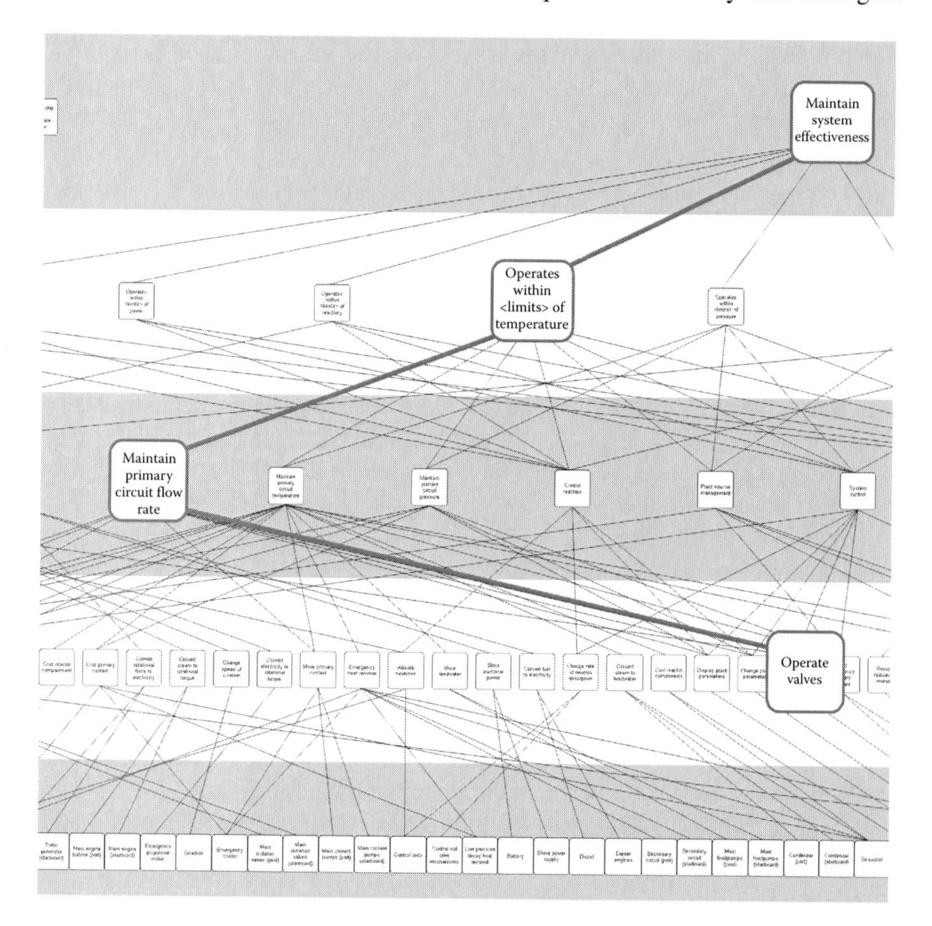

FIGURE 1.35 Means–ends links in the abstraction hierarchy.

previous example, the current state of the valves may be required (open, closed), information may also be needed on the flow rate of the coolant at key points in the system), and various temperature readings may also be required all to inform the questions of can the system provide adequate power while maintaining system effectiveness.

When should the information be displayed? The question of when should the information be displayed can be addressed in the CAT. Here, the functions are plotted against the situation in which they can and typically occur; for example, there is little benefit in presenting information on steam pressure when the ship is alongside before it has been powered-up.

Who should the information be presented to? The various representation used within the SOCA phase can be used inform decisions' related to who should be presented with date. The representations concentrate on the information that could be of use to the role rather than the information that should be. It is possible to rule out representing certain information to specific roles. For example, the TCPO is not required to have knowledge about the flow rate of coolant within the system to perform his or her role.

The remaining questions, 'how should the information be represented?' and 'where should the information be presented?' are less easy to provide examples for they require a complete understanding of the system and the constraints captured in each of the five phases.

1.4 CONCLUSIONS AND RECOMMENDATIONS

In summary, this chapter details the analysis of the propulsion plant onboard a Trafalgar-class submarine in order to assess the utility of CWA. The resultant products of the CWA demonstrated that the approach could be used to analyse the propulsion plant constraints. The seven outputs (namely, AH, ADS, CAT, DL, STA, SOCA and WCA) were then assessed against the criteria, that is, the amount of effort required by the analysts and the SMEs to undertake the analyses, the usefulness of the resultant products of the analyses to inform design decisions (in particular, those identified by EHFA) and the potential for the analyses within CWA to model alternatives and provide comparisons for choosing one alternative over another.

The analyses show that CWA can be applied to the investigation of the propulsion plant. Each of the analyses assessed the various constraints. The appropriateness of the products of CWA was also against a set of criteria, that is, the amount of effort required, the usefulness of the resultant analysis for the human factors design concerns and the possibility of the approach to model alternative systems, informing other EHFAs, hardware and other requirements for conducting CWA using the software tool developed by the HFI-DTC (www.hfidtc.com). The assessment is shown in Table 1.12.

As Table 1.12 shows, CWA supports and contributes to most of the human factors design issues that are of interest to in design of the propulsion plant. CWA supports design of CONOPS, function analysis, normal and emergency operations, allocation of function, training needs analysis, operating procedures specification, design advice and interface design. CWA contributes to the design of TAD, workload assessment, job design and risk assessment. CWA does not support or contribute to the design of operability evaluation, maintainability evaluation and hazard assessment. The time

TABLE 1.12
Assessment of CWA

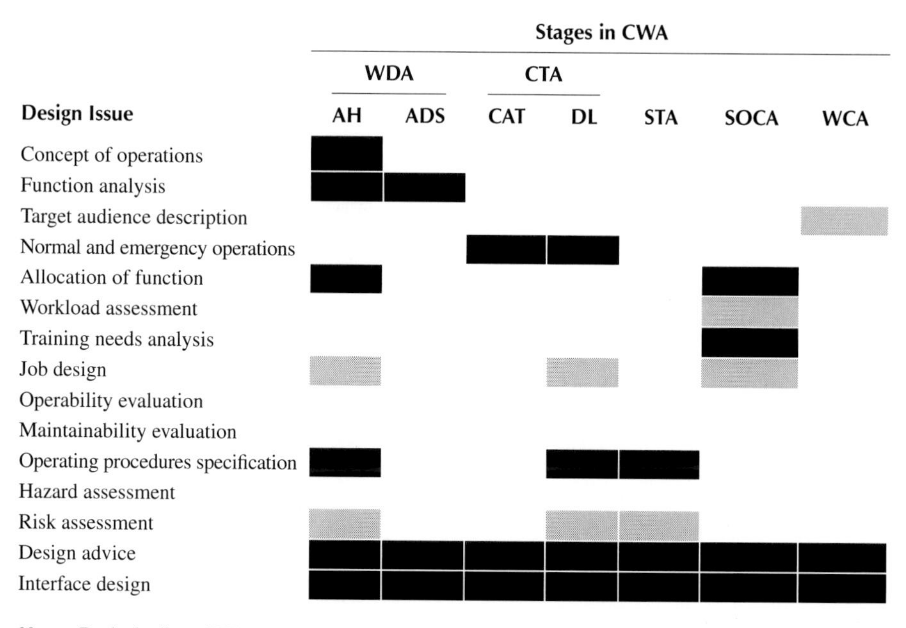

	Stages in CWA						
	WDA		CTA				
Design Issue	**AH**	**ADS**	**CAT**	**DL**	**STA**	**SOCA**	**WCA**
Concept of operations	■						
Function analysis	■	■					
Target audience description							▒
Normal and emergency operations			■	■			
Allocation of function	■					■	
Workload assessment						▒	
Training needs analysis						■	
Job design	▒			▒		▒	
Operability evaluation							
Maintainability evaluation							
Operating procedures specification	■			■	■		
Hazard assessment							
Risk assessment	▒			▒	▒		
Design advice	■	■	■	■	■	■	■
Interface design	■	■	■	■	■	■	■

Note: Dark shading, CWA supports the design issue; light shading, CWA contributes to the design issue but additional forms of analysis are required; no shading, CWA does not support or contribute to the design issue.

taken to undertake the initial CWA is shown at the bottom of the table. This was analyst time only, as additional SME time was also required to help elicit the data and verify the analysis. All of the outputs from CWA can be used to model alternatives and inform EHFA. In the current approach, WCA is rather laborious, but an alternative approach has been proposed, which makes process much easier to conduct.

The outputs from CWA have the potential to inform other human factors analyses being undertaken. This is illustrated in Table 1.13. The human factors methods taxonomy was taken from the text by Stanton et al. (2005, 2013), which identified 10 main classes of method (see Appendix 1B.1). The methods in the taxonomy were then compared with the main outputs of CWA to identify which methods could use the product of CWA or feed their products into CWA. The shaded areas reveal where this would be feasible.

As Table 1.13 shows, each output from CWA may be associated with other human factors methods, either as a potential data source for CWA or as a way of extending the output from CWA. Consideration of the contents of Tables 1.12 and 1.13 together may help to identify which human factors methods may assist in supporting or extending CWA in system design. For example, in developing the TAD (see Table 1.12), WCA may be supported and extended using task analysis and cognitive task analysis methods (see Table 1.13 and Appendix 1B.1). Similarly, in workload

TABLE 1.13
CWA Links to and from Other Human Factors Methods

	Stages in CWA						
	WDA		CTA				
Human Factors Methods Taxonomy	**AH**	**ADS**	**CAT**	**DL**	**STA**	**SOCA**	**WCA**
Task analysis				■	■	■	■
Cognitive task analysis				■	■	■	■
Process charting			■	■	■	■	
Human error identification				■	■		
Situation awareness			■	■		■	
Workload assessment				■	■	■	
Team assessment						■	
Interface analysis	■	■	■	■	■	■	■
Design	■	■	■	■	■	■	■
Performance time prediction				■	■		

assessment, SOCA may be supported and extended with task analysis, cognitive task analysis, process charting, situation awareness and workload assessment methods (see Table 1.13 and Appendix 1B.1). In fact, all of the CWA products may be linked to other human factors methods, as shown in Table 1.13.

It is further suggested that CWA has most benefits when applied in the concept phase of the CADMID (Concept, Analysis, Design, Manufacture, In-Service, Disposal) lifecycle. As a formative approach, CWA can be used to understand the constraints acting on system design. All five phases may be used to explore the constraints and compare alternative designs before they take shape. The pictorial representations enable comparisons to be made, so that human factors guidance may be provided as early as possible. CWA may be continued to be used through the assessment and design phases of CADMID. As the design begins to crystallise, more information may be fed into the CWA modelling. The CWA software supports the process of updating the models, and refining the analysis throughout the design lifecycle.

APPENDIX 1A.1: ABBREVIATIONS AND ACRONYMS

ADS	Abstraction–Decomposition Space
AH	Abstraction Hierarchy
CAT	Contextual Activity Template
ConTA	Control Task Analysis
CONOPS	Concept of Operations
CWA	Cognitive Work Analysis
EHFA	Early Human Factors Analysis
EID	Ecological Interface Design

EPO	Electrical Panel Operator
HFI-DTC	Human Factors Integration Defence Technology Centre
IPT	Integrated Project Team
KBB	Knowledge-Based Behaviour
NCOTW	Nuclear Chief of the Watch
NCW	Network-Centric Warfare
NEC	Network-Enabled Capability
OPS	Operational Performance Statement
PSTAD	Project-Specific Target Audience Description
RBB	Rule-Based Behaviour
RNGTAD	Royal Navy Generic Target Audience Description
RPO	Reactor Panel Operator
RPV	Reactor Pressure Vessel
SBB	Skill-Based Behaviour
SOCA	Social Organisation and Cooperation Analysis
SME	Subject Matter Expert
SRK	Skills–Rules–Knowledge
Stb	Starboard
StrA	Strategies Analysis
TAD	Target Audience Description
TCPO	Throttle Control Panel Operator
TPS	Training Performance Statement
WCA	Worker Competencies Analysis
WDA	Work Domain Analysis

APPENDIX 1B.1: HUMAN FACTORS METHODS TAXONOMY

Task Analysis Methods
Hierarchical Task Analysis (HTA)
Goals, Operators, Methods and Selection Rules (GOMS)
Verbal Protocol Analysis (VPA)
Task Decomposition
The Sub-Goal Template Method (SGT)
Tabular Task Analysis (TTA)

Cognitive Task Analysis Methods
Applied Cognitive Task Analysis (ACTA)
Cognitive Walk-Through
Critical Decision Method (CDM)
Critical Incident Technique (CIT)

Process Charting Methods
Process Charts
Operation Sequence Diagrams (OSD)

Event Tree Analysis (ETA)
Decision Action Diagrams (DAD)
Fault Trees
Murphy Diagrams

Human Error Identification Methods
Systematic Human Error Reduction and Prediction Approach (SHERPA)
Human Error Template (HET)
Technique for the Retrospective and Predictive Analysis of Cognitive Errors (TRACEr)
Task Analysis for Error Identification (TAFEI)
Human Error HAZOP
Technique for Human Error Assessment (THEA)
Human Error Identification in Systems Tool (HEIST)
The Human Error and Recovery Assessment Framework (HERA)
System for Predictive Error Analysis and Reduction (SPEAR)
Human Error Assessment and Reduction Technique (HEART)
The Cognitive Reliability and Error Analysis Method (CREAM)

Situation Awareness Assessment Methods
SA Requirements Analysis
Situation Awareness Global Assessment Technique (SAGAT)
Situation Awareness Rating Technique (SART)
Situation Awareness Subjective Workload Dominance (SA-SWORD)
Situation Awareness of Area Controllers within the Context of Automation (SALSA)
Situation Awareness Control Room Inventory (SACRI)
Situation Awareness Rating Scales (SARS)
Situation Present Assessment Method (SPAM)
SASHA_L and SASHA_Q
Mission Awareness Rating Scale (MARS)
Situation Awareness Behavioural Rating Scale (SABARS)
Crew Awareness Rating Scale (CARS)
Cranfield Situation Awareness Scale (C-SAS)
Propositional Networks

Mental Workload Assessment Methods
Primary and Secondary Task Performance Measures
Physiological Measures
NASA Task Load Index (NASA TLX)
Modified Cooper Harper Scales (MCH)
Subjective Workload Assessment Technique (SWAT)
Subjective Workload Dominance Technique (SWORD)
DRA Workload Scales (DRAWS)
Malvern Capacity Estimate (MACE)

Workload Profile Technique 342
Bedford Scales
Instantaneous Self-Assessment (ISA)
Cognitive Task Load Analysis (CTLA)
Subjective Workload Assessment Technique (SWAT)
Pro-SWORD – Subjective Workload Dominance Technique

Team Assessment Methods
Behavioural Observation Scales (BOS)
Comms Usage Diagram (CUD)
Co-ordination Demand Analysis (CDA)
Decision Requirements Exercise (DRX)
Groupware Task Analysis (GTA)
Hierarchical Task Analysis for Teams (HTA(T))
Team Cognitive Task Analysis (TCTA)
Social Network Analysis (SNA)
Questionnaires for Distributed Assessment of Team Mutual Awareness
Team Task Analysis (TTA)
Team Workload Assessment
Task and Training Requirements Analysis Methodology (TTRAM)

Interface Analysis Methods
Checklists
Heuristic Analysis
Interface Surveys
Link Analysis
Layout Analysis
Questionnaire for User Interface Satisfaction (QUIS)
Repertory Grid Analysis
Software Usability Measurement Inventory (SUMI)
System Usability Scale (SUS)
User Trials
Walk-Through Analysis

Design Methods
Allocation of Function Analysis
Focus Groups
Missions Analysis
Scenario-Based Design
Task-Centred System Design

Performance Time Prediction Methods
Multimodal Critical Path Analysis (CPA)
Keystroke Level Model (KLM)
Timeline Analysis

REFERENCES

Ahlstrom, U. 2005. Work domain analysis for air traffic controller weather displays. *Journal of Safety Research*, 36, 159–169.

Allaway, J. 2004. *Inside the Royal Navy: A Collection of Stunning Cutaway Drawings of Royal Navy Ships, Submarines and Aircraft – Past and Present*. Portsmouth, UK: Navy News (HMSO).

Bertalanffy, L. V. 1950. The theory of open systems in physics and biology. *Science*, 111, 23–29.

Beuscart, J. M. 2005. Napster users between community and clientele: The formation and regulation of a sociotechnical group. *Sociologie du Travail*, 47, S1–S16.

Burns, C. M. and Hajdukiewicz, J. R. 2004. *Ecological Interface Design*. Boca Raton, FL: CRC Press.

Cherns, A. B. 1987. Principles of sociotechnical design revisited. *Human Relations*, 40(3), 153–162.

Cummings, M. L. and Guerlain, S. 2007. Developing operator capacity estimates for supervisory control of autonomous vehicles. *Human Factors*, 49(1), 1–15.

Elix, B. and Naikar, N. 2008. Designing safe and effective future systems: A new approach for modelling decisions in future systems with cognitive work analysis. *Proceedings of the 8th International Symposium of the Australian Aviation Psychology Association*. Sydney, Australia.

Fidel, R. and Pejtersen, A. M. 2005. Cognitive work analysis. In K.E. Fisher, S. Erdelez and E. F. McKechnie (Eds.). *Theories of Information Behavior: A Researcher's Guide*. Medford, NJ: Information Today.

Hackman, J. R. and Oldham, G. 1980. *Work Redesign*. Reading, MA: Addison-Wesley.

Jenkins, D. P., Farmilo, A., Stanton, N. A., Whitworth, I., Salmon, P. M., Hone, G., Bessell, K. and Walker, G. H. 2007a. *The CWA Tool V0.95*. Yeovil, UK: HFI-DTC.

Jenkins, D. P., Stanton, N. A., Salmon, P. M., Walker, G. H., Young, M. S., Farmilo, A., Whitworth, I. and Hone, G. 2007b. The development of a cognitive work analysis tool. In D. Harris (Ed.). *Engin. Psychol. and Cog. Ergonomics*. HCII 2007, LNAI 4562, pp. 504–511.

Kilgore, R. and St-Cyr, O. 2006. SRK inventory: A tool for structuring and capturing a worker companies analysis. *Proceedings of the Human Factors and Ergonomics Society 50th Annual Meeting 2006*. San Francisco, California, pp. 506–509.

Lintern, G. 2005. Work Domain Analysis: Tutorial [On-Line]. http://kn.gd-ais.com/Database/DocMan_Upload/Staging/Tutorial_on_WDA.pdf

Lintern, G., Cone, S., Schenaker, M., Ehlert, J. and Hughes, T. 2004. *Asymmetric Adversary Analysis for Intelligent Preparation of the Battlespace (A3-IPB)*. U.S. Air Force Research Department Report.

MAP-01-011. 2006. Maritime Acquisition Publication No 01-011 Human Factors Integration (HFI) Technical Guide (STGP 11). http://www.aof.mod.uk/content/docs/hfi/map-01-011.pdf

Marsden, P. 1996. Procedures in the nuclear industry. In N. A. Stanton (Ed.). *Human Factors in Nuclear Safety*. London, UK: Taylor and Francis, pp. 99–116.

Naikar, N. 2006a. Beyond interface design: Further applications of cognitive work analysis. *International Journal of Industrial Ergonomics*, 36, 423–438.

Naikar, N. 2006b. An examination of the key concepts of the five phases of cognitive work analysis with examples from a familiar system. *Proceedings of the Human Factors and Ergonomics Society 50th Annual Meeting*. San Francisco, California, pp. 447–451.

Naikar, N. 2008. Modelling activity in network-centric operations with cognitive work analysis: Work situations, work functions, decisions, and strategies. In B. Bolia (Ed.). *Supporting Decision Effectiveness in Network-Centric Operations*. Dayton, OH: Wright Patterson Air Force Base.

Naikar, N., Hopcroft, R. and Moylan, A. 2005. Work Domain Analysis: Theoretical Concepts and Methodology. DSTO-TR-1665.

Naikar, N., Moylan, A. and Pearce, B. 2006. Analysing activity in complex systems with cognitive work analysis: Concepts, guidelines, and case study for control task analysis. *Theoretical Issues in Ergonomics Science*, 7(4), 371–394.

Naikar, N. and Pearce, B. 2003. Analysing activity for future systems. *Proceedings of the 47th Annual Meeting of the Human Factors and Ergonomics Society*. Santa Monica, CA: Human Factors and Ergonomics Society, pp. 1928–1932.

Rasmussen, J. 1974. *The Human Data Processor as a System Component: Bits and Pieces of a Model*. Report no. Risø-M-1722. Roskilde, Denmark: Danish Atomic Energy Commission.

Rasmussen, J. 1983. Skills, rules, knowledge; signals, signs, and symbols, and other distinctions in human performance models. *IEEE Transactions on Systems, Man and Cybernetics*, 13, 257–266.

Rasmussen, J. 1986. *Information Processing and Human-Machine Interaction: An Approach to Cognitive Engineering*. New York: North-Holland. http://www.ischool.washington.edu/chii/portal/literature.html.

Rasmussen, J., Pejtersen, A. and Goodstein, L. P. 1994. *Cognitive Systems Engineering*. New York: Wiley.

Rasmussen, J., Pejtersen, A. M. and Schmidt, K. 1990. *Taxonomy for Cognitive Work Analysis*. Risø National Laboratory Risø-M-2871.

Sanderson, P. M. 2003. Cognitive work analysis across the system life-cycle: Achievements, challenges, and prospects in aviation. In P. Pfister and G. Edkins (Eds.). *Aviation Resource Management* (Vol. 3). Aldershot, UK: Ashgate, pp. 73–85.

Sharp, T. D. and Helmicki, A. J. 1998. The application of the ecological interface design approach to neonatal intensive care medicine. *Proceedings of the Human Factors and Ergonomics Society 42nd Annual Meeting*. Santa Monica, CA: HFES, pp. 350–354.

Stanton, N. A., Salmon, P. M., Rafferty, L. A., Walker, G. H., Baber, C. and Jenkins, D. 2013. *Human Factors Methods: A Practical Guide for Engineering and Design* (second edition). Aldershot, UK: Ashgate.

Stanton, N. A., Salmon, P. M., Walker, G. H., Baber, C. and Jenkins, D. 2005. *Human Factors Methods: A Practical Guide for Engineering and Design* (first edition). Aldershot, UK: Ashgate.

Vicente, K. J. 1999. *Cognitive Work Analysis: Toward Safe, Productive, and Healthy Computer-Based Work*. Mahwah, NJ: Lawrence Erlbaum Associates.

Section II

Requirements Specification

2 Specifying System Requirements Using Cognitive Work Analysis

Neville A. Stanton and Rich C. McIlroy

CONTENTS

2.1 INTRODUCTION

This chapter describes the extent to which cognitive work analysis (CWA) can facilitate the requirements analysis and specification process. The aim is twofold: first, to provide a link between the requirements specification process and CWA (Rasmussen et al., 1994; Vicente, 1999; Jenkins et al., 2009); second, to describe which parts of a CWA can be applied, and how that application should proceed, to specify requirements that lead to usable, functionally relevant, affordable systems that minimise post-deployment redesign, and that do not necessitate extensive training schedules.

Though this second aim may seem, to some, to be ambitious, the application of suitable human factors methods (Stanton et al., 2013), CWA in particular, throughout the requirements development and design phases, can help specify the needs of users and systems such that all users are supported in their performance of system functions.

In order to demonstrate the CWA approach, this chapter uses a real case study based on mission communications planning. Failure to consider human factors requirements in the design specification of new systems can lead to significant deficiencies in the resultant system (Stanton et al., 2009). As a general design principle, the production of new systems should be at least as easy as their previous equivalents. There are concerns that new system design will lead to additional 'emergent' work in terms of both increasing the amount of 'direct' work required and the work associated with operation of the (often digital) tools. Approaches such as CWA can circumvent these problems by helping to ensure consideration of the whole sociotechnical system in specification of design requirements throughout the life cycle of the project.

2.1.1 Cognitive Work Analysis

CWA (Rasmussen et al., 1994; Vicente, 1999; Jenkins et al., 2009) represents a formative approach to the analysis of complex sociotechnical systems. Rather than analysing what a system currently does (*descriptive modelling*) or should do (*normative modelling*), the analysis offers a framework of methods that allow for the in-depth analysis of the properties of the work domain and the workers themselves, therefore defining a set of boundaries that shape activity within the system. The approach leads the analysis into describing how the system *could* perform (*formative modelling*) given its constraints. The analysis comprises five phases: work domain analysis (WDA), control task analysis (ConTA), strategies analysis (StrA), social organisation and cooperation analysis (SOCA) and worker competencies analysis (WCA) (as shown in Figure 2.1). Although each phase builds upon the last, not all of the phases must be used. The framework can be likened to a toolkit of methods, in which the analyst may apply any of the methods individually, or in combination, depending upon the nature and needs of the analysis.

2.1.2 Work Domain Analysis

The first phase of CWA is to develop a WDA. In this phase, the system is described in terms of the environment in which workers operate; the analysis is independent of activity, actors or goals. WDA identifies a fundamental set of constraints that shape activity within the system, providing a foundation for subsequent phases of CWA. The first stage of WDA is to construct an abstraction hierarchy (AH). The AH describes the system at a number of levels of abstraction. At the highest level, the AH describes the reason for the systems' existence – its functional purpose. At the lowest level, the physical objects within the system are described. This is essentially an inventory of the physical resources in the system, both natural and man-made. Each level of the AH can be linked by means–ends relationships through the use of the how–what–why triad. Any node can be taken to answer the question of 'what' it does. In the level below, all of the connected nodes can be used to answer the

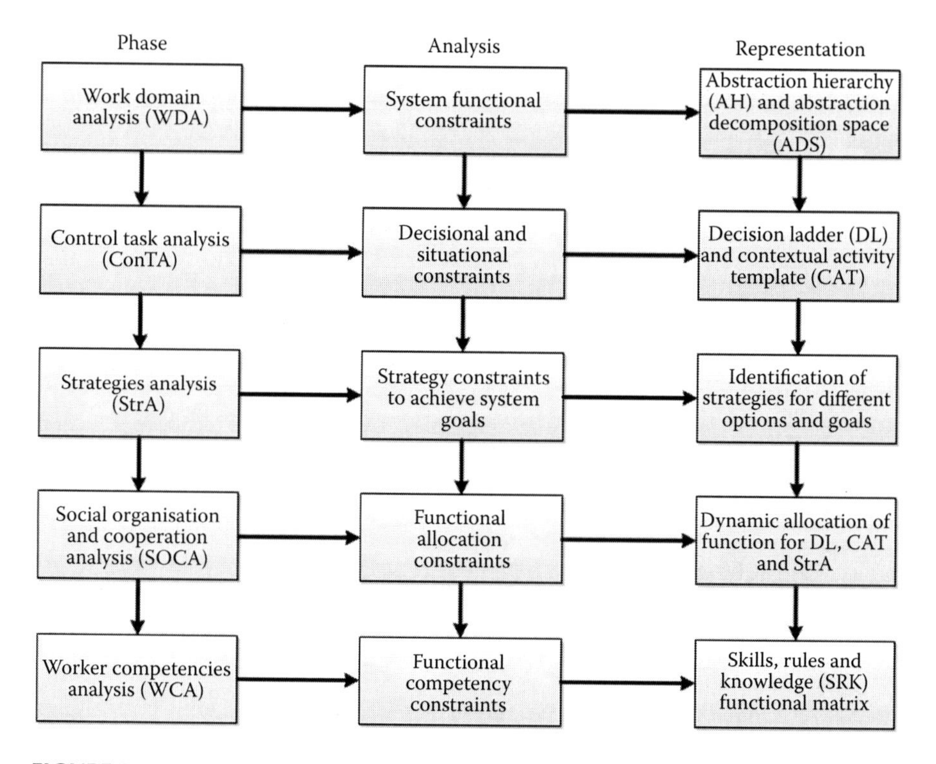

FIGURE 2.1 Five phases of CWA.

question of 'how' this can be achieved. In the level above, the connected nodes are used to answer the question of 'why' it is needed.

2.1.3 Control Task Analysis

The second phase of CWA is ConTA. Building on WDA, this phase considers recurring activities within the system. The analysis focuses on what is to be achieved independent of how the activity is to be conducted and by whom the activity is to be carried out. Naikar et al. (2006) developed the contextual activity template (CAT) for use in this phase of CWA. The template characterises system activity in terms of work functions and work situations (as illustrated in Figure 2.1). These situations may be temporal, spatial or a combination of the two. The activity that occurs is represented in terms of where it is able to be carried out and where it is likely to be carried out (Figure 2.2).

Also developed during the ConTA phase of CWA are the decision ladders (Rasmussen, 1974; Vicente, 1999). These diagrams are used to consider activity in decision-making terms. Decision ladders contain two different types of nodes: the rectangular boxes represent data-processing activities, while the circles represent the resulting state of knowledge. The diagram captures the flow of information processing associated with individual control tasks; it prescribes a sequence of information-processing steps and resultant knowledge states required to fulfil a control task. Although the diagram displays information processing in a linear fashion, different

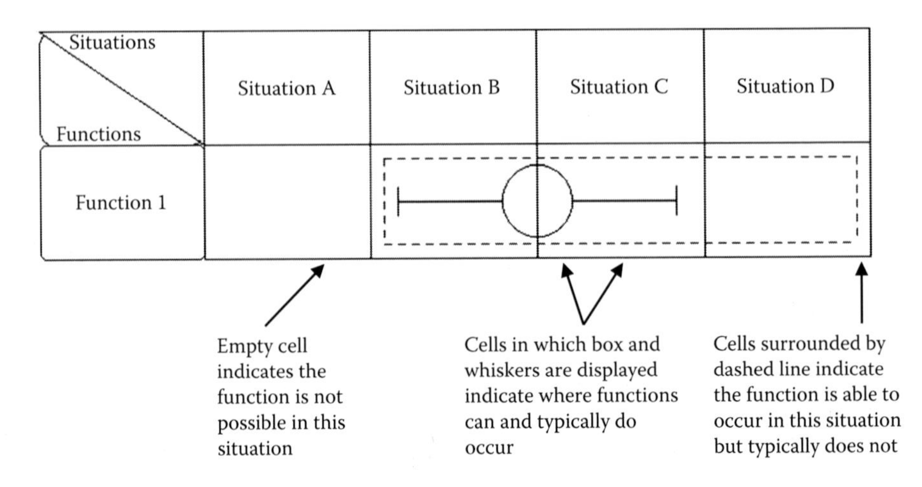

FIGURE 2.2 Explanatory figure of the CAT.

actors are likely to take different routes from the entry point to the end point. More specifically, novice workers are expected to follow the linear sequence while expert actors are often able to take shortcuts. For example, in certain situations, the diagnosis of the system state may lead directly to the execution of a set procedure. There are two types of shortcuts defined in the literature (Vicente, 1999; Jenkins et al., 2009): *shunts* and *leaps*. Shunts connect data-processing activities to non-sequential states of knowledge while leaps connect two states of knowledge.

2.1.4 Strategies Analysis

Strategies analysis (StrA), the third phase of CWA, addresses the constraints governing the way activity is conducted. Where ConTA describes what activities are to be conducted, StrA identifies how these activities are to be performed. It describes different ways of carrying out the same task. It is often the case that there are a number of different ways of achieving the same ends within a system, and the strategies adopted can have considerable variability. Different agents (human or non-human) perform tasks in different ways, and the same agent might perform a task differently in different situations. The flow diagram presentation method was suggested by Ahlstrom (2005) as a suitable method of displaying different strategies and consists of a 'start state' and an 'end state' connected by a number of strategies, each indicating different possible sequences of operations.

2.1.5 Social Organisation and Cooperation Analysis

The SOCA phase of CWA looks at the cooperation between actors within a system. It addresses the constraints imposed by organisational structures or specific actor roles and definitions. The aim of this section of the analysis is to determine the optimal organisation of human and technical factors within a complex sociotechnical system, such that system performance is maximised. The analysis builds upon outputs obtained from previous sections of CWA through the use of shading; different actor roles are each assigned a colour and the various CWA outputs are shaded to indicate where each of the actor groups can conduct tasks. Allocation of function is described in this stage of CWA; different actors are allocated to different functions. The use of differential shading used in this stage of the analysis can be applied to any of the outputs obtained thus far; hence, there are a number of different SOCA views. Its most common application is in colouring the CAT (developed in the ConTA phase) to indicate what activities in which situations can be performed by which system actors.

2.1.6 Worker Competencies Analysis

The final phase of CWA, WCA, is concerned with identifying the competencies required by workers in a system. It takes into account human characteristics, describing the constraints that dictate actor behaviour in different situations. It is the behaviour that is required from actors in order to carry out tasks and fulfil goals, which is the focus of the analysis, and is typically modelled using Rasmussen's (1983) skills–rules–knowledge (SRK) taxonomy. According to Vicente (1999), skill-based behaviour (SBB) consists of automatic actions in response to environmental cues and events. Very little or no conscious effort is required to perform an action of this type. Rule-based behaviour (RBB) is characterised by a set of stored rules and procedures. These rules guide behaviour without the actor having to consider individual system goals and are acquired either through experience or learned from supervisors, former operators and instructional texts. Unlike SBB, the process requires conscious cognitive processing. When more advanced reasoning is required, the actor uses knowledge-based behaviour (KBB). This behaviour is slow and effortful; the actor is required to apply conscious attention, carefully considering the functional principles that govern the system. This type of behaviour is most common in novel, unanticipated events. Although Kilgore and St-Cyr (2006) developed an SRK inventory based on steps on the decision ladders developed in the ConTA phase of CWA, an alternative approach has proposed analyses informed by the object-related processes level of the AH. As with the decision ladder processing steps, the object-related processes can also be considered in relation to the required skill-based, rule-based and knowledge-based understanding. This should be done in order to retain the formative, technology-independent nature of the analysis.

2.2 PLANNING FOR COMMUNICATIONS IN THE MISSION PLANNING SYSTEM FOR APACHE HELICOPTERS

2.2.1 The Mission Planning System

The mission planning system (MPS) (Jenkins et al., 2008) was developed in order to reduce the in-air workload of military pilots. The MPS software provides a number of support tools designed to allow 'crews to plan and execute missions quickly and efficiently and provides information on battlefield data, threat assessment, intervisibility, full deconfliction, engagement zones, communication details, transponder information and Identification Friend or Foe (IFF) settings' (Aerosystems International, 2006). The software is currently used by the U.K. military to plan and assess Apache helicopter single or multiple aircraft sortie missions. Mission plans are developed

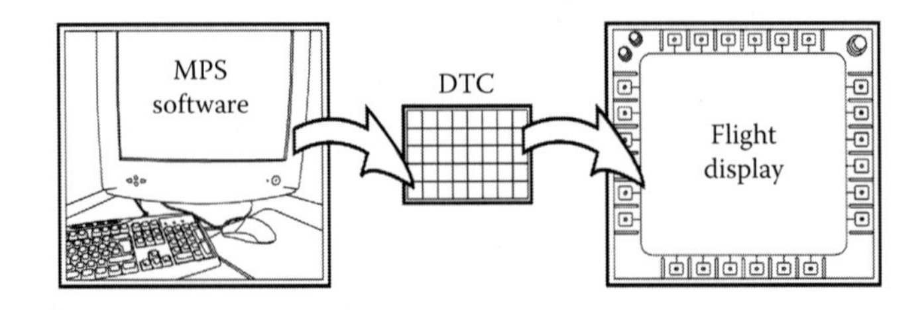

FIGURE 2.3 Process of transferring information between the terminal and the aircraft.

at MPS terminals prior to takeoff and loaded onto a digital storage device called a data transfer cartridge (DTC). The DTC is used to transfer the completed plans to the Apache's onboard data management system (DMS), which supports the mission subsequent to takeoff. This process is graphically represented in Figure 2.3.

Before any mission, military pilots must be aware of a number of communication needs. These include the need to alert authorities on the ground to the presence of any aircraft passing through or travelling in close proximity to any controlled airspace; the need for effective and secure voice communications with other aircraft concurrently airborne; and the need to effectively transfer data between concurrently airborne aircraft, while also maintaining security. Communications with authorities on the ground are dependent upon geographical orientation, while air-to-air communications are dependent upon mission requirements, command chain and the flight times and locality of other aircraft.

The main function of the MPS communications software is to allow pilots to load a collection of radio frequencies such that when airborne, pilots have easy access to all of the frequencies they will require, and that each of these frequencies is properly labelled with regards to where and with whom that frequency is associated. The MPS software contains a visual display of a map of the United Kingdom displaying the boundaries of all major controlled airspaces, including military danger zones and minor and major air fields and airports. By studying the proposed route, marked on the map by a solid black line, pilots must decide what frequencies they will need for their mission. These frequencies must then be looked up in one of the Royal Air Force (RAF) Flight Information Publications, for example, the British Isle and North Atlantic en-route supplement (BINA).

As well as geographical based air-to-ground communications, pilots must plan for air-to-air voice and data communications. These are based on the flight times and locality of other aircraft concurrently airborne. First, the planner must enter the name, call sign and data address for each of the other pilots flying at the same time as them. Second, the planner must enter in a frequency, and the band of this frequency for the voice communication channel. Separate information about frequency and band must also be entered in for the data communication channel. Finally, the planner must take into account the encryption settings and frequency hopping modes for each of these channels. The data entry process involves a number of interfaces through which the information must be inputted, each of which relates to a capability of the aircraft's onboard DMS. Interface organisation in MPS is currently based on the organisation of the DTC (see Figure 2.4; Jenkins et al., 2008).

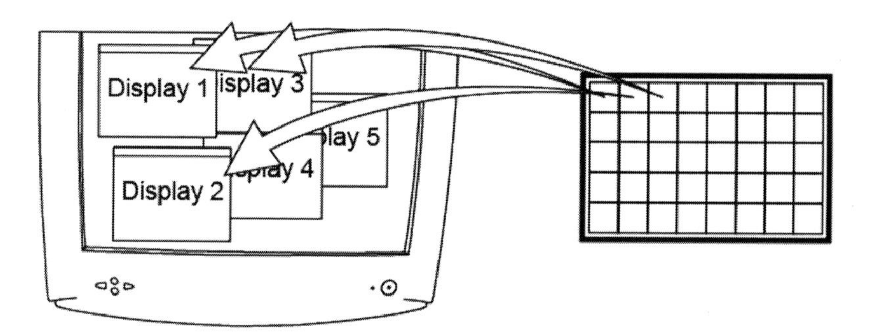

FIGURE 2.4 Current organisation of MPS based on DTC construction.

2.2.2 Cognitive Work Analysis

Outputs from CWA provide a detailed description of the system, in terms of domain, activity, social organisation and worker competency constraints. The formative, constraint-based nature of the analysis allows for a description of the system that is independent of pre-existing technology, therefore supporting revolutionary (as opposed to incremental) design changes to the system.

The AH developed in the first phase of the analysis allows for a new perspective on the system. By developing functional means–ends links across levels of abstraction, it is possible to understand task flow within the system. The AH highlighted the mismatch between the organisation of the task and the organisation of the MPS system. Individual tasks and functions described in the AH require multiple windows to be open in MPS, as currently the software is designed to support DTC data population, rather than to support communications planning. By redesigning the software package to support the communication planning task, it will be possible to reduce this need for multiple windows to be open simultaneously. Additionally, much of the training pilots receive is in how to use the software rather than how to plan communications (Figure 2.5).

By redesigning training and redesigning the MPS interface, it would be possible to focus on training pilots to plan communications instead.

The ConTA phase of CWA is built upon the AH in describing specific recurring activities within the system. The CAT displayed where each of the object-related processes take place in terms of time and place; in this instance, the system situation identified was temporally separated. The diagram revealed that the majority of activity occurs in the final two situations. This indicates an uneven spread of activity across time, with a large requirement for last minute activity; little or no activity is performed in earlier preparatory situations. Two example decision ladders were constructed to describe activity in decision-making terms. The two diagrams, one representing the process of assigning air-to-ground frequencies to the preset channels list and the other representing the construction of data nets for air-to-air data transfer, display the decision-making steps that lead to the completion of the task. Although the diagrams display what decisions are to be made to fulfil a task, they do so without specifying how or by whom. It is expected that

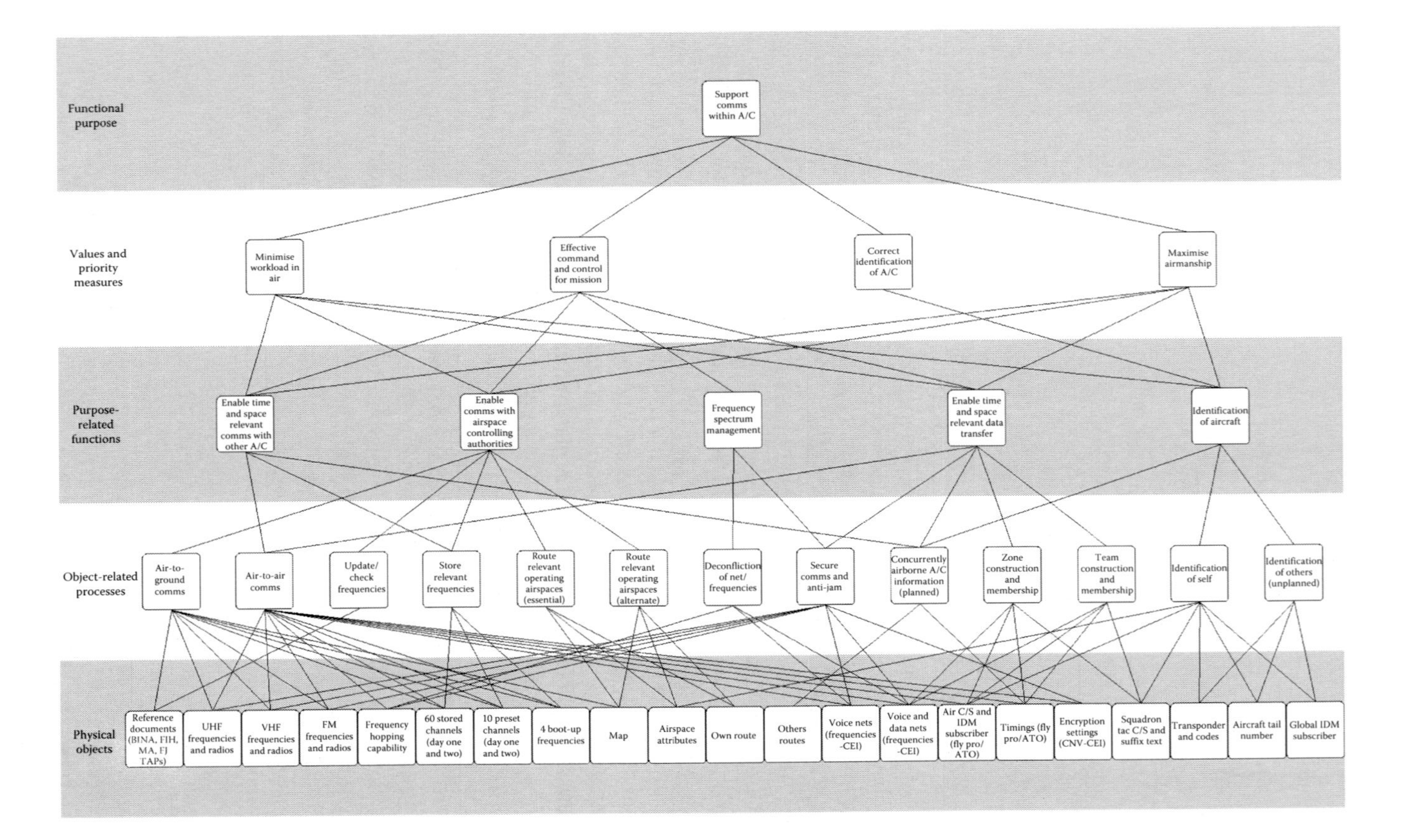

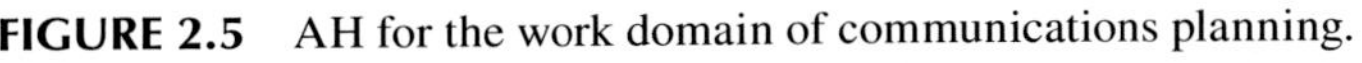

FIGURE 2.5 AH for the work domain of communications planning.

novice users would progress through the sequence of decisions in a linear fashion, while expert users are expected to take shortcuts based on previously learnt rules. The StrA phase of CWA presented two possible strategies for completing the task of populating the preset channels list with air-to-ground communication frequencies. This phase of the analysis was not developed in detail, but was included here for completeness (Figures 2.6 to 2.8).

The SOCA phase of the analysis identified three actors (see Figures 2.9 and 2.10) that are able to carry out the activities presented in the CAT (see Figure 2.11). From this output, it can be seen that the pilot carries out the majority of the activity in the system, and that all of the pilot's activity is carried out in the final two situations. During the final stage of communications planning, activity is carried out solely by the pilot.

The WCA SRK inventory allows for the description of the competencies required by workers to carry out different functions within the system. By describing what is required from workers in this way, it is possible to inform worker selection and training; SBB and RBB may be taught through training; KBBs require more formal education. The knowledge gained from education can be converted into RBB and SBB through practice and experience. The inventory can also be used to guide interface design, in which it can indicate what resources are needed by the worker to engage in effective trouble shooting or problem solving for that particular task or function. In the current research, the SRK inventory was informed by the object-related processes level of the AH, rather than by the processing steps of decision ladders developed in the ConTA phase of CWA. Conducting the analysis in this way allows for the identification of skill gaps at a higher level of abstraction. Comparing current competencies possessed by agents in the system with those competencies required to fulfil complete functions or tasks (rather than single decision-making steps) enables system design and worker recruitment that allow for whole tasks to be carried out by individual agents, thereby supporting task identity and significance (Hackman and Oldman, 1980) – two key characteristics that make a job satisfying, meaningful and engaging (Hackman and Oldman, 1980). Finally, the inventory can be used to inform allocation of function. Each task to be performed carries with it a set of competency requirements; these requirements might be best met by automations, or by less or more qualified human agents. The SRK inventory can help inform system designers which type of agent would be more suited to the task.

In summary, the CWA has highlighted some important issues with the current MPS. It is clear that the design of the interface does not optimally support the communications planning process; rather, it is designed to populate the DTC. Although this goal is a valid one when considering data management and software programming (it is indeed important that all relevant information is fed into the DTC), this is not how the task of communications planning is set out. It is argued that a redesign of the software to support the planning tasks set out in the AH and CAT would reduce the need to have multiple windows open simultaneously. This would, in turn, reduce the chance for omissions, as task-orientated groupings of information would prompt the user to the information they should be considering. The SOCA phase of the analysis highlighted the uneven spread of activity across situations. In the current system, the large majority of the activity is carried out in the latter two situations. If

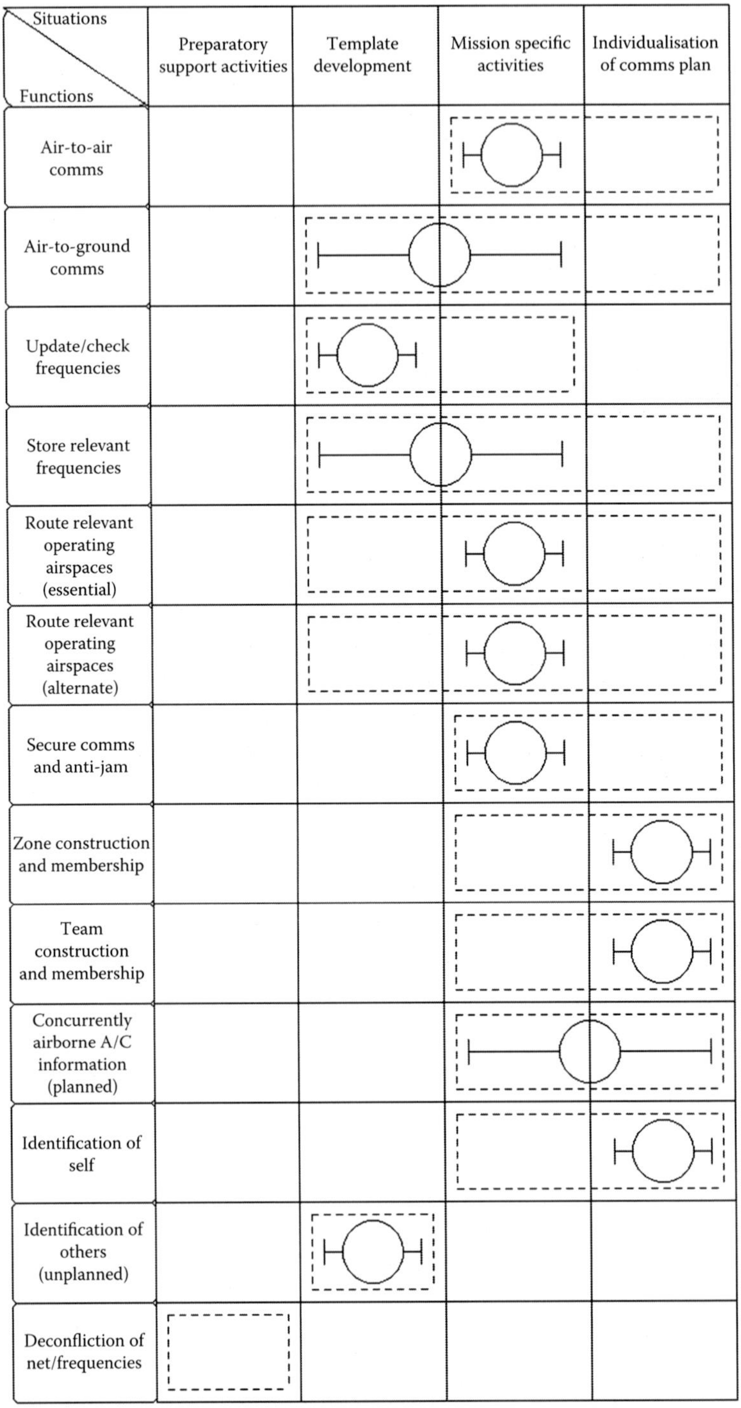

Situations / Functions	Preparatory support activities	Template development	Mission specific activities	Individualisation of comms plan
Air-to-air comms				
Air-to-ground comms				
Update/check frequencies				
Store relevant frequencies				
Route relevant operating airspaces (essential)				
Route relevant operating airspaces (alternate)				
Secure comms and anti-jam				
Zone construction and membership				
Team construction and membership				
Concurrently airborne A/C information (planned)				
Identification of self				
Identification of others (unplanned)				
Deconfliction of net/frequencies				

FIGURE 2.6 CAT with functions and situations.

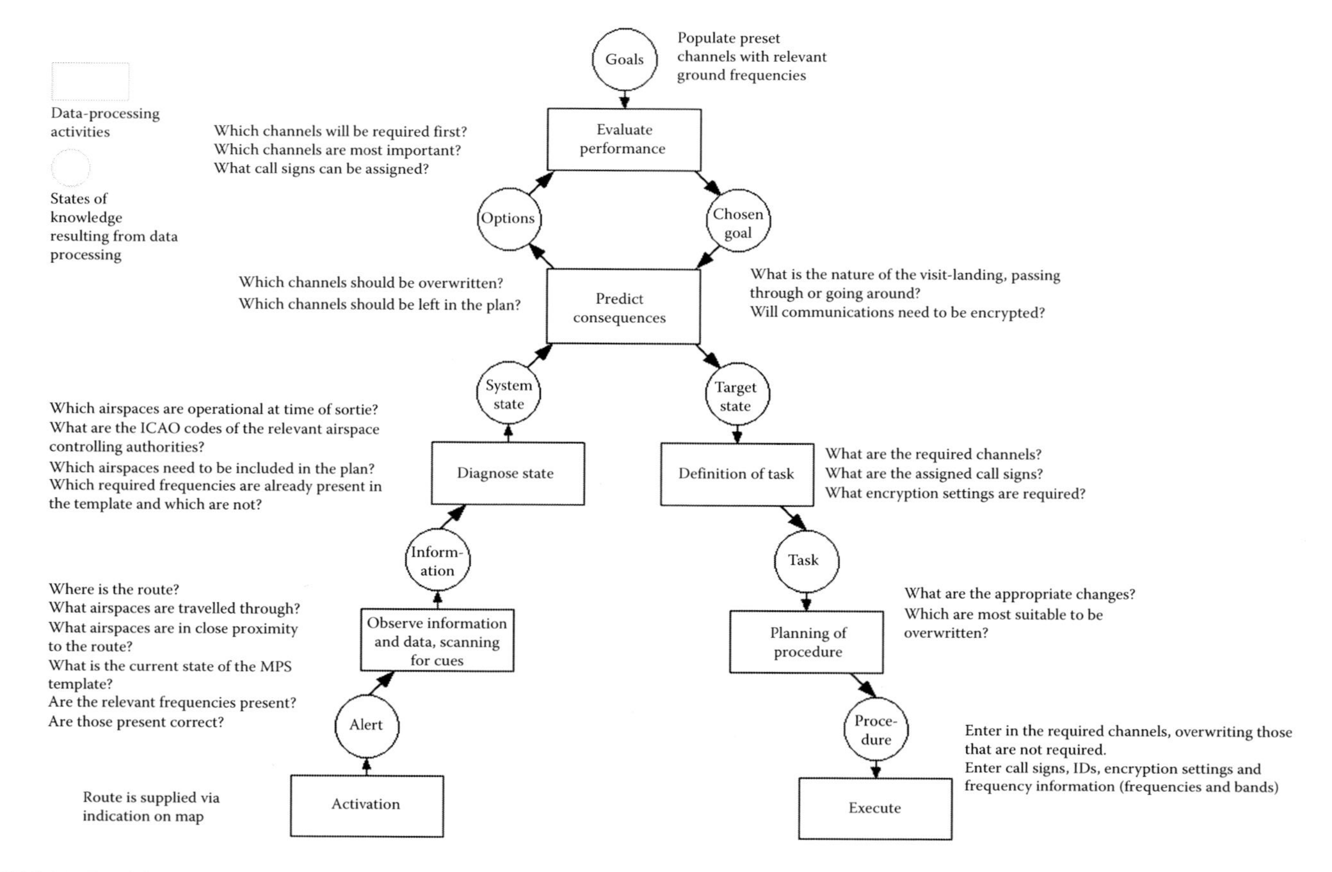

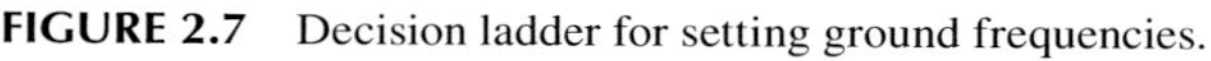
FIGURE 2.7 Decision ladder for setting ground frequencies.

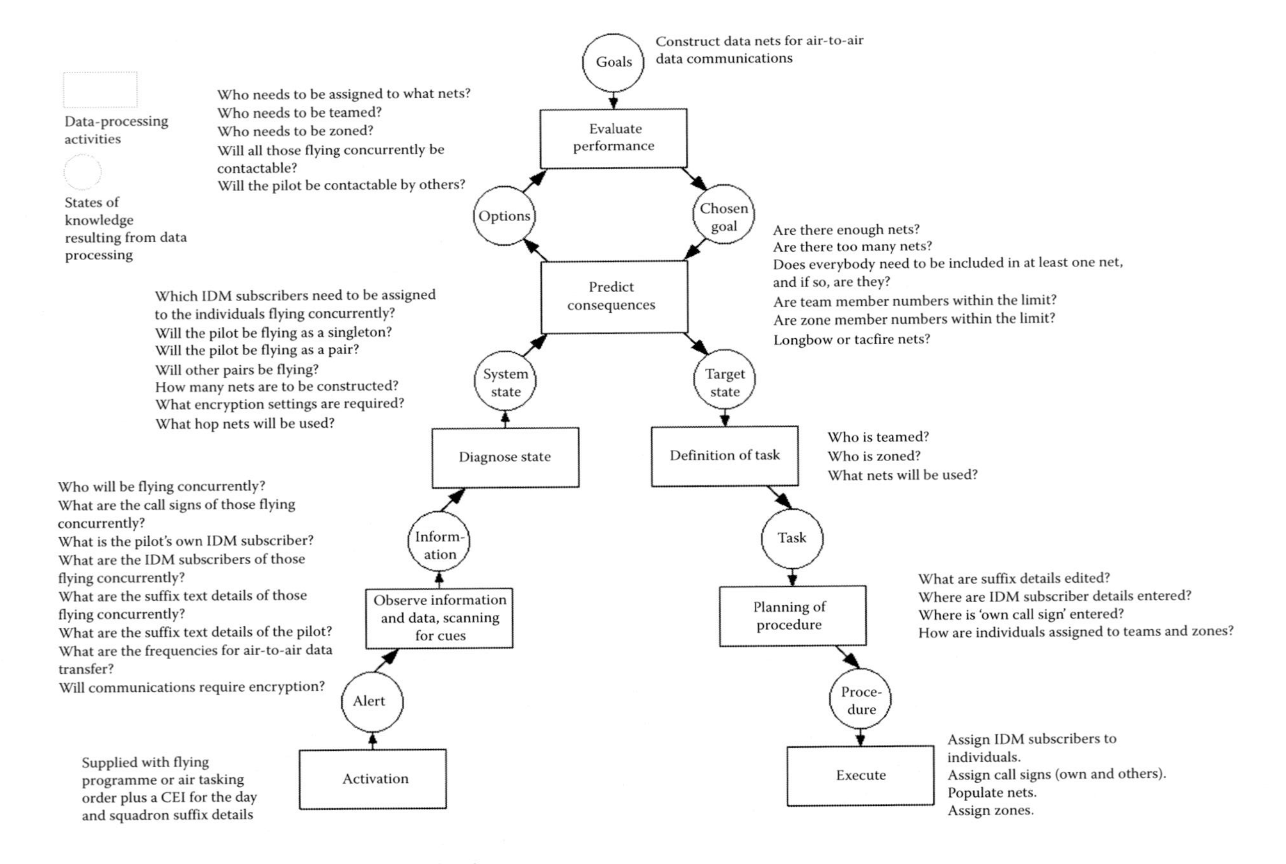

FIGURE 2.8 Decision ladder for air-to-air communications.

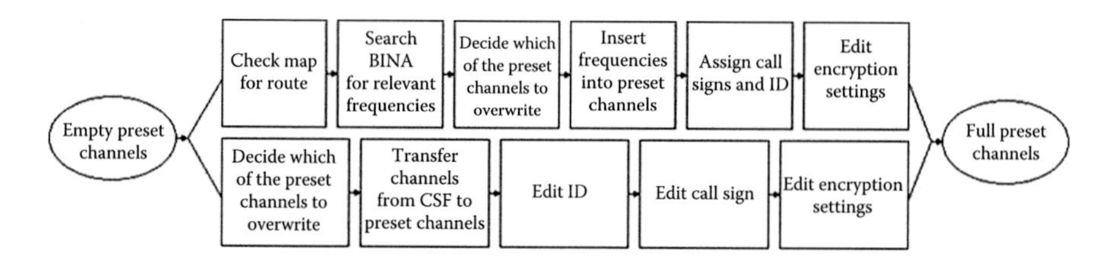

FIGURE 2.9 StrA for completing present channels.

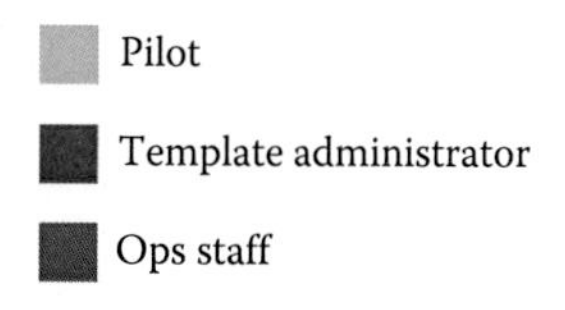

FIGURE 2.10 Key to the coding of the three actors.

more activity could be carried out in the preparatory support activities and template development situations, less time pressure would be put on the pilot immediately before takeoff. In addition, if more of the activity was carried out in the initial three situations, rather than in the individualisation of comms plan situation, the activity could potentially be carried out by a number of different actors, hence dividing workload (Table 2.1).

Redesign of the software package based on task flow would reduce the need for multiple windows to be open simultaneously and reduce the need for extensive training in MPS. The SOCA phase of the analysis is built on the CAT developed in the ConTA phase. The most apparent characteristic of MPS to emerge from this analysis was the requirement for last minute activity. Only a small proportion of the activity carried out within the system happens in the first two situations, namely, preparatory support activities and template development. The large majority of activity occurs in mission-specific activities and individualisation of comms plan. By adding actors to the CAT, it was made clear that not only must the majority of activity be carried out in the latter situations of the planning process, but also that this activity is predominately carried out by the pilot. In summary, the CWA outputs not only highlighted the mismatch between task organisation and system organisation, but also brought to attention the uneven spread of activity across situations and actors.

2.3 CWA FOR REQUIREMENTS SPECIFICATION

2.3.1 Linking Requirements and CWA

While the presentations of CWA and requirements documents differ significantly, the process through which they are constructed bear some important similarities. Most requirements engineers agree that at the outset of any requirements specification process the domain or problem space needs to be defined; it is necessary to define the capability gap, and the domain in which that gap resides, before specifying

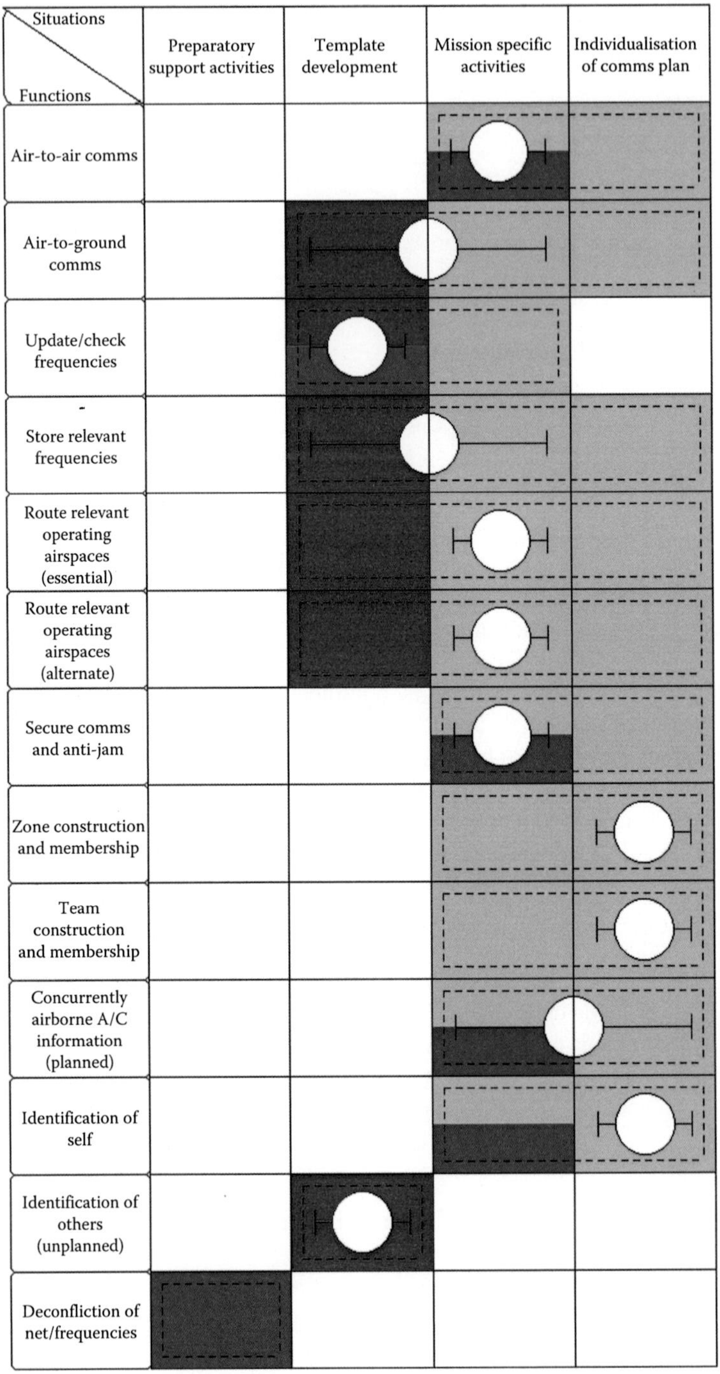

Situations / Functions	Preparatory support activities	Template development	Mission specific activities	Individualisation of comms plan
Air-to-air comms				
Air-to-ground comms				
Update/check frequencies				
Store relevant frequencies				
Route relevant operating airspaces (essential)				
Route relevant operating airspaces (alternate)				
Secure comms and anti-jam				
Zone construction and membership				
Team construction and membership				
Concurrently airborne A/C information (planned)				
Identification of self				
Identification of others (unplanned)				
Deconfliction of net/frequencies				

FIGURE 2.11 SOCA-CAT for communications planning.

TABLE 2.1
WCA for Communications Planning

Object-Related Process	Skill	Rule	Knowledge
Air-to-ground comms	Insert route relevant frequencies	If travelling through an airspace, then insert frequencies for that airspace	Understand what services are required from each airspace authority and the reasons for which communication with that airspace authority is necessary
Air-to-air comms	Insert mission relevant frequencies	From Communications Electronic Instructions, decide which frequencies will be required	Build an understanding of who in the squadron is to be communicated with, why those communications are necessary, and how those communications are enabled
Update/check frequencies	Check frequencies when necessary	Infer from dates on references documents when frequencies must be checked and updated	Understand the need for changing frequencies, when these changes are likely to happen, and how to go about cross-checking them with those stored in mission templates
Store relevant frequencies	Fill call sign frequency list and then present channels list with relevant frequencies	Understand the difference between the different methods of storing frequencies and what information can be contained in each storage mode	Recognise the need for different methods of storing data, which frequencies are most suited to each storage method, and how these frequencies are retrieved in the A/C
Route relevant operating airspaces (essential)	Insert route relevant frequencies	Deduce from the route and reference documents which frequencies will be required	From the route, understand which airspace controlling authorities will need to be contacted. From reference documents understand airspaces' hours of operation and the different rules and regulations related to each type of airspace and airspace boundary
Route relevant operating airspaces (alternate)	Insert frequencies for planned diversions	Deduce from the route and reference documents which frequencies will be required for planned diversions	Understand the need to make plans for possible diversions and the most suitable areas to divert to. From this, have an appreciation of the different airspace operating procedures and times, and how this might affect any possible planned diversion options
Deconfliction of nets/frequencies	Ensure used frequencies/ nets do not interfere with each other	From table, see which frequencies interfere with each other, and avoid those combinations	Know how frequencies may interfere with each other and the consequences of that interference. Ensure this does not happen by selecting frequencies that are of a sufficient separation

(*Continued*)

TABLE 2.1 ***(Continued)***
WCA for Communications Planning

Object-Related Process	Skill	Rule	Knowledge
Secure comms and anti-jam	Ensure correct SATURN or HaveQuick net and crypto net variables	Insert correct SATURN or HaveQuick nets and crypto net variables based on the Communications Electronic Instructions	Understand the need for secure comms and the capabilities of SATURN and HaveQuick radios. Understand the need for correct nets and crypto net variables when communicating with other military organisations or individuals
Concurrently airborne A/C information (planned)	Select required information based on own and others' routes	Check flying programme or Air Tasking Order to determine who will be flying concurrently. Insert the relevant A/C information into the planner	Understand the need for different types of information based on the flying times of those in the squadron, what information is required, and what this information will enable the pilot to do
Zone construction and membership	Select individuals for zone membership	Assign Improved Data Modem subscribers to those flying concurrently, and assign the relevant number of zone members	Understand what zone membership entails and what information can be sent to and from zone members. An understanding of the importance of Improved Data Modem subscribers and suffix text, the limit on number of zone members, and the connection between channels and zones
Team construction and membership	Select individuals for team membership	Assign Improved Data Modem subscribers to those flying concurrently, and assign the relevant number of team members	Understand what team membership entails and what information can be sent to and from team members. An understanding of the importance of Improved Data Modem subscribers and suffix text, the limit on number of team members and relevance of initial Improved Data Modem nets
Identification of self	Insert relevant transponder codes	Insert transponder codes based on A/C details, mission details, and pilot information	Understand the importance of the different transponder modes, how they relate to A/C tail number, mission requirements and individual pilot information. Understand the importance of being able to identify oneself
Identification of others (unplanned)	Assign global Improved Data Modem subscriber	Assign global Improved Data Modem subscriber and the relevant Improved Data Modem net, based on relevant information	Understand the need to identify others encountered en-route that are not originally planned for. Understand the ability to acquire information from others without the need for voice communications, and how the global Improved Data Modem subscriber and nets enable this

any systems aiming to fill that gap (Jackson and Zave, 1993). This is a central tenet of the first stage of CWA – WDA. In this stage of the analysis, an AH is constructed that describes the system across five levels of abstraction (Jenkins et al., 2009). The description is based not on activities or goals specific to actors, like some other human factors methods (e.g. cognitive systems engineering [Hollnagel and Woods, 1999] or cognitive task analysis [Schraagen et al., 2000]), rather it is focussed on abstract functions and domain structure. The analysis does not focus on how activities are to be achieved but on what needs to be fulfilled within the constraints of the domain. This concept of describing *what* needs to achieved, independent of *how* it is to be achieved, is central to both user requirement document (URD) and system requirement document (SRD) (AOF, 2010a). Both CWA and requirements specification processes are, or at least should be, problem or domain driven, independent of technology (Myers, 1985; Jenkins et al., 2009). The concept that domain understanding must come before consideration of specific, technologically focussed activity is elegantly described by Ernst et al. (2006) in the context of software requirements engineering:

> This is akin to understanding the terrain before understanding what paths one can take therein. (p. 3)

With both requirements engineering and CWA come the need for prior acquisition of information. It is not possible to analyse or describe any system without understanding the domain in which it is to be situated. It is crucial that a solid foundation of information regarding the domain and the actors is laid before specifying any system to be used in that domain, and by those actors. There are a number of requirements frameworks that make this point explicit (e.g. the Knowledge Acquisition in Automated Specification [KAOS] methodology [Dardenne et al., 1991] and the i* framework [Yu, 1997]). Indeed, in a review of requirements engineering programmes in software development, Curtis et al. (1988) asserted that a poor understanding of the domain was the primary cause of project failure.

2.3.2 WDA for Requirements

2.3.2.1 Structuring and Communicating Information

CWA and requirements engineering frameworks both stipulate the need to gather information about the domain; however, it is in the structural organisation of that information that CWA can provide significant support to the requirements process. This is particularly true when considering large, complex systems and those systems in which the interaction between technology and human governs success, namely, sociotechnical systems. As Yu (1997) notes, as the complexity of a domain increases, so does the need for a tool to assist in the representation of that complexity. Verbal protocol analysis (Ericsson and Simon, 1980, 1993; McIlroy and Stanton, 2011a; McIlroy et al., 2012) and rich pictures (Checkland, 1981; Stanton and McIlroy, 2012) are two methods to gather information about the task of planning for communications in MPS. The information gathered from these activities served to inform CWA (McIlroy and Stanton, 2011b), and in turn CWA served to structure and organise the information gathered.

The benefit provided by CWA in structuring information first comes in the earliest stage of the analysis – WDA. It is in this phase that the AH is developed. The AH provides a representation of the system that displays the underlying organisation of the work domain and of the system. Not only can this representation facilitate the requirements specification process (described in detail in Section 2.3.2.2) but it also has the added benefit of being easily interpreted. Hence, it offers a means for communicating this information in a format more likely to be understood, and to be understood more quickly, than if the system description was presented in a textual format (Walker et al., 2010). This point refers to the adage 'a picture is worth a thousand words'; it would likely require many pages of text to describe in equal detail the information that the AH provides about the structure of a system. This has benefits not only for the human factors researcher (in developing system understanding) and the requirements engineer (in saving time and effort trying to convey the information in text), but also for the individual tasked with designing the envisioned system. Though success in writing requirements is critical if the envisioned system is to provide all required capabilities, it is only one half of the process; the requirements must also be interpreted.

2.3.2.2 Functions, Connections and Conceptual Model

As described above, the AH is a representation of the work domain that focuses on abstract functions and functional connections; it is independent of actors and specific activities. Indeed, it is the authors contention that it is the AH that provides the most useful and applicable tool for informing the requirements specification process as it is WDA that helps think about the reasons for a system's existence (Sanderson et al., 1999). Furthermore, the AH specifies the domain objectives and functions that must be available and satisfied if the system is to achieve its goals(s). The goal (or goals in the case of systems with more than one overriding purpose) at which functions and objectives are aimed at satisfying is described on the topmost level of the AH (Vicente and Rasmussen, 1992); it is the system's reason for existence that is described. As described above, this is akin to the single statement of need in which it provides a short description of the needs of the capability, that is, the goal, or purpose of the system. Hence, developing an AH for a domain in which there is a capability gap will provide the purpose of goal of a required system; this can then be used to inform, if not provide, the single statement of need.

In the next level down from the functional purpose, the values and priority measures of the system are described. These provide two benefits when considering requirements and system design. First, though the functional purposes can be used to set a benchmark against which to judge system performance, they do not cover all aspects of the system. The values and priority measures, on the other hand, encompass the system in its entirety; they therefore offer a means for measuring the effectiveness of a system as a whole and as such may be used as a form of high-level acceptance criteria. Second, and more importantly for the actual design of the requirements, they describe the high-level constraints in the system. Only by satisfying these constraints can the overall system purpose be satisfied. It is for these reasons that the values and priority measures should be considered when designing a new or updated system. Not considering the nodes at this level of the hierarchy would present two significant risks: a system may be developed that cannot be validated (there would be no way to measure how well the system is achieving its purpose) or a system may be developed that does not perform within the constraints of the work domain or environment (hence does not perform at all).

The approach considered so far is one of a top-down nature, insofar as the purpose is first considered, then the constraints within which that purpose must be fulfilled. It is also possible, however, for the AH to inform requirements specification from a bottom-up perspective. The bottom level of the AH lists the physical objects that comprise the system, while the second level up, the object-related processes, describes the affordances of those physical objects. When considering whether or not a system satisfies the requirements described in the URD and SRD, there is a tendency to focus on technical specifications, that is to say the focus is on assessing whether or not each specific component performs the individual function for which it was designed (Naikar and Sanderson, 2001). It is the case, however, that though requirements can be met at the physical level this does not necessarily mean that those physical requirements will combine to fulfil or achieve a higher-level function or purpose. Where the AH provides benefit is in the description of the linkages between lower-level system components and higher-level functions and goals. It is therefore possible not only to assess the level of component performance, but also to consider the additive effect of all object-related processes, and hence to assess system performance in terms of higher-order functions and purposes.

Through the combination of bottom-up and top-down considerations of system design and requirements specification, it is possible to develop a functional model of a system. As previously described, CWA provides a means for structuring and organising information gathered from experts and end users; it can therefore be used to represent, in terms of functions and linkages, the conceptual models of experts. A conceptual model is an internal, mental representation, based on previous experience, of how something works (Craik, 1943). Individuals use conceptual models to direct their interactions with the world. In a work context, individuals use conceptual models to enable them to perform activity and operate systems (Wilson and Rutherford, 1989). Rouse et al. (1992) suggested that the closer the parallel between system and user models, the less cognitive effort is required from the user to interpret data, solve the problems and perform tasks.

There is, however, no mention of the owner of this conceptual model (one representative user? multiple users? domain experts?) and no description of how to go about obtaining information regarding this conceptual model. The first issue can only be addressed by indicating to whose conceptual model the statement refers. The second issue can be addressed using the AH; it is possible to use the AH as a tool to present information in an organised fashion. Given the AH is constructed through the collaborative effort of analysts, experts and representative users, it follows that the system model, if it is to be effective and valid, must be composed of the functional nodes and relationships intrinsic to the work domain as described in the hierarchy (Elm et al., 2003). Considering and developing the AH in this manner results in a system representation that can be used as a tool to directly influence the requirements specification process such that not only is it possible to specify the requirement for a system to match the users' conceptual models, but also allows for a description of what these conceptual models consist.

2.3.2.3 The AH for Training System Requirement

Naikar (2006) describes the use of CWA in the specification of training system requirements for the Australian Air Force's F/A-18 fighter aircraft (Naikar and Sanderson, 2001; Naikar, 2006). As Naikar describes, there has been a tendency in the past for organisations to assume expense equates to quality; expensive systems must, by their very nature of being expensive, be worthwhile. This is not, however, necessarily the case (Lintern and Naikar, 1998). There is still a significant requirement not only for an analysis of the training needs, but also of the specification of the system requirements to meet those needs. To address this need, Naikar (2006) suggests using the AH, offering an explanation of the utility of each level in terms of the specification of requirements. Although the analysis was performed to specify training system requirements, the process is equally applicable to the specification of requirements for systems across platforms and domains. Naikar's (2006) derivations of training system requirements from each level of the AH are presented in Table 2.2.

An important point to make here is that the training requirements described by Naikar (2006) are slightly different in nature to the requirements set out in the URD and SRD. This was an analysis of a domain aimed at providing the functional understanding of the requirements for training; that is to say, Naikar (2006) used CWA to describe the domain in such a way as to inform for which functions and activities a training system needs to train. Though this process is different from that of constructing the requirements for a system intending to fill a capability gap, the concepts remain useful. Naikar investigated a system to inform training requirements; we propose that CWA can be used to investigate a domain or environment to inform system requirements. An analysis of an operational system results in a detailed understanding of its functions and purposes, hence allows for an understanding of the functions and purposes on which training should be focussed. An analysis of a required

TABLE 2.2
Training System Requirements Relating to Each of the Five AH Levels

Functional purposes	The training system must be capable of satisfying the training objectives of the work domain
Values and priority measures	The training system must be capable of collecting data relating to the measures of performance in the work domain
Purpose-related functions	The training system must be capable of generating scenarios for training the fundamental functions of the work domain
Object-related processes	The training system must be capable of simulating the functionality of physical devices and significant environmental conditions in the work domain
Physical objects	The training system must be capable of recreating the functionally relevant properties of physical devices and significant features of the environment in the work domain

Source: Adapted from Naikar, N. 2006. *International Journal of Industrial Ergonomics*, 36, 423–438.

system, and the environment in which that required system will be situated, results in a detailed understanding of the functions that will need to be performed, hence an understanding of design needs; requirements are necessarily informed.

2.3.2.4 The AH for Interface Design Requirement

CWA has often been posited as a method for informing interface design (e.g. Vicente and Rasmussen, 1992; Burns and Hajdukiewicz, 2004); however, it is most commonly considered in the latter stages of the design cycle, after user and system requirements have already been specified. It is argued, however, that the AH can be used much earlier in the system's life cycle, namely, during the requirements specification stage. Although it may be premature to determine the specific layout of the interface at this stage of the process, it is possible to use the structure defined in the AH to provide some suggestions for interface organisation.

For example, in the investigation of the communications planning suite of the MPS (McIlroy and Stanton, 2011b; Stanton and McIlroy, 2012), it was determined that to perform specific functions (described on the second level from the bottom of the AH) multiple windows had to be open simultaneously. This equated to a need to switch between various windows when performing a single function. To address this issue, it was suggested that for each function on the object-related processes level of the AH, all of the connected physical objects should be available to the user simultaneously (Figure 2.12). The key concept is that an interface should be organised such that the user may perform individual functions described in the AH without the need to switch between windows or modes. Though this analysis was carried out on an existing system, with the physical objects in place, we contend that the principle remains equally applicable to systems that have yet to be defined to that level of specificity. By stipulating the requirement for the functional organisation of information and options presented to the end user, it will be possible to help ensure a system that supports task-orientated groupings of information, which would not only prompt the user as to the information they should be entering, but also support

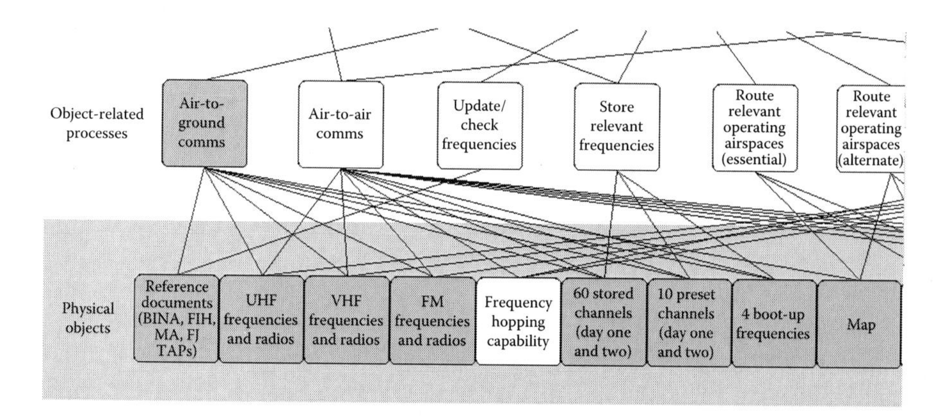

FIGURE 2.12 Function of planning for air-to-ground communications, with all connected physical objects shaded grey. (Adapted from McIlroy, R. C. and Stanton, N. A. 2011. *Applied Ergonomics*, 42(2), 358–370.)

task engagement and minimise the cognitive load associated with switching between interfaces (Eggleston and Whitaker, 2002).

The use of the AH in specifying requirements would be most suitable when constructing the SRD as it considers the functional systems architecture in facilitating and clarifying the allocation of required functionality to individual systems or components (AOF, 2010b). In this instance, the windows, modes or option screens made available to the user can be considered as separate components of the system. This does not describe *how* information is to be presented (an obligation of all requirements documents, military or otherwise); rather, the focus is on *what* information is to be presented in concert. This provides the benefit of allowing for the specification of interface structure without providing direct guidance on the aesthetic properties of the interface; this is a task for the solution provider. Note, though, that the authors are of the opinion that this process is not ideal. Handing off requirements documents to a solution provider and waiting for a system to be tested makes the assumption that the solution provider will themselves apply human factors methods in design, supplying an interface that is optimally designed to support the task.

2.3.2.5 Function-Based WCA

The focus so far has been on the use of the AH to inform requirements; however, this is not to say that the remaining phases are not of use. Of particular use in the specification of requirements is WCA, the final stage of CWA. WCA addresses the constraints dictating actor behaviour for different system activities (Jenkins et al., 2009) and is commonly modelled using Rasmussen's (1983) SRK taxonomy. Previous analyses of worker competencies have commonly adopted a decision ladder approach; that is, an SRK inventory is constructed for each control task, with each row describing the different types of behaviour required for individual steps on decision ladders for that control task (Kilgore and St-Cyr, 2006; Kilgore et al., 2009). We contend that although this process is indeed useful in the latter stages of the design process, it is not a method that can be applied in requirements specification. Describing competencies in this way requires the system to have had activities described and analysed. It is not possible to construct decision ladders for systems in which no specific activities, and strategies therein, have yet to be described. This does not, however, prevent this form of analysis providing an input in early stages of requirements analysis; rather, the method must be adapted. McIlroy and Stanton (2011b) provide a description of the SRK inventory informed by nodes taken from the AH rather than decision ladder steps. In their argument, they contend that using the SRK inventory in this way allows for a description of the competencies required from system actors before a particular solution has been defined.

The process of constructing the SRK inventory in this manner requires a considerable volume of expert and user input. Once this input is supplied, however, converting the information presented in the inventory into system requirements incurs minimal time and effort. The resulting requirements can be used not only to inform the design of the interface, but also to inform personnel recruitment requirements and training requirements (should the skills and knowledge be unavailable in the current personnel base). As an example, consider the function 'deconfliction of

TABLE 2.3
SBB, RBB and KBB Required to Perform the Function 'Deconfliction of Net/Frequencies'

Purpose-Related Functions	SBB	RBB	KBB
Deconfliction of nets/frequencies	Ensure used frequencies/nets do not interfere with each other	From table, see which frequencies interfere with each other, and avoid those combinations	Know how frequencies may interfere with each other and the consequences of that interference. Ensure this does not happen by selecting frequencies that are of a sufficient separation

Source: Adapted from McIlroy, R. C. and Stanton, N. A. 2011. *Applied Ergonomics*, 42(2), 358–370.

nets/frequencies' in the WCA carried out in the investigation of the communications suite in MPS (McIlroy and Stanton, 2011b) (Table 2.3).

In terms of interface design, for the SBB, the requirement specified may be worded as such: 'The user shall always be able to ensure used frequencies and nets do not interfere with each other'. It is necessary to explicitly state that this must always be possible as it is a requirement of all users, novice to expert. For the remaining two behaviours, however, it should be explicitly stated that information should only be available upon user request; having this information continually presented would not provide any benefit to the expert user as it is predominantly skill they use to guide actions (see Rasmussen, 1983, for a discussion on the types of behaviour governing actions of expert and novice actors). Only less experienced users would require this information; therefore, wording of requirements should make this explicit. For example, the requirement specified to support RBB may be as follows: 'The system shall provide a table, upon user request, that provides information displaying the frequencies and nets that interfere with each other'. Similarly, to support KBB, 'The system shall be provide information, upon user request, regarding the confliction of nets and frequencies, for why these conflictions arise, and offer selection options that are of sufficient separation as to avoid the problem of confliction.' Specifying the interface in this way avoids this issue of determining *how* an interface is to be designed by simply stating *what* information is to be presented, thereby adhering to the fundamental principle of specific solution-independency (once again, however, the authors would like to make their position clear that they believe this process to be suboptimal).

This method of specifying requirements is, in effect, allowing for the description of the information an actor needs to perform a task. This type of information requirements specification can be contained within SRDs, though often is not and even when it is, it is rarely specified to the same level of detail. Of the requirements documents viewed by the analysts, only one had a sufficiently extensive section on human factors principles; however, though the principles contained in the section were beneficial, the requirements did not, in the analysts' judgement, go into enough detail when describing interface design requirements. The example presented below represents the most specific of all the requirements analysed with regards to information requiring presentation:

> The information presented and the style of its display should reflect what the operator needs to perform the task.

This requirement assumes that the solution provider has a sufficiently detailed understanding of the users' tasks to provide an interface that optimally supports work. As the solution provider is rarely, if ever, the end user, there should be no reason to make this assumption. Without an understanding of the task, it will be impossible to design a system that optimally supports it. Furthermore, the ill-defined, somewhat vague nature of this requirement is inherently disadvantageous in two primary ways: (1) it does not constrain the solution provider in a concrete way and so may have no effect on interface design and (2) without the addition of an explicit description of the task (which is outside the scope of the requirements documents) it is immeasurable and unfalsifiable, therefore providing no benefit and offering no means for judging the efficacy or success of the acquired system.

2.3.3 The Importance of Process

Although CWA provides pictorial and graphical representations of the system that can, on their own, be used to inform requirements specification, it is the process of performing the analysis that provides the greatest benefit. In order to have a solid understanding of the system under consideration, it is crucial that the customer is involved throughout the process. Involvement must include not only the customer, but also human factors specialists, domain experts and representative users of the envisioned system. This collaborative process between the customer and industry is not strictly participatory design (e.g. Schuler and Namioka, 1993), but a participatory analysis of need, that is to say the needs of the users and the system. Furthermore, it is critical that each party is involved at every stage; it is not sufficient to create a CWA output in one session and expect that to be satisfactory and lasting. The process is inherently iterative, requiring a number of rounds of development and evaluation of analyses. Moreover, any changes made to any of the outputs will likely have an effect in other parts of the analysis, therefore requiring validation by all involved. This point is succinctly described in Elm et al. (2003):

> The artefacts serve as a post hoc mechanism to record the results of the design thinking and as stepping stones for the subsequent step of the process. Each intermediate artefact also provides an opportunity to evaluate the completeness and quality of the analysis/design effort, enabling modifications to be made early in the process. The linkage between artefacts also ensures an integrative process; changes in one artefact cascades along the design thread necessitating changes to all. (p. 7)

Admittedly, it is not always practical to conduct a full CWA on an envisioned system in the earliest stages of its life cycle. If a system is still in the concept stage, the activities to be performed may not yet have been described; hence, the ConTA, the StrA and the SOCA stages may be quite effortful to construct, and as such incur more time and cost than the customer may be willing to offer. It is argued, then, that in these instances a CWA should be conducted on the system that is currently in place, that is to say the system that will be replaced, or on systems performing similar purposes. The descriptions of activity, strategies and social organisation will still be applicable to the proposed or envisioned system, therefore providing a basis for development of the new system, unless the system under acquisition is intending to provide a capability that is currently unavailable, involving completely new roles and activities. As this is rarely the case, the full CWA can be applied in most instances. In the few instances where this is the case, it is proposed that WDA and WCA will still be applicable and beneficial in relation to requirements specification.

Reducing customer involvement after requirements documents have been passed off to industry cannot ensure a system that is optimally designed. It is argued that the full CWA process should be used a support tool throughout the system's life cycle, with the AH and WCA providing considerable benefit in the requirements stages, and the remaining three phases applicable throughout design and manufacture. It is argued that CWA, applied correctly by individuals of sufficient expertise in human factors, can be used to structure and organise the collaborative and iterative process of design and manufacture as long as the customer and a representative sample of domain experts and prospective end users remain involved. The AH provides a graphical representation that summarises the system in an easily interpreted format. This can also be said for the other CWA outputs; each provides a different representation of the system that, if replicated in text, would require an extensive written account to describe. As with the AH, not only can the representations offered by CWA facilitate system understanding, they can also provide a means for communicating that information to those responsible for designing, and ultimately building (hardware), or coding for (software) that system.

2.4 CONCLUSIONS

The primary conclusions of the exploration of CWA as a tool for informing requirements specification are that not only does CWA bear significant resemblance to some requirements engineering frameworks, but that CWA can offer a number of ways of informing, if not supplying, user and system requirements. Although the CWA process comprises five stages, it is argued that the AH constructed in the WDA stage and the AH informed SRK inventory developed in the worker competencies stage are the most applicable. Despite the minimal discussion of the other stages of CWA, namely, ConTA, StrA and SOCA, it is argued that they are equally important when designing new systems. It is merely contended that they are not as applicable in the requirements specification stage of the system's life cycle; rather, they are of more benefit during the actual design process. Though the requirements documents aim to remain solution independent, it is the authors' contention that the customer should retain direct involvement throughout the design process, applying human factors methods, throughout. This is not the current standard practice; it is not uncommon that once requirements documents are constructed and passed down to industry, the customer ends its involvement until some pre-defined testing of prototype systems. At this stage, it is too late to provide design guidance; it is much easier and much cheaper to change something when it is in a paper- or computer-based design rather than something half-built in a factory.

In summary, this discussion of CWA and requirements has three major conclusions in terms of requirements specification, and has three major conclusions in terms of the issues with the current requirements and acquisition process. In terms of the specification of requirements, the three key points are as follows: (1) CWA's AH can be used to help understand the domain of interest in terms of its constraints, the opportunities it affords, and the conceptual models held by domain experts and prospective end users; (2) the communication of information through a graphical or pictorial format provides significant benefit, in terms of ease of understanding, over the use of textual descriptions of systems and (3) the AH and WCA can inform requirements specification, and, in some cases, can directly provide requirements that need only minimal adaptation. In terms of the issues with the current process, the three central conclusions are as follows: (1) the governmental procurement organisation does not always determine its own success criteria, rather they leave it to industry to regulate themselves. (2) The iterative process of requirements analysis, specification and design is not always adhered to, with human factors analyses only applied during the earliest stages, if at all. Although it is critical that human factors play a role at this early stage, it is of equal importance that the process is carried throughout the design stages; the process is inherently iterative. (3) The governmental procurement organisation does not always retain involvement throughout the requirements or design phases; rather, it is common for the governmental procurement organisation to take a backseat once requirements documents have been handed to industry and only become involved once more when a testable prototype has been produced. Indeed, it is the author's experience in talking to those involved in the process (across visits to prime contractors in industry) that the governmental procurement organisation does not even always have this much involvement, with many of the requirements documents being written by the companies paid to produce the solution. Though this approach may be the most cost-effective in the short term, it is clear that this system of requirements production and system procurement is not performing at an acceptable level of expense, in terms of both time and money (Gray, 2009).

REFERENCES

Aerosystems International. 2006. *Mission Planning System Flyer.* Yeovil, UK: AeI.

Ahlstrom, U. 2005. Work domain analysis for air traffic control weather displays. *Journal of Safety Research*, 36, 159–169.

AOF (Acquisition Operating Framework). 2010a. *Requirements and Acceptance.* http://www.aof.mod.uk/aofcontent/tactical/randa/index.htm. Accessed 24 May 2010.

AOF (Acquisition Operating Framework). 2010b. *Defining System Requirements.* http://www.aof.mod.uk/aofcontent/tactical/randa/content/srdosanddonts.htm. Accessed 25 May 2010.

Burns, C. M. and Hajdukiewicz, J. R. 2004. *Ecological Interface Design.* Boca Raton, FL: CRC Press.

Checkland, J. 1981. *Systems Thinking, Systems Practice.* Chichester, UK: John Wiley.

Craik, K. 1943. *The Nature of Explanation.* Cambridge: Cambridge University Press.

Curtis, B., Krasner, H. and Iscoe, N. 1988. A field study of the software design process for large systems. *Communications of the ACM*, 31, 1268–1287.

Dardenne, A., Fickas, S. and van Lamsweerde, A. 1991. Goal-directed concept acquisition in requirements elicitation. *Proceedings of the IWSSD-6 – 6th International Workshop on Software Specification and Design*. Como, Italy, pp. 14–21.

Eggleston, R. G. and Whitaker, R. D. 2002. Work-centred support systems design: Using organizing frames to reduce work complexity. *Proceedings of the Human Factors and Ergonomics Society 46th Annual Meeting*, Baltimore, Maryland. Human Factors and Ergonomics Society, pp. 265–269.

Elm, W. C., Potter, S. S., Gualtieri, J. W., Roth, E. M. and Easter, J. R. 2003. Applied cognitive work analysis: A pragmatic methodology for designing revolutionary cognitive affordances. In E. Hollnagel (Ed.). *Handbook for Cognitive Task Design*. London: Lawrence Erlbaum Associates.

Ericsson, K. A. and Simon, H. A. 1980. Verbal reports as data. *Psychological Review*, 87, 215–251.

Ericsson, K. A. and Simon, H. A. 1993. *Protocol Analysis; Verbal Reports As Data*. Cambridge, MA: MIT Press.

Ernst, N. A., Jamieson, G. A. and Mylopoulos, J. 2006. Integrating requirements engineering and cognitive work analysis: A case study. *Presented at the Fourth Conference on Systems Engineering Research*. Los Angeles, CA.

Gray, B. 2009. *Review of Acquisition for the Secretary of State: An Independent Report by Bernard Gray*. London: The Stationary Office.

Hackman, J. R. and Oldman, G. R. 1980. *Work Redesign*. Reading, MA: Addison-Wesley.

Hollnagel, E. and Woods D. D. 1999. Cognitive systems engineering: New wine in new bottles. *International Journal of Human-Computer Studies*, 51, 339–356.

Jackson, M. and Zave, P. 1993. Domain descriptions. *International Symposium on Requirements Engineering*. San Diego, CA, pp. 56–64.

Jenkins, D. P., Stanton, N. A., Salmon, P. M. and Walker, G. H. 2009. *Cognitive Work Analysis: Coping with Complexity*. Farnham, UK: Ashgate Publishing Limited.

Jenkins, D. P., Stanton, N. A., Salmon, P. M., Walker, G. H. and Young, M. S. 2008. Using cognitive work analysis to explore activity allocation within military domains. *Ergonomics*, 51, 798–815.

Kilgore, R. and St-Cyr, O. 2006. SRK inventory: A tool for structuring and capturing worker competencies analysis. *Proceedings of the Human Factors and Ergonomics Society 50th Annual Meeting 2006*. pp. 506–509.

Kilgore, R., St-Cyr, O. and Jamieson, G. A. 2009. From work domains to worker competencies: A five-phase CWA. In A. M. Bisantz and C. M. Burns (Eds.). *Applications of Cognitive Work Analysis*. Boca Raton, FL: CRC Press, pp. 15–47.

Lintern, G. and Naikar, N. 1998. Cognitive work analysis for training system design. *Proceedings of the Australasian Computer Human Interaction Conference (OzCHI'98)*. Los Alamitos, CA: IEEE Computer Society Press, pp. 252–259.

McIlroy, R. C. and Stanton, N. A. 2011a. Observing the observer: Non-intrusive verbalisations using the concurrent observer narrative technique. *Cognition, Technology and Work*, 13(2), 135–149.

McIlroy, R. C. and Stanton, N. A. 2011b. Getting past first base: Going all the way with cognitive work analysis. *Applied Ergonomics*, 42(2), 358–370.

McIlroy, R. C., Stanton, N. A. and Remington, R. E. 2012. Developing expertise in military communications planning: Do verbal reports change with experience? *Behaviour and Information Technology*, 31(6), 617–629.

Myers, W. 1985. MCC: Planning the revolution in software. *IEEE Software*, 2, 68–73.

Naikar, N. 2006. Beyond interface design: Further applications of cognitive work analysis. *International Journal of Industrial Ergonomics*, 36, 423–438.

Naikar, N. and Sanderson, P. 2001. Evaluating design proposals for complex systems with work domain analysis. *Human Factors*, 43, 529–542.

Naikar, N., Moylan, A., and Pearce, B. 2006. Analysing activity in complex systems with cognitive work analysis: Concepts, guidelines, and case study for control task analysis. *Theoretical Issues in Ergonomics Science*, 7(4), 371–394.

Rasmussen, J. 1974. *The human data processor as a system component: Bits and pieces of a model*. Report no. Risø-M-1722. Roskilde, Denmark: Danish Atomic Energy Commission.

Rasmussen, J. 1983. Skills, rules, knowledge; signals, signs, and symbols, and other distinctions in human performance models. *IEEE Transactions on Systems, Man and Cybernetics*, 15, 234–243.

Rasmussen, J., Pejtersen, A. and Goodstein, L. P. 1994. *Cognitive Systems Engineering*. New York: Wiley.

Rouse, W. B., Cannon-Bowers, J. A. and Salas, E. 1992. The role of mental models in team performance in complex systems. *IEEE Transactions on Systems, Man, and Cybernetics*, 22(6), 1296–1308.

Sanderson, P., Naikar, N., Lintern, G. and Goss, S. 1999. Use of cognitive work analysis across the system life cycle: From requirements to decommission. *Proceedings of the 43rd Annual Meeting of the Human Factors and Ergonomics Society*. Santa Monica, CA: Human Factors and Ergonomics Society, pp. 318–322.

Schraagen, J. M., Chipman, S. F. and Shalin, V. J. 2000. *Cognitive Task Analysis. Expertise: Research and Application Series*. Mahwah, NJ: Lawrence Erlbaum Associates.

Schuler, D. and Namioka, A. 1993. *Participatory Design: Principles and Practices*. Mahwah, NJ: Lawrence Erlbaum Associates.

Stanton, N. A. and McIlroy, R. C. 2012. Designing mission communication planning: The role of rich pictures and cognitive work analysis. *Theoretical Issues in Ergonomics Science*, 13(2), 146–168.

Stanton, N. A., Salmon, P. M., Rafferty, L. A., Walker, G. H., Baber, C. and Jenkins, D. P. 2013. *Human Factors Methods: A Practical Guide for Engineering and Design* (second edition). Aldershot, UK: Ashgate.

Stanton, N. A., Walker, G. H., Jenkins, D. P., Salmon, P. M., Revell, K. and Rafferty, L. 2009. *Digitising Command and Control: Human Factors and Ergonomics Analysis of Mission Planning and Battlespace Management*. Aldershot, UK: Ashgate.

Vicente, K. J. 1999. *Cognitive Work Analysis: Toward Safe, Productive, and Healthy Computer-Based Work*. Mahwah, NJ: Lawrence Erlbaum Associates.

Vicente, K. J. and Rasmussen, J. 1992. Ecological interface design: Theoretical foundations. *IEEE Transactions of Systems, Man and Cybernetics*, 22, 589–606.

Walker, G. H., Stanton, N. A., Baber, C., Wells, L., Gibson, H., Salmon, P. and Jenkins, D. 2010. From ethnography to the EAST method: A tractable approach to representing distributed cognition in Air Traffic Control. *Ergonomics*, 53, 184–197.

Wilson, J. R. and Rutherford, A. 1989. Mental models: Theory and application in human factors. *Human Factors*, 31, 617–634.

Yu, E. S. 1997. Towards modelling and reasoning support for early-phase requirements engineering. *International Symposium on Requirements Engineering*. Annapolis, MD, pp. 226–235.

3 Using the Abstraction Hierarchy to Create More Innovative Specifications

Daniel P. Jenkins

CONTENTS

3.1 INTRODUCTION

Since its conception in the nuclear power domain, cognitive work analysis (CWA) has been applied to a variety of other safety critical complex environments. It is fair to say that comparatively, the framework is yet to be widely applied in domains that demand less rigour or regulation. The purpose of this chapter is twofold: (1) to highlight the utility of the framework in domains other than those that fall under the category of complex safety critical and (2) to provide a description of how, by modelling constraints, CWA can be used to inspire more innovative design.

3.1.1 WHAT ARE DESIGN SPECIFICATIONS?

Regardless if you are designing a train or a toothbrush, specifications can be used to provide explicit requirements to guide the design process. This often comes in the form of physical dimensions (e.g. maximum size, location and size of interfacing parts), environmental factors (e.g. temperature and humidity to withstand), ergonomic factors (e.g. description of the target population), aesthetic or sensory factors (e.g. requirement to represent brand values), cost (e.g. material, manufacture, purchase), maintenance that will be needed, quality and safety.

In regulated industries, such as transportation or medical devices, regulations and widely adopted standards often form the cornerstone of a design specification. Taking the example of door controls within trains, they prescribe acceptable locations (e.g. maximum and minimum height above floor level), arrangements (open above close) and actuation forces (e.g. maximum force), along with specifications for reliability and robustness.

Detailed specifications are something that most engineers are very comfortable with, particularly in the later stages of the design process. When described appropriately, they provide measurable and testable requirements for an artefact – creating a clear description of what the product is and how it should perform. Conversely, the prescriptive nature of a specification can also be viewed as stifling, particularly by designers in the early concept generation stages of a project. Tightly defined physical features and functions can limit the scope for lateral thinking. As such, it could reasonably be argued that design processes that are reliant on a product specification are better placed for evolutionary, as opposed to truly innovative products.

One approach used to mitigate the constraining nature of specifications is to discount, or dramatically restrict, the specification at the initial stages of the design process – delaying its introduction until after a series of concepts has been created. The classic example of this would be a 'concept car' that explores a new design direction without being overly concerned by details such as the construction techniques required, the material costs or its impact on fuel performance. While this approach of ignoring constraints in the early stages of the process can have clear advantages for creativity, it can reinforce an 'over the wall culture' between design and engineering, playing up to role stereotypes. Furthermore, it could also be argued that an examination of constraints can direct creativity to develop new ways of managing these constraints.

Few would argue with the notion that detailed specifications can stifle creativity; however, their value to the design process, in terms of efficiency and focus, is also clearly evident. The natural question then becomes, how can the system constraints be managed to ensure that the process retains the advantages of a prescriptive specification, without constraining the process of innovation? In short, as this chapter will explain, it is contended that the consideration of constraints and a specification can, and should, exist throughout the design process. However, the level of detail, and means of presentation, should change to meet the requirement of the design activity at hand.

3.1.2 The Role of Specifications in Teams

The importance of a design specification is particularly evident in projects with sizeable development teams and supply chains. A clear and auditable development process is of paramount importance, and the design specification plays a critical role in this. 'Ownership' for specific requirements can be assigned to individuals, regardless of whether they are in the core project team or cascaded to the supply chain. Likewise, where staged development processes are adopted, structured test plans can be employed to ensure that the design is compliant before project gateways are passed.

The reductional nature of a design specification is one of its great strengths in allowing the roles and responsibilities to be shared. However, without some form of systemic oversight, there is the very real danger that components, or constituent parts, may be designed to be compliant to the identified sections of the design specification; however, they may fail to adequately meet the purpose or values of the system. This is particularly relevant in cases where purposes are not cascaded with the requirements.

Where large organisations are ostensibly producing variants of the same core product (e.g. automotive companies, white goods producers), a highly structured process is particularly valued. The development process can remain common and be honed and refined based on previous practical experience. At the start of the project, a template can be used to form the base specification. Input from different roles in the company such as marketing, ergonomics and benchmarking teams can be used to set specific values within the specification. These can describe the target audience, along with their requirements. The resulting high level of consistency between these specifications has clear advantages for working with ambitious timescales. The specification provides focus and direction across the wider team and substantially reduces the duplication of effort and excessive exploration. From a management perspective, this also has a number of advantages. According to Klein (2014), organisations value predictability because they like projects to run smoothly. Companies like to plan the steps that will take each project from start to finish, the resources for each step, and the schedule. That way, managers can quickly notice perturbations and make the necessary adjustments. The converse to this is that where time pressures are critical, it can be far quicker and less risky to develop a variant of a proven product than to strive for true innovation.

3.2 METHOD

At the risk of oversimplifying things, the design process can be captured by the relationship between the following three words.

The specification should sit central to this; it describes in detail exactly *what* the artefact, which is being developed, should do. As the development process progresses, the design team adds detail to the design to explain *how* each of these requirements can be met. What can be missing in this process is the explicit link up to *why* the requirement or even the product exists.

On the basis of the description thus far, there are a number of shortfalls with the classic design specification. Many design specifications could be improved by the following:

1. Creating an explicit link that describes *why* a requirement is needed
2. Allowing specifications to be viewed at differing levels of abstraction (i.e. what should it achieve at a physical level, what impact should it have on the end user's life)
3. Describing the inter-relationships between components in a system that influence how requirements are to be met

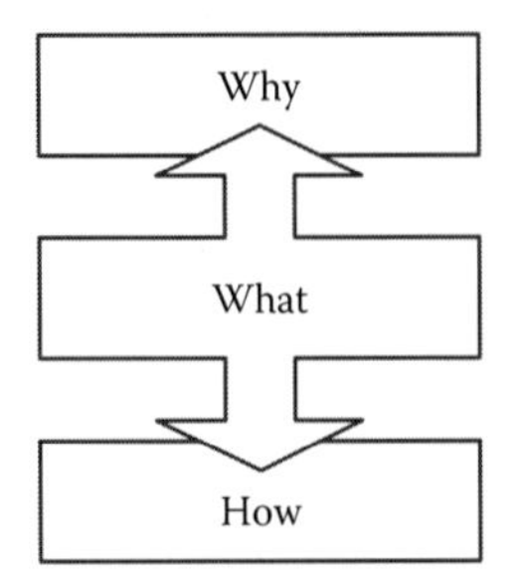

FIGURE 3.1 Why–what–how triad.

One means of addressing these challenges is to adopt the abstraction hierarchy (Rasmussen, 1986; see Figure 3.3) as a means of describing the system that the artefact inhabits and the constraints around it. As the title suggests, the technique and the resulting diagram describe a system at a number of levels of abstraction (typically five). At each level, the model can be used to describe what the system should do at a different level of abstraction. Moreover, each of the nodes in the model can be used to explore the 'what–why–how' triad introduced in Figure 3.1. A given node answers the question of what is required, while the linked nodes below describe how the system can achieve this, whereas the linked nodes above can be used to answer the question why. These upward links are particularly valuable as they provide a rationale for the system. Where there are multiple connections from a single node, the diagram also describes the interrelationships between components and how the same affordances can be achieved by different physical objects.

3.3 CASE STUDY: A THERMOSTAT

The utility of the abstraction hierarchy as part of product specification is perhaps best explained with an example – in this case, a humble thermostat. A relatively simple example has been chosen to illustrate the utility of the approach and also to highlight that CWA can be effectively applied in non-safety critical domains.

The thermostat we will be initially focussing on is deliberately simple – in this case, a product designed for Honeywell by Henry Dreyfuss and launched in 1953 (see Figure 3.2). At the risk of oversimplification, the product can be reduced to four core components:

- A temperature sensor – to measure ambient temperature
- A switch – to toggle the state of the heating control unit (on/off)
- A connection to the heating control unit – to send messages to the heating control unit
- A rotary dial – to capture user inputs and display the set temperature

These physical objects and their functions can be described at the base of the abstraction hierarchy (see Figure 3.3). The diagram lists the physical objects at the

FIGURE 3.2 'Round' thermostat designed by Henry Dreyfuss for Honeywell (c. 1953).

very base, while their direct functions are listed in the row above. A link is made between them to show which objects relate to which functions.

The abstraction hierarchy can also be constructed top-down by considering the overall purpose (or purposes) of the product – in this case, to optimise the heating system and the user experience (see top row of Figure 3.3). Unlike goals, these domain purposes do not change with time or as a result of different events, but remain fixed. The level below, the domain values, provides high-order measures of performance used to determine whether or not the functional purposes are being achieved – in this case, to maximise thermal comfort and convenience, while minimising environmental impact and energy usage.

Different users of the system may place different priorities on these values. Some users may value thermal comfort above all else and have little regard for the energy usage or the environmental impact. Likewise, other users may be incredibly motivated by cost saving, or by minimising their environmental footprint, and prioritise this over their thermal comfort. The majority of users, however, will at some point place some priority on each of these values (this can be explicitly explored in the social organisation and cooperation analysis phase of CWA). Likewise, the priority placed on these values is also likely to change in different situations (this can be explicitly explored in the activities analysis phase of CWA using a contextual activity template). While many of the values may be complementary, for example, reducing cost and reducing environmental impact, it is also highly likely that there will be times where these values are in conflict, for example, maximising thermal comfort and minimising energy usage. As such, users may have to make decisions as to which value to prioritise in a given situation (this can be explored in the activity analysis in decision-making terms phase of CWA using decision ladders).

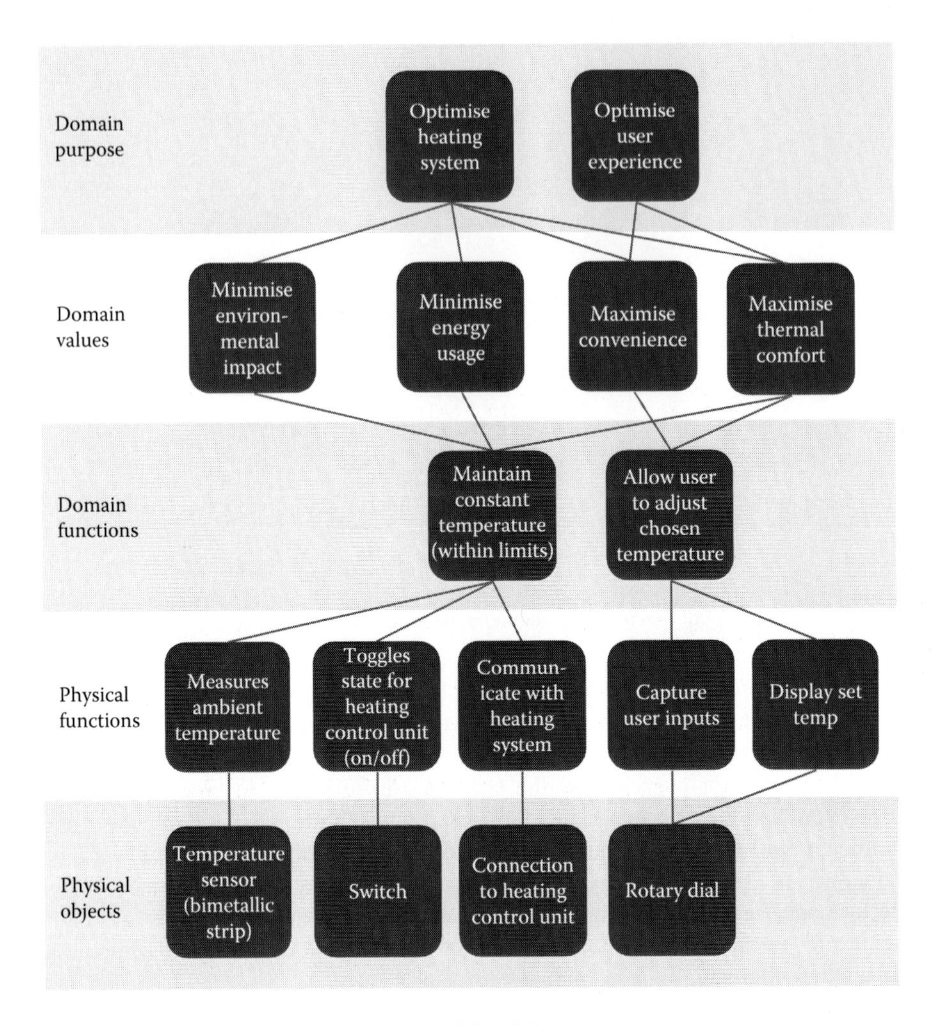

FIGURE 3.3 Abstraction hierarchy for traditional thermostat.

The row in the centre of Figure 3.3, the domain functions, links the diagram together. These are the functions that need to take place to meet the purpose of the system – in this case, 'maintaining a constant temperature' and 'allowing the users to adjust the chosen temperature'. Applying the what–why–how triad to the first of these (maintaining a constant temperature), the *why question* can be answered by following the links up (to minimise energy and environmental impact and to maximise thermal comfort), whereas the how question can be answered by chasing the links down – by measuring temperature and toggling the heating on/off via the connection to the control unit.

Together, the diagram creates an explicit link between the functions of the physical object at the base of the hierarchy and the system values (often user values) and purposes at the top. The complexity of the linking between the nodes provides an indication of how these different needs can be met by a combination of physical objects.

One of the useful things we can do with the abstraction hierarchy model is to explore new or different ways of meeting the high-order domain values. By considering the system in a top-down fashion, each of the domain values can be taken in turn and questioned to explore if there are alternative ways of how they could be achieved. Taking 'minimise the cost and environmental impact' as an example, it may be possible to reduce energy consumption by reducing the temperature while the occupants of the home are bed, tucked up under thick duvets or blankets. This example is highlighted in a new version of the abstraction hierarchy shown in Figure 3.4. Following the link down, one means of determining when the occupants are in bed would simply be to use a clock (assuming the occupants sleep at similar times of the day every day).

An alternative approach for reducing the energy consumption would be to determine when the home is unoccupied (this example is also illustrated in Figure 3.4). Following the links down, this could be achieved with some form of motion sensor (such as a passive infrared sensor [PIR]); alternatively, it could be linked to door locks, or user fobs or smartphones.

Approaching the system top-down has a number of advantages; it encourages the design team to start with a focus on the purpose of the system and its high-order values. This can help to avoid so-called 'function creep' where new features are added to a system because it is technically viable to do so, rather than because they offer an enhancement to the performance of the system.

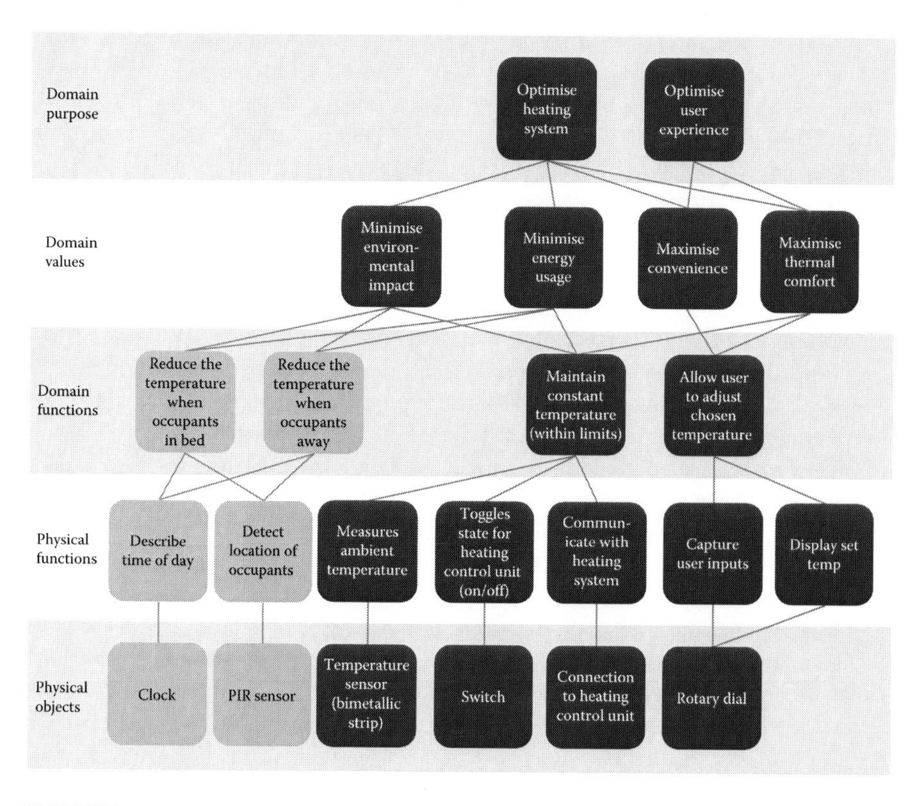

FIGURE 3.4 Abstraction hierarchy for thermostat with added features.

3.3.1 A Connected Thermostat

The other way of considering the system is to approach it bottom-up and assess how new technologies (physical objects) could have an impact on the system.

While the idea behind the Internet of things (IoT) has been around for quite some time, it is recently receiving unprecedented levels of interest. Ostensibly at least, the IoT involves creating a connection with everyday household objects (e.g. heating systems, locks, doors and windows, fans, lights) to allow them to be controlled remotely or automatically in response to events (e.g. a time of day, a change in environmental conditions, a message from a human). The Nest Learning Thermostat is used by many as the de facto standard example when describing the IoT and connected devices. Due to its familiarity, it has been adopted here for the purpose of explaining the approach.

A revised version of the abstraction hierarchy has been created (as shown in Figure 3.5) to represent a system with a Nest thermostat. The diagram has been coded to highlight what the Nest adds over and above a traditional heating system at different levels of abstraction. As such, the impact of the introduction of new components (physical objects) can be considered. For example, following the links up the model, the Wi-Fi card provides a connection between the heating unit and the Internet. This allows remote control of the temperature either within the home or remotely. This has a positive impact on convenience.

Likewise, the introduction of humidity sensor allows ambient humidity to be measured; this allows the temperature to be adjusted to compensate for humidity, which, in turn, maximises thermal comfort while minimising energy usage and environmental impact. This has the potential to positively impact both purposes of reducing energy consumption and maximising user experience. Explicitly considering these high-order impacts can help to reduce 'function creep'.

3.3.2 Systems Thinking

In addition to the Nest thermostat, other components in the wider system have also been included to the abstraction hierarchy in Figure 3.5, such as smart phones and apps that may be required to interface with the product. Other artefacts that may have overlapping affordances have also been captured (such as utility bills). The functions of each of these objects are described in the physical functions row. Describing these functions in a generic way encourages the analyst to consider how functions and physical object can be used in different ways.

At the base level of the hierarchy (physical objects), each link can be taken in turn and questioned to establish if there are other components already in the systems that can perform the same functions. It may be that other components can perform a given function instead, allowing a component to be removed (a cost saving). Likewise, the function may be better serviced by a different component (greater systems performance). Alternatively, more than one object may offer the same function; in this case, multiple objects can provide greater flexibility or resilience.

Similarly, other functions or affordances of the existing components can be considered to see if they have the ability to have a positive impact on the overall system

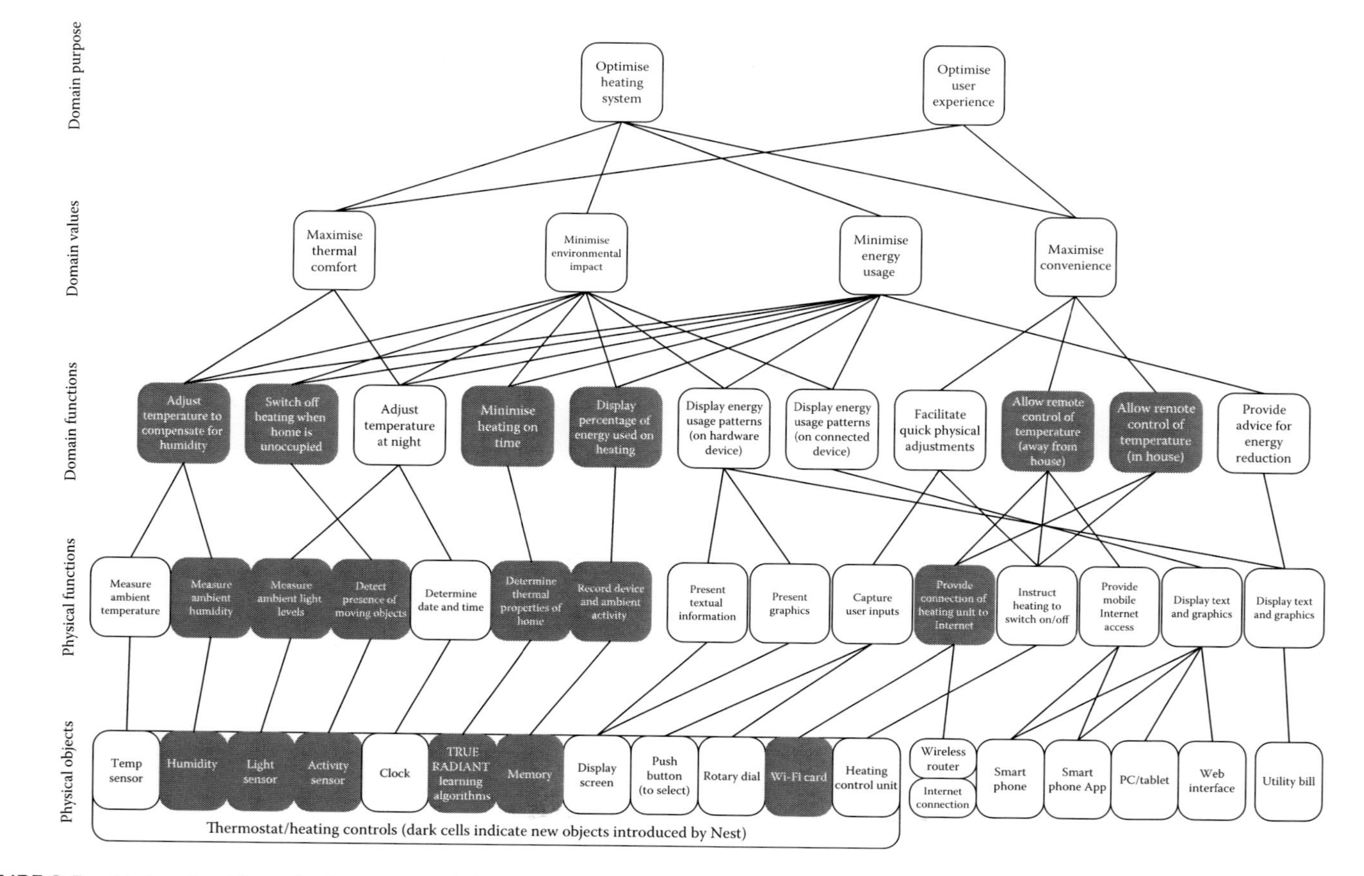

FIGURE 3.5 Abstraction hierarchy for connected thermostat.

goals. The boundaries of the system can also be expanded to consider other products and components and how functions can be shared between these. This is of particular interest in the IoT where numerous devices are to be connected.

One of the key advantages of including additional components (such as smart meters and utility bills) is that the compatibility of the product under analysis can be explored. In some cases, this may mean that functions performed by other artefacts can be subsumed by the product in question; likewise, it may mean that functions should be left to other artefacts or that the product should be designed to be complementary to them.

3.4 DISCUSSION AND CONCLUSIONS

The abstraction hierarchy provides a description of the system constraints at a high level encouraging the design team to focus on how the product or artefact is meeting the values of the system that it inhabits. Furthermore, it encourages the analyst to consider the relationship between the base-level physical objects in the systems and the high-order purpose of the system. As a stand-alone representation, the abstraction hierarchy lacks the detail of a traditional product specification. As such, the approach described is in no way intended to replace the standard specification, rather it aims to support it.

In most cases, the design process would start with the abstraction hierarchy while the overall system architecture is being decided. Once the system architecture is agreed, the abstraction hierarchy would inform the development of a more detailed design specification. This would include all of the traditional constraints such as relevant legislation, cost, size, safety, context restraints and time. Explicit links can be made between these two representations (the abstraction hierarchy and the specification) allowing a clear audit trail and also allowing the specification to be updated in line with changes to the abstraction hierarchy should constraints or assumptions be modified.

3.4.1 Is the Abstraction Hierarchy Really Necessary in the Design Process?

The challenge of generating a connected device can, of course, be viewed solely at a physical level by augmenting the current thermostat system with a means of communication. An existing detailed design specification from a legacy product could simply be modified to add a new requirement for connectivity. Accordingly, this connectivity can be viewed as a 'bolt-on' function – and the resultant focus would be on the decision of which type of communication protocol to adopt (e.g. Wi-Fi or low energy radio communications) to reduce cost and increase reliability. Indeed, an approach similar to this is likely to have been adopted by a number of manufacturers prior to the release of the Nest and deployed with varying level of success and adoption.

It is perhaps interesting though that it was a small independent, albeit well-funded, company that designed and developed the product, which is now viewed as the most innovative. It was the Nest Learning Thermostat, not a product from one of the large established suppliers, that decided to move beyond a technical innovation, to one that seeks to actively engage the end user to meet the higher-order domain values. A number of the incumbent companies subsequently produced competitors to the Nest, offering products with similar components and functions at a physical level. However, by this point, benchmarking the competition would have allowed them to develop a detailed specification of what these products should do and the individual requirements for each of the components.

It would, of course, be naïve to think that the reason this level of innovation came from a start-up, and not a large organisation, was solely down to large organisations focusing on the physical and functional level of a project – brought about by over-reliance on design specifications. However, what is clear is that such a reliance on a 'traditional' product development process is not embracing opportunities for innovation.

The abstraction hierarchy is, again of course, not unique – there are a wide range of tools and techniques that encourage the design team to focus on user and stakeholder needs and seek to find innovative solutions to identified markets and problems (it is not clear what approach was used to develop the Nest product). Likewise, the approach described in the chapter is certainly not proposed as a silver bullet allowing a perfect balance to be struck between project efficiency and innovation. However, the described approach is proposed to help inform these trade-offs, to encourage debate early on in the project of the purpose of the system.

REFERENCES

Klein, G. 2014. No, your organisation really won't innovate. *Wired.* http://www.wired.co.uk/magazine/archive/2014/05/ideas-bank/gary-klein. Accessed 05 September 2014.

Rasmussen, J. 1986. *Information Processing and Human-Machine Interaction: An Approach to Cognitive Engineering.* Amsterdam, the Netherlands: North-Holland.

4 Using the Decision Ladder to Reach Interface Requirements

Daniel P. Jenkins

CONTENTS

4.1 INTRODUCTION

The link between the quality of a user interface and system performance is now almost universally accepted. High-profile incidents, such as the Three-Mile-Island accident, brought the consequences of poor interface design into the public consciousness. As a result, it is a topic that has received considerable attention from a range of disciplines. The challenge of interface design is often tackled by software engineers, human factors specialists, usability experience (UX) specialists or graphic designers, or, in cases, a combination of these disciplines.

For very simple interactions, such as an alarm clock app for a mobile phone, developing an interface may be an intuitive and straightforward process. The adoption of style guide and consideration of a set of heuristics (e.g. Ravden and Johnson, 1989; Nielsen and Molich, 1990) may be enough to ensure a useable design. However, the challenge is proportional to the complexity of the product or service being designed. The consequence of system failure is also an important consideration in the approach adopted; while the failure of an alarm clock may result in missed appointments or even flights, it is unlikely to cause a fatality. Conversely, in safety critical environments, the cost of failure may be much higher. The Therac-25 radiation machine is a frequently cited case (see Leveson and Turner, 1993) of where interface failure tragically contributed to multiple fatalities. Six incidents were reported between 1985 and 1987 where patients were given significant overdoses of radiation.

The Therac-25 machine had two modes of operation: direct electron-beam therapy (E) and megavolt x-ray beam therapy (X). When operating in the direct electron therapy mode (E), the machine delivers shallow lower-power beams. Conversely in x-ray mode (X), a deeper focused beam is delivered, which is flattened with an attenuating filter that is positioned in the path of the beam.

One case occurred in Tyler, Texas, in 1986 after 2 years of use and nearly 500 treatments. The operator was performing a routine electron treatment (22 MeV) of 180 rad to the patient's upper back as part of a 6000 rad treatment programme spread over six and a half weeks. As per usual, the patient was set up, face down, on the table and the operator left the room. At the control terminal, the operator manually entered the treatment details as follows:

1. Operator entered prescription data (treatment type and dose).
2. Operator confirmed settings noting that she had mistyped X (for x-ray mode) rather than E (for an electron treatment).
3. Operator used the up arrow on the keypad to move the cursor up over the X to edit it.
4. Operator typed E to overwrite X (within 8 s of the first entry).
5. Operator typed B for beam on.
6. An unfamiliar error was presented on the screen '*Malfunction 54*'; however, no information was provided on the details of this error. The operator manual supplied with machine did not explain or address the malfunction codes, nor did it give any indication that these malfunctions could place a patient at risk.
7. As system errors were a relatively common occurrence and routinely accepted, the operator typed P for proceed.

At a much later date, it became apparent that the interface had not recognised the change in beam type (X to E) as a piece of legacy code ignored changes made within 8 s of the initial entry. Not only was the wrong treatment delivered, but also a dose designed to penetrate the thick attenuating filter was delivered, without the filter in place. As a result, the patient received a massive overdose (16,000 rad instead of 180 rad). Previous versions of the machine had a mechanical interlock in place to prevent this situation; however, this was not present in the Therac-25.

As with most complex systems, it is difficult and, more importantly, inappropriate to seek a single root cause for this failure; however, the user interface played a critical role. The product lacked the required interlocks and the interface lacked the feedback to communicate the situation and its criticality. As discussed by Leveson (1995), there is a common conflict between making a system easier and more efficient to use and safety goals.

4.1.1 Complex Thinking for Complex Systems

For more complex systems, a structured approach is needed to ensure, first, that all the required information elements are considered and, second, that they are included in the optimal way to ensure an appropriate balance of system values (e.g. safety, efficacy, efficiency, usability and resilience).

Most interface designs start by first establishing the information requirements. More often than not, these information requirements are communicated as a text-based document. The resultant document typically forms the bridge between systems architects, or engineers and the interface designers. Perhaps, unsurprisingly, the quality of these information requirements has a direct relationship on the quality of the resultant interface and the performance of the systems in which they are used. Thus, in order to ensure the safety, efficacy, efficiency, usability and resilience of products and services, it is important that the process for developing information requirements is fit for purpose.

Thus, ostensibly at least, the foundation for a well-designed interface design lies in establishing the following:

1. What information is required?
2. When it needs to be displayed?
3. Where it should be displayed?
4. Whom it should be displayed to?
5. How – in what format?

4.1.2 Top-Up and Bottom-Down Thinking

When formulating a system's information requirements, it is often useful to think of the system both from a technical and an intentional perspective. In other words, what the components of the system can physically do, and what the system is intended to achieve (the systems' higher-order purposes). Both approaches can reveal rich sources of information.

The performance of a single pump (flow rate, power consumption) in a large system may be an important information source; likewise, its relationship to the safe and efficient performance of the wider system is also likely to be important. For most systems, an abstraction hierarchy is an excellent foundation for system understanding. It provides an explicit link between the physical elements in the system and the higher-order stakeholder values. These stakeholder values can be used as a starting point for establishing user goals during different situations (this will be explored in a little more detail shortly).

4.1.3 Decisions, Decisions, Decisions ...

The main underlying assumption for the approach discussed in this chapter is that the interface being developed is intended to enable some form of control of a system. A second assumption is that the control task is distributed between actors. In the case of computer interfaces, this distribution is expected to include some form of computer automation and at least one human actor.

Ostensibly, decision-making is at the heart of all control tasks. There have been many attempts to model the decision-making activity. Most involve some form of observation of information, orientation to the current situation, a choice as to which action to adopt, and finally an action (see the Observe, Orientate, Decide, Act [OODA] loop). The decision ladder (Rasmussen, 1974) is the tool most commonly used within cognitive work analysis to describe decision-making activity. Unlike some of its counterparts, its focus is on the entire decision-making activity, rather than the moment of selection between options. It is not specific to any single actor; rather, it represents the decision-making process of the combined work system. In many cases, the decision-making process may be collaborative, distributed between a range of human and technical decision-makers.

As illustrated in Figure 4.1, the ladder contains two different types of nodes: the rectangular boxes represent data-processing activities, while the circles represent resultant states of knowledge. Novice users (to the situation) are expected to follow the decision ladder in a linear fashion. When followed linearly, the process starts with some form of activation (on the bottom of the left leg). Observations are then made and information elements are collected. These information elements are then combined to determine a system state. Options are then formulated and evaluated against goals. On the basis of an understanding of the situation, a context-specific goal is chosen and target state is selected. A task is then defined and broken down into a procedure, which is finally executed. Thus, the left side of the decision ladder represents the observation of the current system state, whereas the right side of the decision ladder represents the planning and execution of tasks and procedures to achieve a target system state.

Expert users, familiar with the situation, are expected to link the two halves of the ladder by shortcuts. In familiar situations, users are expected to recognise an appropriate task after collecting information from the system, negating the need to explicitly consider different options and evaluate them against goals (illustrated as a shortcut in Figure 4.1).

4.1.4 Applicability

It is contended that the approach described here applies to a wide range of applications, and could include the control of bank current accounts, an unmanned aerial vehicle (Elix and Naikar, 2008; Jenkins, 2012), a military command and control system (Jenkins et al., 2010), a tank training simulator (Jenkins et al., 2011b), an

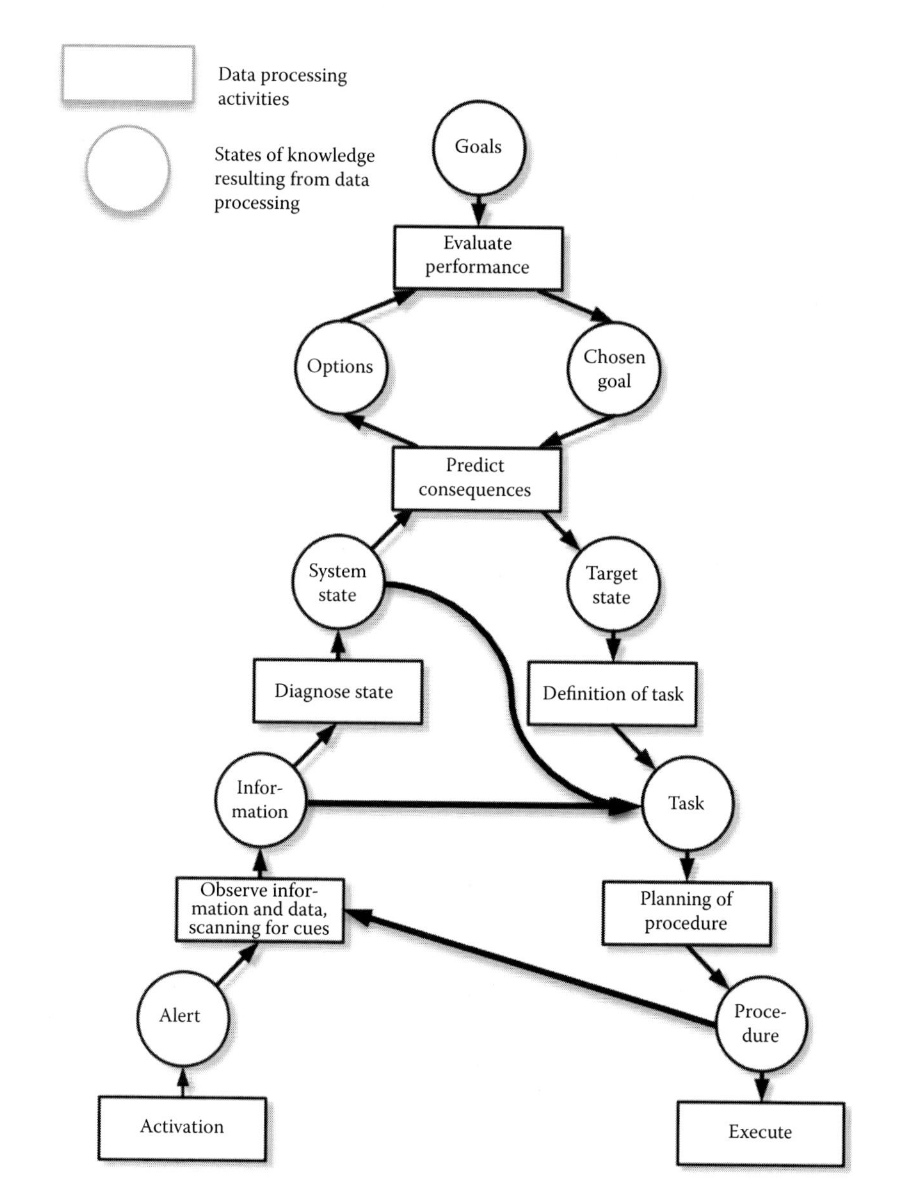

FIGURE 4.1 Decision ladder template.

automotive interface (McIlroy and Stanton, 2015), a policing system (Jenkins et al., 2011a) or a medical device.

4.2 METHOD

The approach starts with a list of the information requirements that could be needed by the system. This list is then coded to provide additional detail and constraints, such as when, where, to whom and how information should be displayed.

4.2.1 Eliciting Information Requirements: What Information Is Required?

The approach for eliciting the systemic information requirements is based on a series of semi-structured interviews with system experts and/or stakeholders. These interviews are structured around a template with a decision ladder (see Figure 4.1) at its centre.

The process involves capturing the questions that decision-makers pose themselves and the system at each stage of the decision-making process. A separate model should be created for each of the key situations. These are typically identified through a contextual activity template or a hierarchical task analysis (HTA). The process is explored in greater detail in Table 4.1.

4.2.1.1 Practical Tips

The activity can be conducted on paper templates or entered directly into a digital model. Given the iterative nature of the task, it is useful to create the models electronically. This makes it easier to reclassify elements and reorder them. Another advantage of creating them electronically is that there is often a level of duplication between tasks. Models from a similar task can often be edited reducing the workload. As it is a collaborative exercise between the analyst(s) and the domain experts(s), it is useful to work on a large display that all parties can see.

Most text and graphical software packages lend themselves to this task. MS PowerPoint or MS Visio are often favoured as they can be distributed to stakeholders.

4.2.2 Analysing the Models: When Should Information Be Displayed?

The level of consistency between the decision ladder models can be a very useful cue to design. Information elements can often be divided into two groups: (1) persistent and (2) context specific. As the name suggests, persistent information elements should be presented regardless of the situation or task, while context-specific information elements should only be displayed during the tasks or situations where they are relevant. For interfaces that predominately contain persistent data, an argument could be made for showing all information elements as this reduces the complexity of a moded display.

A table can then be created and the list of information elements can be coded to show which elements are present in which situations or tasks.

4.2.3 Defining the Interface Type: Where Should It Be Displayed?

The where question can be addressed in two ways; first, a decision needs to be made on where information should be displayed in the environment. This may mean the following:

- Different sites (e.g. maybe in different countries)
- Different rooms within a site (e.g. control room, plant room, treatment room)
- Different locations within a room (e.g. wall-mounted display, equipment display, indicator lamp, hard-copy manual, whiteboard, poster)

TABLE 4.1
Interview Process

Stage 0 – Define Task Steps
Prior to starting the interviews, the activity should be decomposed into separate parts. The optimal method of decomposition will vary depending on the system. Activities that can easily be delineated into a series of notably different task steps are best deconstructed using a task analysis technique such as HTA. Activities that are defined by more environmental conditions, such as location, are better deconstructed in a contextual activity template.
A separate template should be produced for each task step or situation, it is also wise to have plenty of spare templates should a different breakdown become evident during the interview.
Stage 1 – Determining the Goal
The first stage of the interview process for each model is to structure the goal of the system. The expert should be asked to provide a high order goal, along with a number of constraints affecting it. The expert should be reassured that the constraints could possibly be in conflict. The information works well placed in the format 'To (insert goal) (insert constraints)'.
In many cases, the abstraction hierarchy is very useful for populating the constraints. The second level from the top, the domain values, can often be imported directly.
Stage 2 – Alert
The expert should be asked to begin the walk-through at the chronological start of the process. Alerts capture the events that first draw them to the need to make a decision.
Stage 3 – Information
The expert is asked to list the information elements they would use to gain an understanding of the situation. Experts often discuss a combination of system states (see below) and information elements together at this stage of the process. The key here is not to stop the expert but to capture both and classify them at a later stage.
Stage 4 – System State
The system states represent a perceived understanding of the work system based upon the interpretation of a number of information elements. The key distinction between an information element and a system state is that system states are formed of more than one quantifiably different elements of information. In short, information elements are processed and fused to form system states.
Stage 5 – Options
The options within the ladder can be described as the opportunities for changing the system state in an attempt to satisfy the overall goal. The points are structured as questions in the form: 'is it possible to (...)?' The number and type of options available will be informed by the system state. It is anticipated that in certain situations there may be only one option available.
Stage 6 – Chosen Goal
The chosen goal, at any one time, is determined by selecting which of the constraints receives the highest priority. In this case, a priority needs to be placed on the domain values. This does not have to be an absolute choice per se, rather, one takes a higher priority than the other does in the given situation.

(*Continued*)

Table 4.1 (*Continued*)
Interview Process

Stage 7 – Target State
The target states mirror the option available; once a particular option is selected, it becomes the target state. The options are rephrased in the form 'Should (option) take place?'
Stage 8 – Task
The listed task questions relate to the tasks required for achieving the target state while maintaining the overall goal.
Stage 9 – Procedure
The procedure lists questions that will inform the choice of task procedure.
Stage 10 – Analysing the Models
Once a decision ladder has been created for each task or scenario, the variability between the models can be compared. At this stage, it is useful to give each element in the model a unique identifier, for example, Alerts (AL001), Information elements (IE001), System states (SS001), Options (OP001), Goals (GL001), Target states (TS001), Tasks (TA001) and Procedures (PR001). Where appropriate, similar elements can often be combined and reworded to reduce the total number of elements.

The second way of addressing the question is to consider the arrangement within each of these locations (e.g. the location on the poster or the screen). There are a number of applicable approaches for grouping information elements. These may be the following:

- Spatial – arranged in relation to their location in the physical world
- Process based – arranged in a sequence based on a process flow
- Functional grouping – grouped based on the physical functions
- Critical – arranged hierarchically based on importance

The output of the analysis approach provides a useful means of structuring the interface. By explicitly mapping which information elements relate to which systems states. By adding a column to the table for each location the information elements can be coded to indicate the relationship.

4.2.4 Whom It Should Be Displayed To?

In a similar way, different actors in the system may need different access to information, which may include the following:

- Digital agents
- Operators
- Supervisors
- Administrators
- Maintenance staff

The matrix of information elements and system states can also be coded to indicate which actors the information should be displayed to. This can help to inform and document decisions relating to whether separate system views are required and whether actor types can be combined to reduce the number of views required.

4.2.5 How Should the Information Be Displayed

The decision as to how information should be displayed will be informed by a consideration of the factors above. Once the information elements for each situation, location and actor have been defined, the decision on representation needs to also consider the appropriateness of the representation. Key questions for single displays are as follows:

- Is it binary/multistage/analogue information?
- Is it between limits?
- Are the limits fixed or variable?
- Is it critical to monitor?
- Does the trend need to be tracked?
- Does the rate of change need to be tracked?
- Do values need to be tracked in relation to other elements?

4.3 CASE STUDY: RADIOGRAPHY EQUIPMENT

The following case study is based on a generalised description of a radiography machine. The high-level task flow for the machine operation is described in Table 4.2. In this case, the activity is well delineated by the task step; hence, an HTA is deemed to be an appropriate means of dividing up the process.

4.3.1 What Information Is Required

To illustrate the process, Task 5 – deliver radiation – will be explored in greater detail as detailed in Figure 4.2 and Table 4.3. This process would be repeated for each of the task steps.

TABLE 4.2
High-Level Task Description

Task Step	Description	Operator's Location
1	Patient check-in	Control room
2	Machine preparation	Treatment room
3	Patient set-up	Treatment room
4	Machine set-up	Treatment room
5	Deliver radiation	Control room
6	Unload patient	Treatment room
7	Clean machine	Treatment room

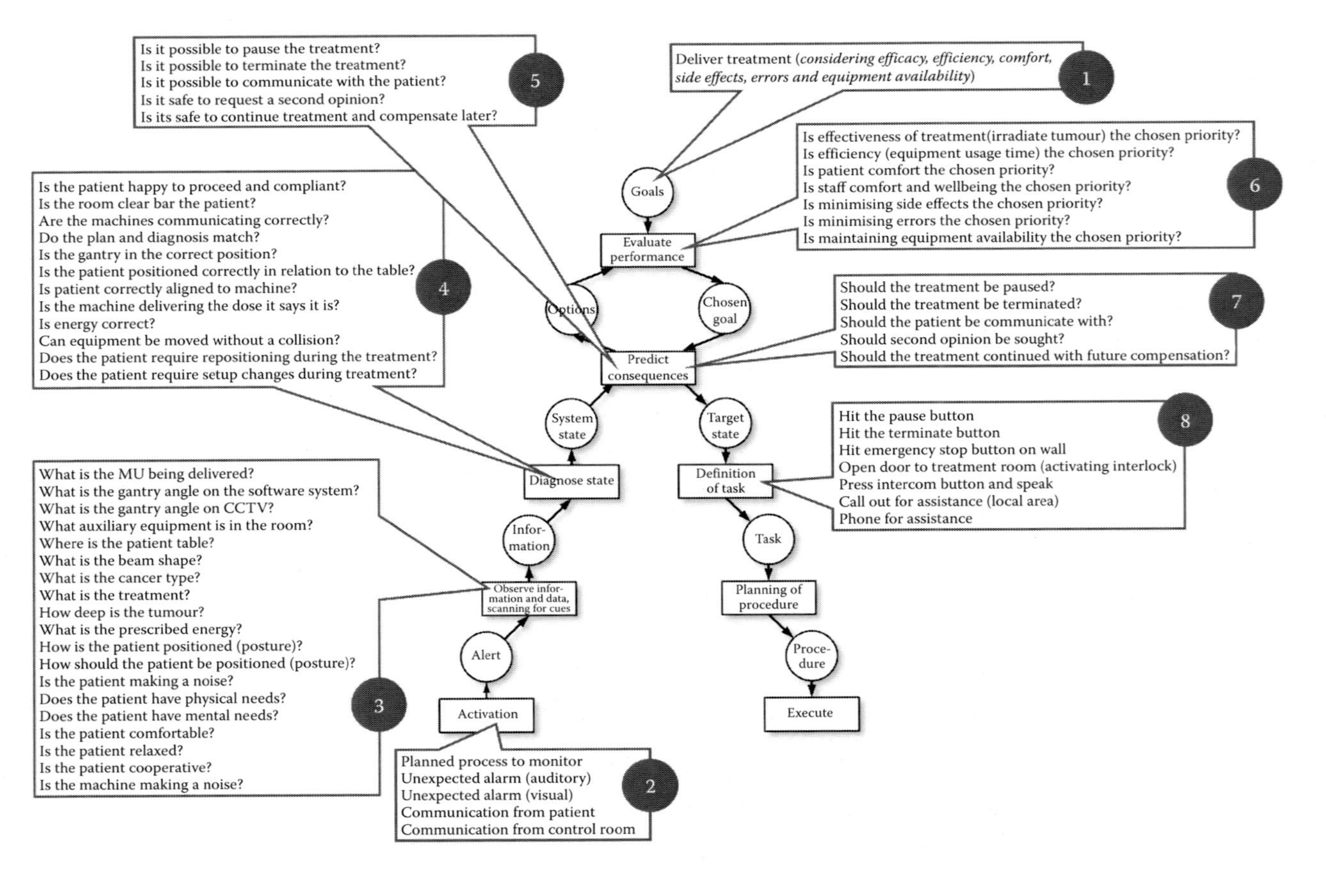

FIGURE 4.2 Decision ladder template for Task 5 – deliver radiation.

TABLE 4.3
Data Collection Process for the Case Study

Stage 1 – Determining the Goal
The process starts at the top of Figure 4.2. The goal at this stage of the process is simply to 'deliver the treatment'; the caveat is that it must also consider the system values of efficacy, efficiency, comfort, side effects, error and equipment availability.
Stage 2 – Alert
Moving to the base of Figure 4.2, during the delivery process alerts to a system state change include monitoring the process, alarms (visual and auditory), communications from the patient and communications from other members of staff in the control room.
Stage 3 – Information
The information elements are the nuggets of information that can be brought together to understand the state of the system. In this case, they include information about the physical equipment (e.g. the gantry angle, the equipment in the room, the position of the table) along with information from the human machine interface (HMI) (e.g. the dose being delivered, the beam shape), information from patient records (e.g. treatment type and location) and information from the patient (e.g. are they comfortable, relaxed).
Stage 4 – System State
The system states represent a perceived understanding of the system based upon the interpretation of a number of information elements. Questions such as '*is the patient positioned correctly?*' can be assessed by considering the treatment type and the current position of the patient.
Stage 5 – Options
At a high level, during treatment there are five main options available to the operator: to pause the treatment, terminate the treatments, communicate with the patient, request a second opinion and continue treatment and compensate later.
Stage 6 – Chosen Goal
The chosen goal, at any one time, is determined by selecting which of the constraints receives the highest priority. In the case of the Therac-25, the operator had to decide which aspect of the goal to focus on based on the information presented. The system values of efficacy, efficiency, comfort, side effects and error and equipment availability were in this case in conflict. Most notably, the efficiency and equipment availability were in conflict with safety and side effects.
Stage 7 – Target State
The target states mirror the option available.
Stage 8 – Task
The tasks relate to the actions required (e.g. hit the pause button, press intercom button and speak).

4.3.2 When Is Information Required

A matrix can be created listing each of the alerts, information element, system states, option, goals, target states, tasks and procedures. These are collected from models for each of the task steps described in Table 4.2. To illustrate the process, a small subset of these is presented in Table 4.4. The matrix can be coded to show

TABLE 4.4
Example Elements Coded by Task Step

ID	Description	1. Patient Check-in	2. Machine Prep	3. Patient Set-Up	4. Machine Set-Up	5. Deliver Radiation	6. Unload Patient	7. Clean Machine
AL01	Patient appears agitated							
AL02	Unexpected alarm (auditory)							
AL03	Communication from patient							
IE01	What is the name of the patient?							
IE02	What is the weight of the patient?							
IE03	What is the size of the patient?							
IE04	Does the patient have physical needs?							
IE05	What is the MU being delivered?							
IE06	What is the cancer type?							
IE07	Does the patient have multiple appointments?							
SS01	Is the patient happy to proceed?							

Note: Dark grey cells indicate typically needed; light grey cells indicate may be needed.

when the elements are typically needed (dark grey cell) and when they may be needed (light grey cell). As illustrated in Table 4.4, some of the information elements may be required all of the time such as the name of the patient, whereas other elements are required only in specific situations (e.g. the dose being delivered during Stage 5).

4.3.3 Where Is Information Required

The same matrix that was introduced in Table 4.4 can be coded to indicate the location of the elements. An example of this is presented in Table 4.5. This is only a small subset of the elements from the full model; however, it shows that there is a clear mix between information that is collected from the physical world and information that exists in documents. For systems that have most information available via machine feedback or documents, it may be appropriate to control them from a screen where physical environment information may be supplied by CCTV or sensors.

For systems that are dominated by information in the physical environment, it may be appropriate to explore how information elements from documents and machine feedback can be presented in the physical world. This could be in the format of indicator lamps, wall-mounted displays, embedded displays or auditory feedback. Other technologies such as projectors or wearables (fobs around the neck, bracelets, glasses) may also be appropriate.

4.3.4 Whom Should Information Be Displayed To

Just as with the other questions, the matrix can be extended to explore which of the actors typically (dark cells) or may need (light cells) access to the different elements. An example of this is illustrated in Table 4.6. As would be expected, the information requirements of the porter are quite different to those of the radiotherapist. The value of this approach, however, is that it forces the analyst to explicitly question which information elements are provided to which actor type. This can be used to directly inform the design of equipment in terms of displays and indicators, as well as the design of printouts or apps that are provided to different individuals.

4.3.5 How Should the Information Be Displayed

The last question 'how should the information be displayed?' is perhaps the one that involves the most experience and 'black art'. That being said, there are a set of heuristics that can be applied to quickly arrive upon the first iteration of a design. As discussed in Section 4.2.5, attributes of the information element such as whether it is between limits or if trend is important inform this. Table 4.7 provides examples for each of the information elements.

TABLE 4.5
Example Elements Coded by Location

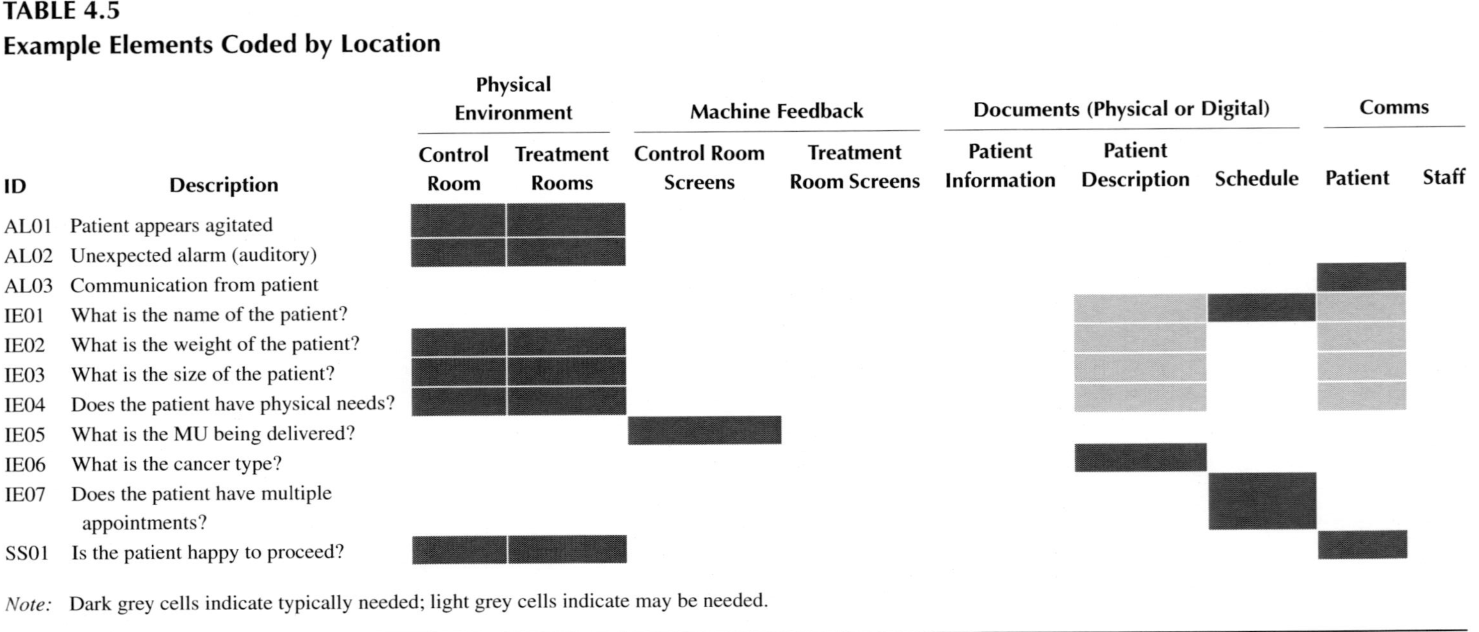

ID	Description	Physical Environment		Machine Feedback		Documents (Physical or Digital)			Comms	
		Control Room	Treatment Rooms	Control Room Screens	Treatment Room Screens	Patient Information	Patient Description	Schedule	Patient	Staff
AL01	Patient appears agitated									
AL02	Unexpected alarm (auditory)									
AL03	Communication from patient									
IE01	What is the name of the patient?									
IE02	What is the weight of the patient?									
IE03	What is the size of the patient?									
IE04	Does the patient have physical needs?									
IE05	What is the MU being delivered?									
IE06	What is the cancer type?									
IE07	Does the patient have multiple appointments?									
SS01	Is the patient happy to proceed?									

Note: Dark grey cells indicate typically needed; light grey cells indicate may be needed.

TABLE 4.6
Example Elements Coded by Actor

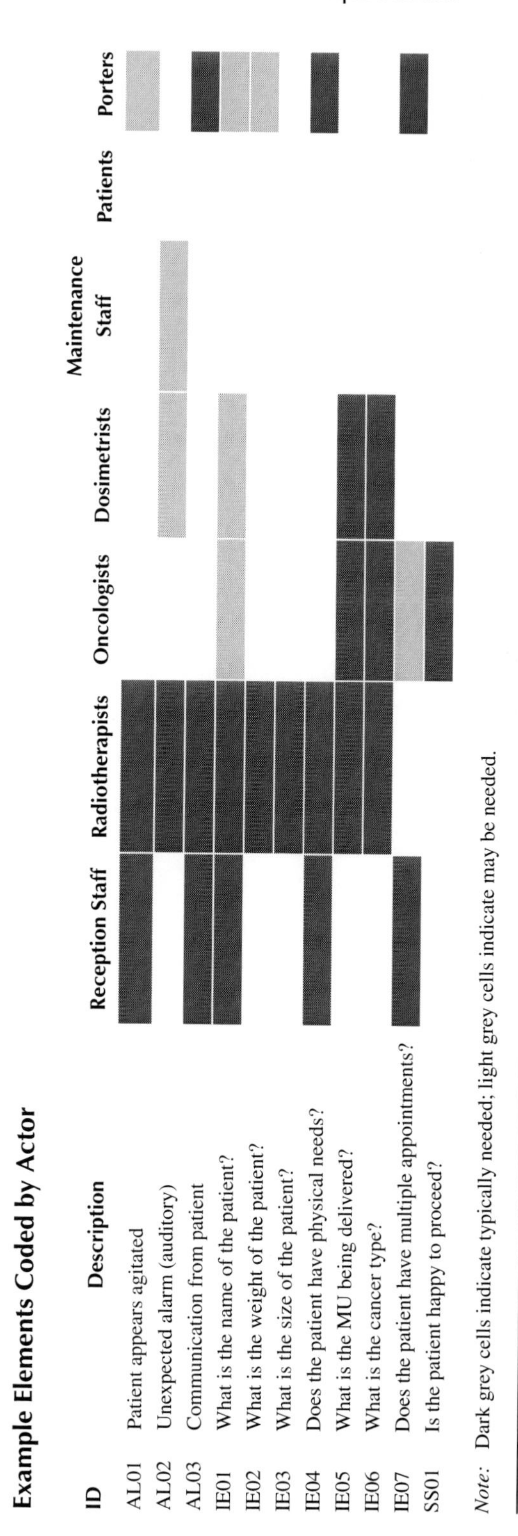

ID	Description	Reception Staff	Radiotherapists	Oncologists	Dosimetrists	Maintenance Staff	Patients	Porters
AL01	Patient appears agitated	Dark grey	Dark grey					Light grey
AL02	Unexpected alarm (auditory)		Dark grey		Light grey	Light grey		
AL03	Communication from patient	Dark grey	Dark grey					Dark grey
IE01	What is the name of the patient?	Dark grey	Dark grey	Light grey	Light grey			Light grey
IE02	What is the weight of the patient?		Dark grey					Light grey
IE03	What is the size of the patient?		Dark grey					
IE04	Does the patient have physical needs?	Dark grey	Dark grey					Dark grey
IE05	What is the MU being delivered?		Dark grey	Dark grey	Dark grey			
IE06	What is the cancer type?		Dark grey	Dark grey	Dark grey			
IE07	Does the patient have multiple appointments?	Dark grey		Light grey				Dark grey
SS01	Is the patient happy to proceed?			Dark grey				

Note: Dark grey cells indicate typically needed; light grey cells indicate may be needed.

TABLE 4.7
Example Elements Coded by Format

ID	Description	Format
AL01	Patient appears agitated	High-quality image of patient
AL02	Unexpected alarm (auditory)	Unique sounding alarm louder than background
AL03	Communication from patient	High-quality audio
IE01	What is the name of the patient?	Text
IE02	What is the weight of the patient?	Numerical with units
IE03	What is the size of the patient?	Numerical with units
IE04	Does the patient have physical needs?	Text field
IE05	What is the MU being delivered?	Numerical with units
IE06	What is the cancer type?	Text field/map of body
IE07	Does the patient have multiple appointments?	Schedule
SS01	Is the patient happy to proceed?	High-quality image of patient

4.4 DISCUSSION AND CONCLUSIONS

As discussed in the introduction (see Section 4.1.4), the approach described in this paper has proved to be effective in a wide range of situations. It provides welcome structure to the process of eliciting and structuring information requirements that focus on end users and stakeholders.

One of the clear strengths of the approach is that it provides a very explicit link between the data collection, the analysis and the design. Returning to the criteria raised in the introduction:

- What information is required? – this is explicitly captured in the decision ladder models.
- When it needs to be displayed? – this is captured in the summary table (each information element is coded to indicate which situation or task it is relevant to).
- Where it should be displayed? – this is coded in the matrix.
- Whom it should be displayed to? – the matrix can be coded to indicate which actors should receive each of the information element.
- How, in what format? – by determining the data type (e.g. between limits, binary, multistate), it is possible to determine the best solution for each information element.

4.4.1 Conclusions

As discussed in the introduction, decision-making is fundamental to safe and effective system control. There is a long established connection between the quality of decision-making and the information available to decision makers. This does not necessarily mean presenting more information; on the contrary, too much information can be as detrimental as too little. Rather, to optimise system performance (e.g.

safety, efficacy, efficiency, resilience), effective decision-making must be supported by the right information, at the right time, in the right place, to the right actors, in a format that can be readily understood.

It is certainly not claimed that the structured approach described in this paper would prevent all cases of error in isolation; however, as part of a suite of error prevention measures, it is contended that it would lead to the development of safer, more considered systems. Furthermore, owing to the focus on user information requirements, it is contended that it results in more useable interfaces that will have a positive impact on multiple system performance metrics (e.g. efficacy, efficiency, resilience).

REFERENCES

Elix, B. and Naikar, N. 2008. Designing safe and effective future systems: A new approach for modelling decisions in future systems with cognitive work analysis. *Proceedings of the 8th International Symposium of the Australian Aviation Psychology Association.*

Jenkins, D. P. 2012. Using cognitive work analysis to describe the role of UAVs in military operations. *Theoretical Issues in Ergonomics Science*, 13(3), 335–357.

Jenkins, D. P., Salmon, P. M., Stanton, N. A., Walker, G. H. and Rafferty, L. 2011a. What could they have been thinking? How sociotechnical system design influences cognition: A case study of the stockwell shooting. *Ergonomics*, 54(2), 103–119.

Jenkins, D. P., Stanton, N. A., Salmon, P. M. and Walker, G. 2011b. A formative approach to developing synthetic environment fidelity requirements for decision-making training. *Applied Ergonomics*, 42(5), 757–769.

Jenkins, D. P., Stanton, N. A., Salmon, P. M., Walker, G. H. and Rafferty, L. 2010. Using the decision ladder to add a formative element to naturalistic decision-making research. *International Journal of Human-Computer Interaction*, 26(2–3), 132–146.

Leveson, N. G. 1995. Medical devices: The Therac-25 (Appendix A). In N. G. Leveson (Ed.). *Software: Systems Safety and Computers*. Boston, MA: Addison Wesley.

Leveson, N. G. and Turner, C. S. 1993. An investigation of the Therac-25 accidents. *IEEE Computer*, 26(7), 18–41.

McIlroy, R. C. and Stanton, N. A. 2015. A decision ladder analysis of eco-driving: The first step towards fuel-efficient driving behaviour. *Ergonomics*, 58(6), 866–882.

Nielsen, J. and Molich, R. 1990. Heuristic evaluation of user interfaces. *Proc. ACM CHI'90 Conference*. Seattle, WA, 1–5 April, pp. 249–256.

Rasmussen, J. 1974. The human data processor as a system component: Bits and pieces of a model (Report No. Risø-M-1722). Roskilde, Denmark: Danish Atomic Energy Commission.

Ravden, S. J. and Johnson, G. I. 1989. *Evaluating Usability of Human-Computer Interfaces: A Practical Method*. Chichester, UK: Wiley.

5 From Cognitive Work Analysis to Software Engineering

Anandhi Dhukuram and Chris Baber

CONTENTS

5.1 INTRODUCTION

In this chapter, our concern is with the relationship between cognitive work analysis (CWA) and software engineering. The reason for this is to address the gap between the description that CWA produces and a specification that can be used as the basis of design. As Lintern (2005) has pointed out, CWA (like the majority of human factors methods) is able to make recommendations for the design of a system but does not translate these into a form that a designer is able to use. In some instances, the 'designer' would be the person conducting the analysis. More often, the design is created by a software engineer who will interpret 'requirements' from the CWA to produce a prototype. However, CWA is not a requirements analysis tool. Rather, it describes the elements of a system and allows analysts to consider how these elements might be configured. Converting from CWA to a design, therefore, involves close collaboration between analysts and designers. We interpret this as an invitation to explore ways of translating CWA into system requirements. Part of the problem lies in establishing a means of communication between human factors engineers who conduct the analysis and the software engineers who will design and code the application (Handley and Smillie, 2010). Bruseberg et al. (2008) points out that 'Human Factors practitioners and SE [Software Engineering] practitioners often find that there are communication difficulties' (p. 220). We propose that there might be a potential synergy between the views that software engineering specification techniques such as Unified Modeling Language (UML) employ and the different 'views' of a system that CWA provides. If it is possible to make simple translations between these, then it could be possible to generate views that can be read as UML but interpretable as CWA. This provides a means by which software engineers can work on the development of an application while having direct reference to the CWA output. It also means that any mismatch in information captured in the views (or needed to make the views coherent) can form the basis for discussion in the design team. For example, there might need to be a specific function in the UML view to allow information to pass between objects, but this function might not be in any CWA description. So, this means either that the CWA description is incomplete or that there will be a change to the way in which a given task is performed. In either case, raising this as an issue during the design stages can help to ensure matching between the software engineer and human factors engineer in their understanding of what is being developed and what activity it will support.

5.2 BACKGROUND AND CONTEXT

The purpose of UML is to provide a shared means of communicating the formal specification of a 'system' to everyone involved in the design process (Booch 2005). The approach combines several views that are brought together ('unified') under a common umbrella and that allow designers to describe and model a system using formalised drawing tools. UML, developed from a collection of object-oriented software design approaches, is intended to function as an industry standard for software design. In this chapter, we show how CWA can be translated into a subset of the different views used by UML. These views are the following:

- Use case diagrams: to represent the interactions between the system and the users or other external systems; useful for mapping the functional requirements or user needs for the system.
- Class diagrams: 'classes' are used in object-oriented programming languages such as C++ or Java or C#, and define the relationships, attributes and operations of objects.
- Package diagrams: for grouping classes and interfaces.
- Sequence diagrams: used to specify the type and order of messages passed between objects during execution.
- State diagrams: used to describe system states and transitions to be triggered for state changes.

It is worth pointing out at this stage that UML does *not* provide any support for the design of user interfaces. Of course, CWA does have an approach to user interface design through its theory and practice of ecological interface design (EID) (Vicente and Rasmussen, 1992; Burns and Hajdukiewicz, 2013). The first step in EID is to derive information requirements from the CWA and then propose that these are represented in the user interface. For example, the values and priorities in the abstraction hierarchy (AH) represent a trade-off space in which the system seeks to balance the achievement of these functions. The object-oriented functions (in the AH and control task analysis) could represent the individual tasks that the user of the system needs to perform and that need to be supported. This gives a flavour for what needs to be displayed on the user interface. In much the same way, the classes in UML could be used to define which objects might need to be displayed on the user interface (but not how to do this). However, it is fair to say that EID does not lend itself to checking of the design (Jamieson and Vicente, 2001; Mendoza et al., 2011) and, as such, UML offers advantages in terms of auditing and checking.

5.3 AIMS AND RESEARCH QUESTIONS

The research question for this chapter is, how can CWA be converted into UML as the basis for system design?

5.4 CASE STUDY DESCRIPTION

In order to illustrate the approach taken in this work, we adapt the analysis reported by Jenkins et al. (2012) of the Apple iPod 80 GB. The aim in this chapter is to rework an existing CWA (rather than attempt to redesign the iPod). By taking an analysis that has been performed for other purposes, we run the risk of losing information (because it might not have collected for the purpose we have in mind). However, the benefit of doing this is that the process can be directly related to an extant CWA and this means that we do not run the risk of conflating our need to produce UML diagrams with the need to produce CWA. This provides, we feel, a clean break between the human factors engineering and the software engineering, while also illustrating how these disciplines are bridged in this process.

5.4.1 Abstraction Hierarchy

In their analysis of the iPod 80 GB, Jenkins et al. (2012) focus on the physical components and the functions that they support. The ‘functional purpose’ of the device is to provide access to a collection of music while on the move. As Jenkins et al. (2012) note, there are additional purposes at play in the deployment of the iPod, not least is to provide a revenue stream for Apple. For the purposes of the analysis in this chapter, we concentrate on playing music.

The object-related functions from Figure 5.1 are used to construct the social organisation cooperation analysis (SOCA-CAT) (Figure 5.2). In Figure 5.2, these functions are considered in a range of situations and allocated to different ‘actors’ in the system. These actors can be human, that is, the user, or can take the form of automation, that is, iPod, ICE, power supply, PC or hi-fi.

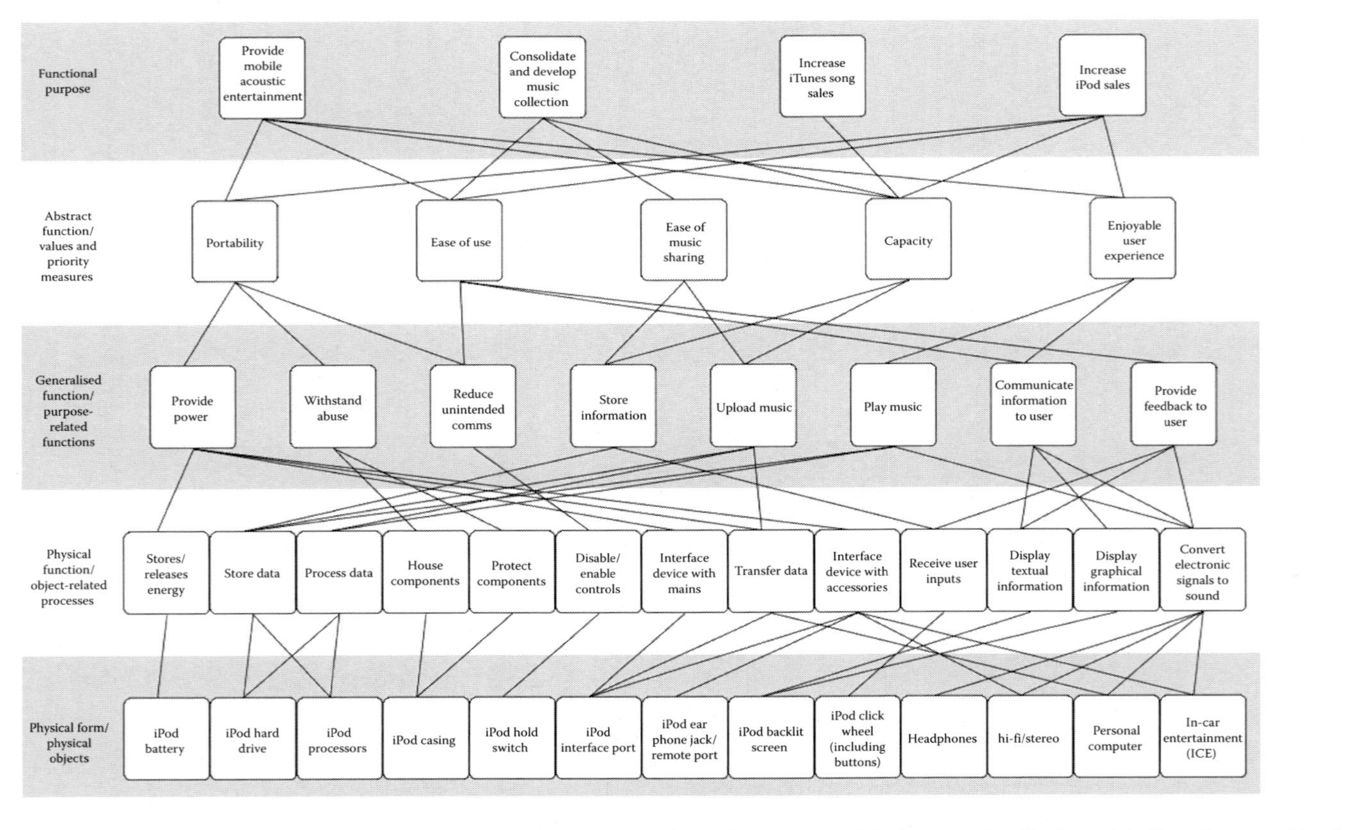

FIGURE 5.1 AH for Apple iPod 80 GB. (Adapted from Jenkins, D. P. et al. 2012. *Cognitive Work Analysis: Coping with Complexity*. Aldershot, United Kingdom: Avebury Ashgate.)

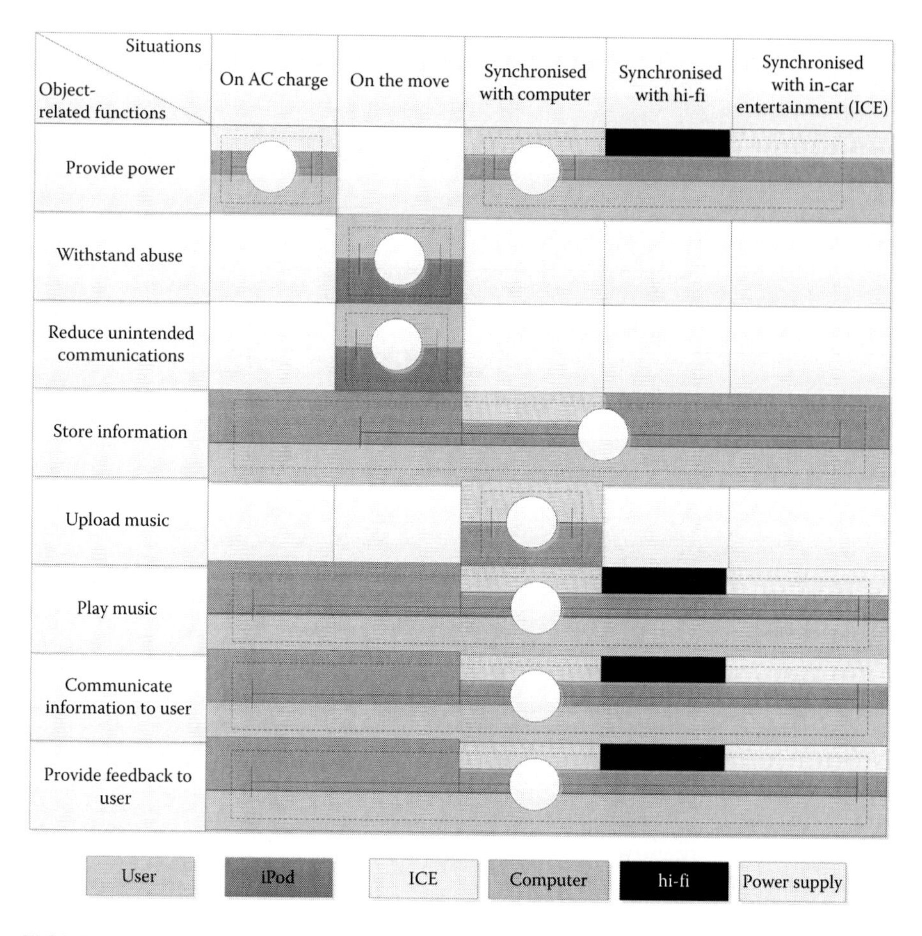

FIGURE 5.2 SOCA-CAT derived from Figure 5.1.

In order to achieve a particular purpose-related function (see Figure 5.1), actors can perform object-related functions in a variety of sequences. These define the strategies for the system, and examples of these are illustrated in Figure 5.3.

5.5 TRANSLATING CWA TO UML

In this section, we provide examples and instructions to convert from the CWA diagrams presented in the previous section to the subset of UML diagrams considered in this chapter. The challenge is to capture the content of the CWA in sufficient detail to not lose or distort this content while presenting it in a format that has the appearance of UML. We would expect any of the diagrams we develop in this section to be treated as initial sketches, which would then be reworked as part of the design process. In other words, just as CWA can be seen as an iterative process of description and redescription (Lintern, 2012) so we would see the process of translating to UML as iterative. Both processes continue until there is consensus in the design team.

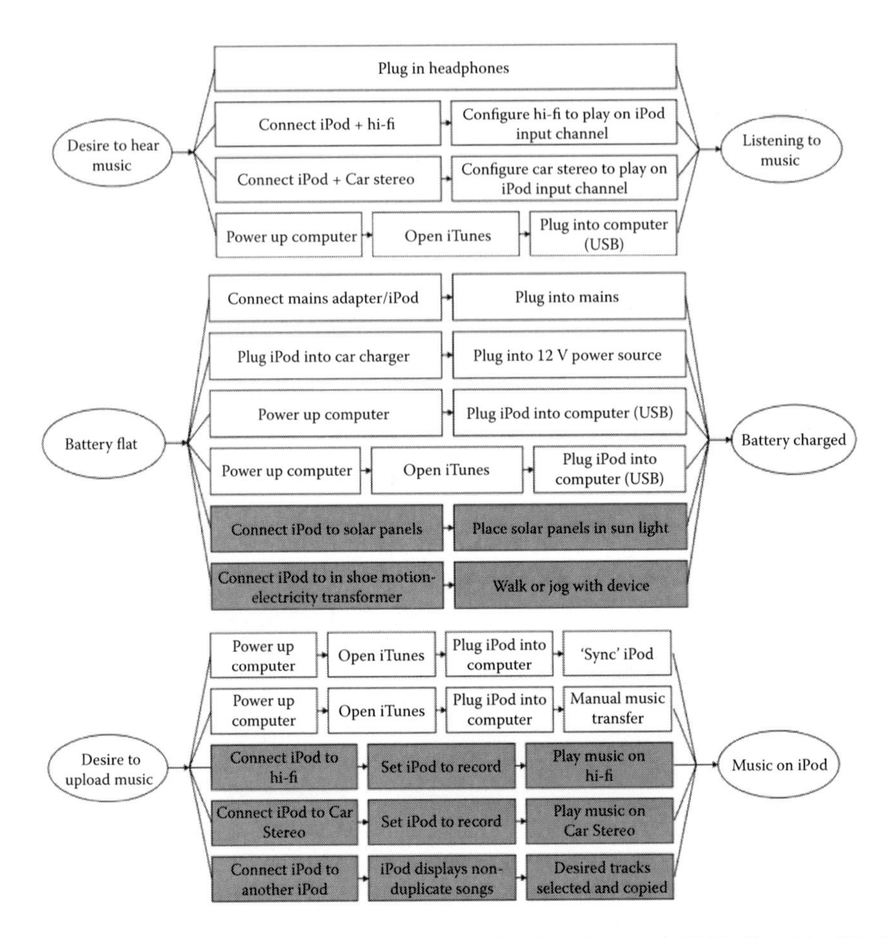

FIGURE 5.3 Strategies analysis. (Adapted from Jenkins, D. P. et al. 2012. *Cognitive Work Analysis: Coping with Complexity.* Aldershot, United Kingdom: Avebury Ashgate.)

5.5.1 Creating a Use Case Diagram

In UML, the use case diagram graphically presents the behaviour of the system. It shows the relationship between actors in a system and the actions (use cases) that they are able to perform. In addition to representing the behaviour of the system, use case diagrams can also be used to imply the structure of the system, in terms of the associations between use cases and between actors. The concept of actors can be related directly to the SOCA-CAT and to the objects in the AH. Use cases can be considered in terms of object-related functions in the AH, and the relationships between elements come from the lower level of the AH. Thus, Figure 5.4 shows the collection of use cases that relate to the functional purpose of 'listen to music' (through headphones). If we wished to consider listening to music through hi-fi or in-car entertainment system or personal computer, then we just replace the 'headphones' actor. We have shaded the heads of some of the actors because these are all part of the iPod system (hence the border around these (Table 5.1).

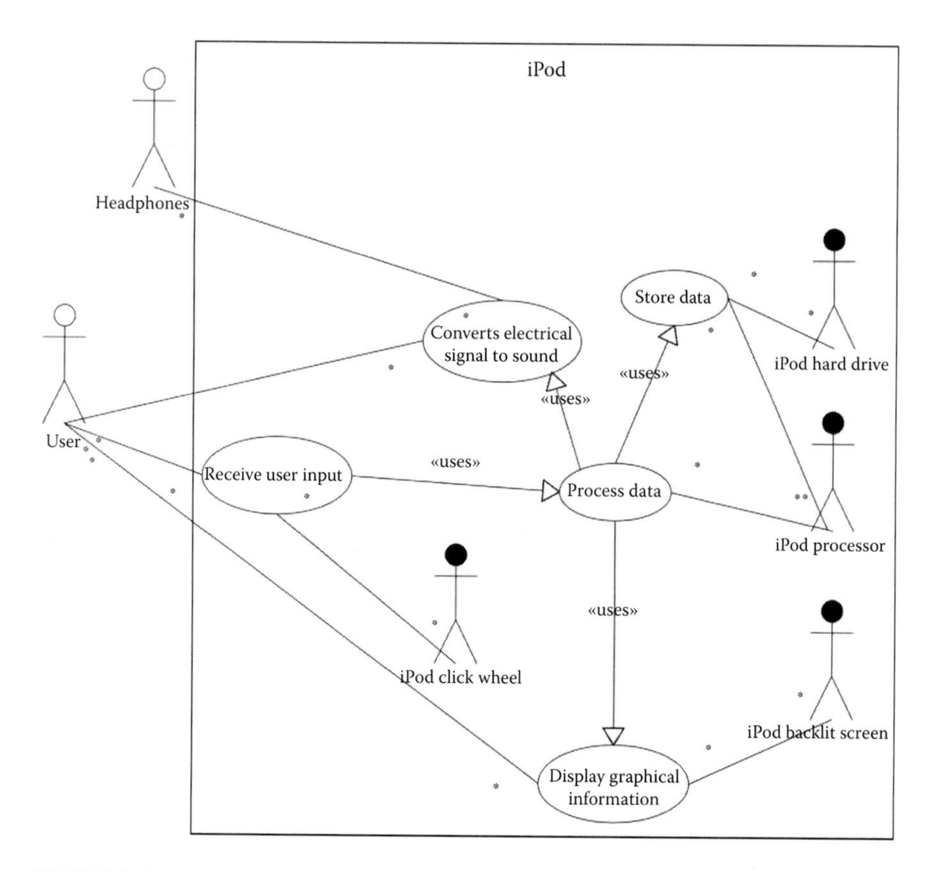

FIGURE 5.4 Use case for iPod listening to music.

TABLE 5.1
Use Case Elements and Their CWA Source

Elements of UML Use Case Diagram	Derived from the Following CWA Diagrams
1. Actors	SOCA-CAT
2. Use cases	AH's 'purpose-related functions'
3. Relationships	AH's 'object-related functions' and AH's 'object-related processes'

5.5.2 Creating a Class Diagram

In UML, a class diagram represents the architecture of the system to be designed. It shows the discrete functions that the system will perform and the attributes of these functions in terms of what information they require and the operations that these functions are able to perform (Figure 5.5).

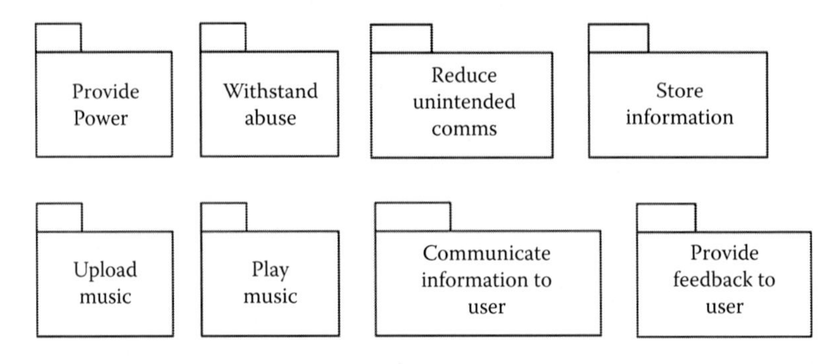

FIGURE 5.5 Derived packages.

TABLE 5.2
Class Diagram Elements and Their CWA Source

Elements of Class Diagram	Derived from the Following CWA Diagrams
1. Class name	AH's 'object-related processes'
2. Attributes	AH's 'object-related processes' and AH's 'physical objects'
3. Operations	AH's 'physical objects'
4. Class relationships	Determined by the designer

Table 5.2 defines the elements required and the CWA source, and Figure 5.6 presents the class diagram derived from these.

5.5.3 Creating Packages

In UML, a package is a collection of elements, with a unique name. Each package can be thought of as a container with elements that relate to each other (Table 5.3).

5.5.4 Creating Sequence Diagrams

In UML, a sequence diagram is used to illustrate the dependencies between processes. It highlights how one process leads to another and what information might be required to ensure that each process complete efficiently (Table 5.4).

The elements that are shown within angular brackets in Figure 5.7 are needed to make the diagram clear, but are not explicitly stated in the original AH (Figure 5.1). These are (we feel) implicit in the AH and do not necessarily mean that it should be modified. However, one can see how the addition of elements in the UML diagrams could lead to a need to modify the CWA diagrams, in order to ensure consistency between them (Table 5.5).

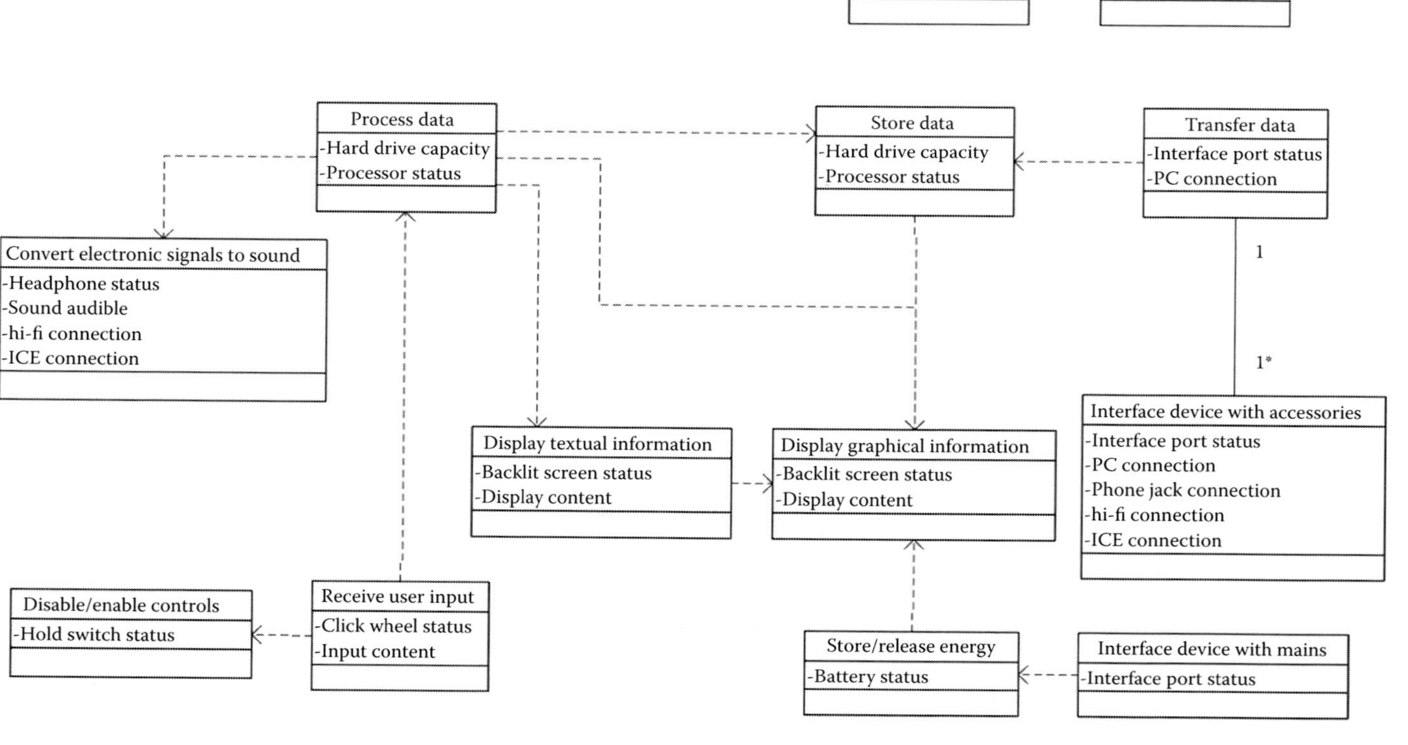

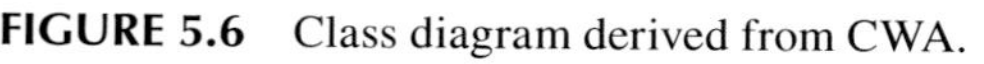
FIGURE 5.6 Class diagram derived from CWA.

TABLE 5.3
UML Packages and Their CWA Source

Information for Construction	CWA
1. Identify packages	*AH*
Identify the packages required	AH's 'purpose-related functions'
2. Identify class groups	*AH*
Organise classes into logically related groups	AH mappings between 'purpose-related functions' and 'object-related processes' to group the classes

TABLE 5.4
Sequence Diagram Elements

Information for Construction	CWA
1. Identify actors	*AH/use case*
Identify the actors that interact with the system	Use cases derived from AH show the users of the system
2. Identify user interface fragment	*Decision ladder*
User interface to help in user interaction	Look at decision ladder's 'data processing'
3. Identify fragments	*Decision ladder*
Fragments are represented as a box to enclose the interactive portions	By looking at the decision ladder's 'state of knowledge' and its use case diagrams
4. Identify database interactions	*AH/use case*
This fragment is optional depending on the requirement	Shows the interaction of the system

5.5.5 Statecharts

Figure 5.8 is derived from the strategies analysis (Figure 5.3) and the AH (Figure 5.1). Rather than defining actions, Figure 5.8 shows the state of the system once an action has been performed. Thus, 'track selected' and 'play pressed' relate to the object-related processes 'display textual information' and 'receive user inputs', and 'music playing' relates to 'convert electrical signals into sound' (Figure 5.9).

5.5.6 Combining the Two Approaches

The different views in CWA and UML complement each other. Indeed, in our applications we have found that the questions raised when completing the UML have led us to question and refine the CWA. This supports an iterative approach to specifying the user interface. As CWA analysis represents a high-level view of the system, it lacks lot of details necessary for the development, as shown in Figure 5.8. Therefore, by applying UML modelling, it is easy to bridge the gaps for systems engineering. Our approach for this integration is a minimalist, streamlined approach that focuses on the core set of UML diagrams that will be needed to do most of the modelling work to capture CWA analysis for coding.

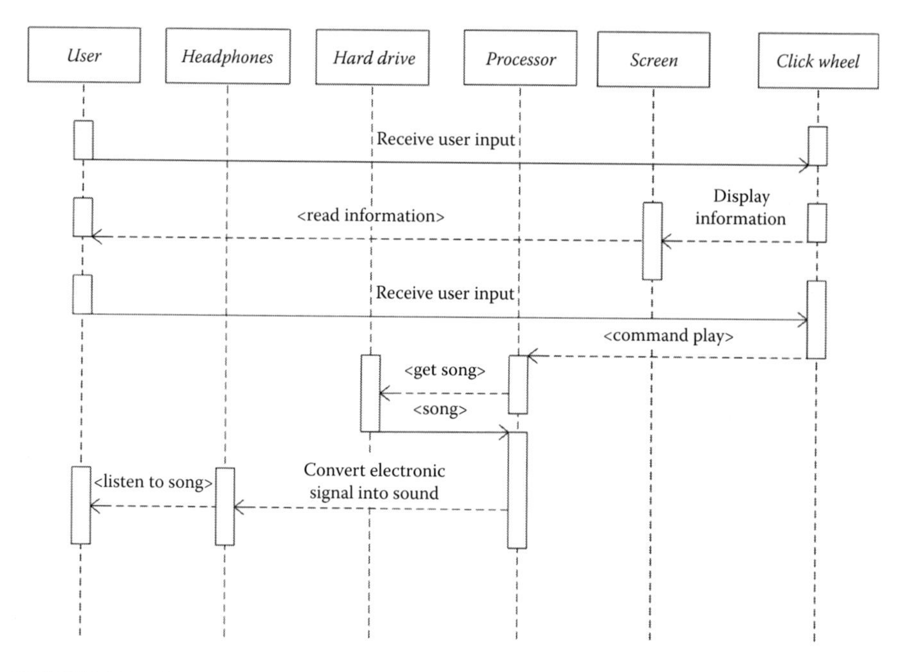

FIGURE 5.7 Sequence diagram.

TABLE 5.5
Statechart Elements

Information for Construction	CWA
1. Initial and final nodes Represents the start and end of the activity	*Strategies analysis* This is derived using the start and end of strategies analysis course of action
2. Object nodes Represents the flow of data through an activity	*Strategies analysis* Identified using the steps involved in each course of action
3. Actions Important steps that take place in the activity	*AH* Looking at the AH mappings for the object nodes

5.6 FROM SPECIFICATION TO USER INTERFACE

While the combination of CWA and UML results in identifying what needs to be put on the user interface, neither approach informs the decision of *how* this information should be displayed. The question of how to present information on the user interface and how this relates to CWA (or UML) remains very much an art. However, an interesting approach to supporting design decisions is offered by Upton and Doherty (2008) who describe a simple, systematic process that allows the requirements derived from an AH to be mapped into a visual form. This process is outlined in Figure 5.10.

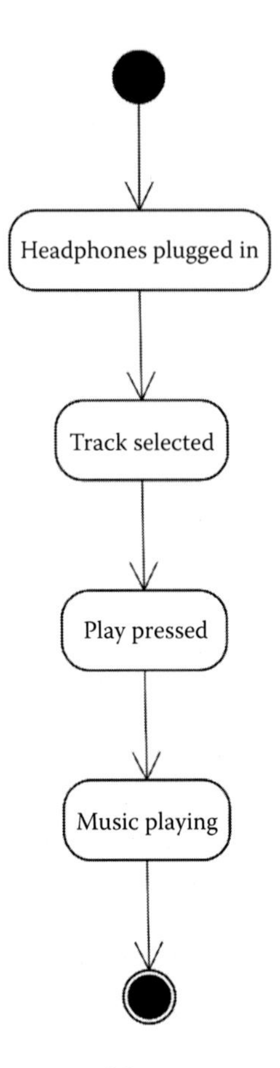

FIGURE 5.8 Statechart for 'listen to music'.

In terms of the design phase, this methodology begins with data-scale analysis. This involves deciding what format to use to present the data on the user interface. Following Stevens (1946), it is assumed that data can be categorised into one of these four scales: nominal, ordinal, interval and ratio. Furthermore, Bertin and Borg (2011) has suggested that graphic displays have a set of dimensions that can be mapped onto the data scales in terms of different forms of perception. The dimensions include the following: position, size, tone, texture, orientation, shape and hue. Bertin and Borg's (2011) forms of perception are as follows:

- *Associative*: A nominal category (say a square icon) can be perceived in the presence of distractor shapes and colours.
- *Selective*: A nominal category 'pops' out of a display; for example, all red dots in a scatterplot might appear to the viewer to form a shape.
- *Ordinal*: Data can be placed on an ordered scale, perhaps in terms of size or in terms of spatial arrangement.
- *Quantitative*: The viewer can judge the degree of difference between values.

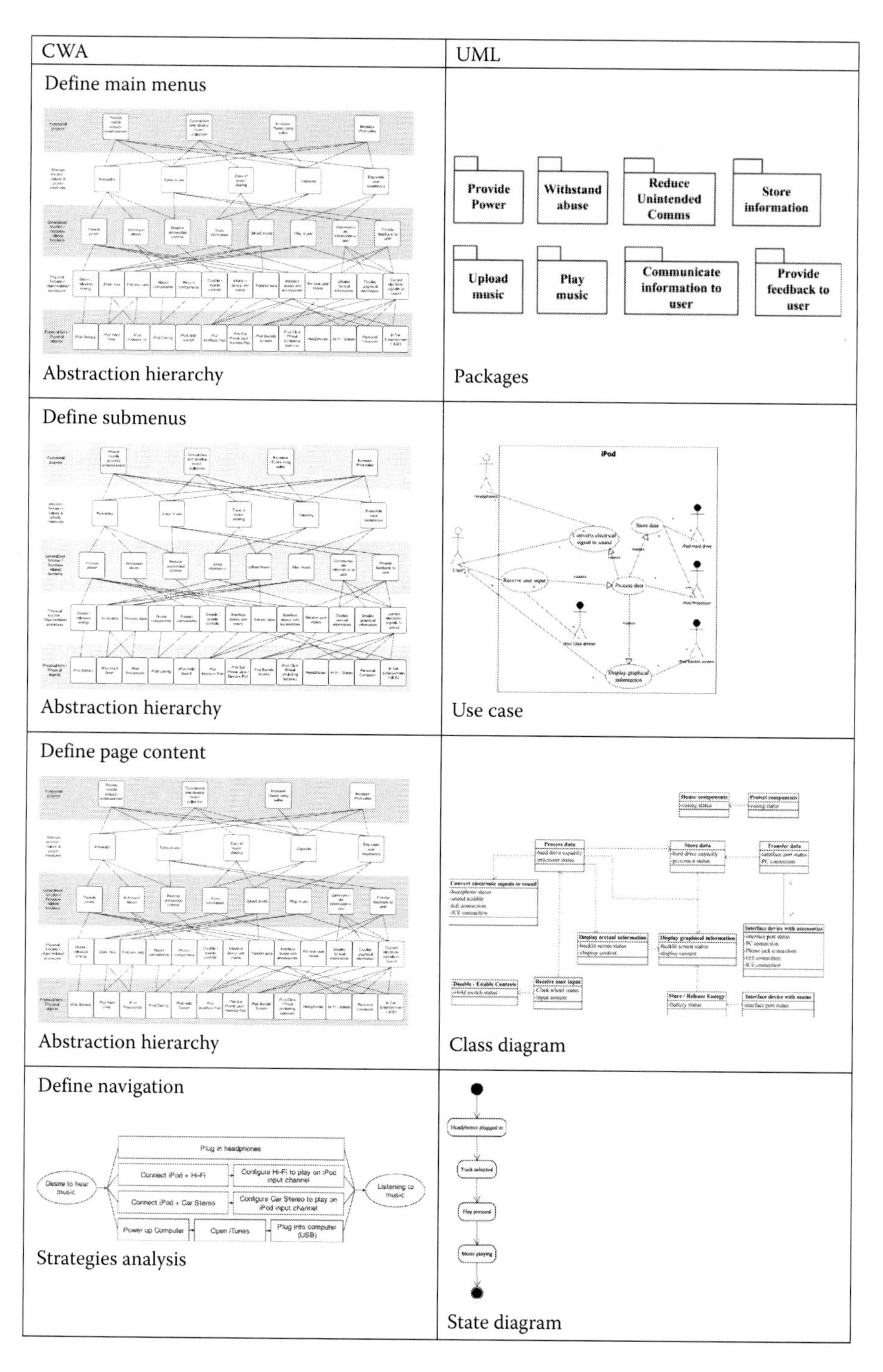

FIGURE 5.9 Relating CWA to UML views.

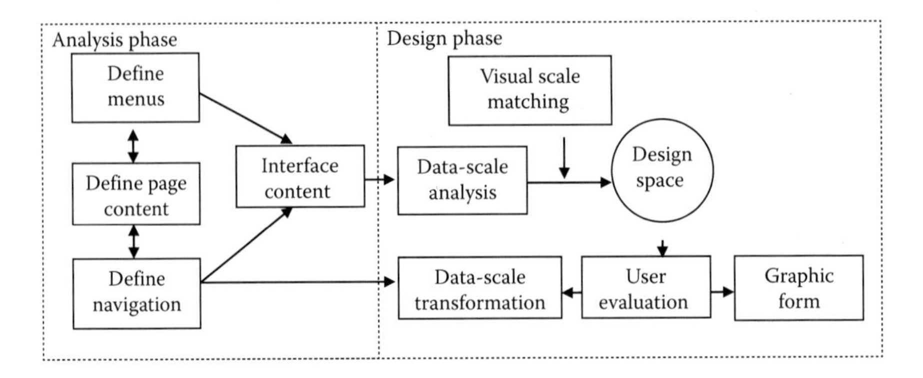

FIGURE 5.10 Visual design process. (Modified from Upton, C. and Doherty, G. 2008. *International Journal of Human-Computer Studies*, 66, 271–286.)

Relating these forms of perception to the skills, rules, knowledge (SRK) framework (Rasmussen, 1983; Chapter 1), one can see how associative and selective are likely to be skill-based (i.e. automatic) behaviours while ordinal is likely to be rule-based behaviour and quantitative is a combination of rule-based and knowledge-based behaviours. In other words, the effort required to extract information using the associative and selective forms of perception will be much lower than the quantitative form.

Relating this to the notion of main and submenus (Figure 5.8), we might say that the main menu could involve associative perception, say in terms of an icon, while the submenu could involve ordinal perception, say in terms of a list of topics. The page content would involve quantitative perception relating to reading the content. For warnings, alerts and advisories, one might expect associative and selective perception. Relying on the skill-based forms of perception means that the viewer is likely to trade-off accurate interpretation and understanding of data for a looser 'feel' for what the information conveys. For the iPod example used in this chapter, indicating battery or volume levels could involve associative perception, in the form of colours or shapes on the display. It might even be useful to show power (energy use) and volume, to indicate that playing at a higher volume drains the battery quickly.

5.7 CONCLUSIONS

This chapter has addressed a specific issue of extracting the CWA analysis to UML for the design and implementation through the definition of a step-by-step process and its illustration through a case study. The idea that UML is, to some extent, agnostic as to the ways in which objects are instantiated which captures the flexibility of CWAs focus on how the system might look. We suspect that the flexibility was at root of original EID work and the context in which this work was conducted resulted in mass–energy balance as an obvious solution. Subsequent developments have used this visualisation as a template (rather than the context-specific solution it was) and played down the need to reflect the flexibility of activity. It is proposed that the use of 'objects' returns us to that state of affairs, while also providing a basis for sharing designs with software engineering.

The approach is not meant to design approaches using EID (Vicente and Rasmussen, 1992; Burns and Hajdukiewicz, 2013). However, it is felt that it does offer some advantages in comparison with such approaches. CWA helps shift the focus from the user per se to the system in which activity occurs. This offers the potential to capture the richness of the domain in ways that focusing design on the user's tasks and goals could miss. The use of UML defines 'objects' in the software. The benefit of this approach (in comparison with EID) is that the relations between objects are not assumed to follow the mass–energy balance but could have other relations. Moreover, creating UML models using CWA helps not only to verify requirements, but also to make sure that CWA analysis is not missed in the design process. As UML models are developed, designers can feed information and changes back into the CWA, fleshing it out with details to refine the system.

REFERENCES

Bertin, J. and Berg, W. J. 2011. *Semiology of Graphics: Diagrams, Networks, Maps*. ESRI Press.

Booch, G. 2005. *The Unified Modeling Language User Guide*. New York: Pearson Education.

Bruseberg, A. 2008. Human views for MODAF as a bridge between human factors integration and systems engineering. *Journal of Cognitive Engineering and Decision Making*, 2, 220–248.

Burns, C. M. and Hajdukiewicz, J. 2013. *Ecological Interface Design*. Boca Raton, FL: CRC Press.

Handley, H. A. and Smillie, R. J. 2010. Human view dynamics – The NATO approach. *Systems Engineering*, 13, 72–79.

Jamieson, G. A. and Vicente, K. J. 2001. Ecological interface design for petrochemical applications: Supporting operator adaptation, continuous learning, and distributed, collaborative work. *Computers & Chemical Engineering*, 25, 1055–1074.

Jenkins, D. P., Walker, G. H., Stanton, N. A. and Salmon, P. M. 2012. *Cognitive Work Analysis: Coping with Complexity*. Aldershot, United Kingdom: Avebury Ashgate.

Lintern, G. 2005. Integration of cognitive requirements into system design. *Proceedings of the Human Factors and Ergonomics Society 49th Annual Meeting*. Santa Monica, CA: HFES, pp. 239–243.

Lintern, G. 2012. Work-focused analysis and design. *Cognition, Technology & Work*, 14, 71–81.

Mendoza, P. A., Angelelli, A. and Lindgren, A. 2011. Ecological interface design inspired human machine interface for advanced driver assistance. *IET Intelligent Transport Systems*, 5, 53–59.

Rasmussen, J. 1983. Skills, rules, and knowledge; signals, signs, and symbols, and other distinctions in human performance models. *IEEE Transactions on Systems, Man and Cybernetics*, 13, 257–266.

Upton, C. and Doherty, G. 2008. Extending ecological interface design principles: A manufacturing case study. *International Journal of Human-Computer Studies*, 66, 271–286.

Vicente, K. J. and Rasmussen, J. 1992. Ecological interface design: Theoretical foundations. *IEEE Transactions on Systems, Man and Cybernetics*, 22, 589–606.

energy management systems (EMSs). Electricity supply and demand are linked by transmission and balanced second-by-second by market mechanisms, supervisory manual interventions and automatic feedback control. Grid operations have periods of high temporal demand – where operators need to quickly and accurately extract information from a variety of sources – and quieter periods where operators inspect system statuses, plan upcoming procedures or assess events from the previous shift (Obradovich, 2011).

Power grid operation is a challenging domain because of how strongly network elements are coupled together (EPRI, 2009). Faults propagate in milliseconds, so operators must anticipate how automated circuit breaker systems will behave and which network configurations could occur. These domain characteristics make it difficult to pre-analyse the grid into summary parts. Power grids are open systems affected by uncontrollable factors such as weather, natural disasters or even negligently operated construction equipment. Longer-term societal trends are increasing uncertainty and variability in power grid operations: deregulation has opened the electricity market to more, smaller participants; intermittent renewable wind and solar power are displacing easily controllable fossil fuel generation; power grids are being further interconnected across jurisdictions, increasing grid monitoring scope (Schneiders et al., 2012); and national security priorities are increasing requirements to defend against physical or computer attacks (Kröger, 2008).

Electric blackouts are a severe risk to the welfare and security of households, health and emergency services, and more. Due to the critical nature of power grid infrastructure and the growing sources of system uncertainty, operators need appropriate tools to support a complete and accurate understanding of the state and trajectory of these rapidly evolving complex systems.

6.1.2 Aims and Research Questions

We partnered with the Independent Electricity System Operator (IESO), the Reliability Coordinator responsible for managing the power grid for the province of Ontario. The project goals were to evaluate opportunities for control room improvement and develop design concepts for (1) monitoring and controlling power grid elements in a multi-jurisdictional scope (also known as wide-area monitoring); and (2) integrating analyses of operating limit constraints. We undertook the work domain analysis (WDA) described in this article to help us understand the work demands of power grid operation.

6.2 METHOD

We developed our understanding of power grid operations through literature review, field observations, interviews, focus groups and a questionnaire. We first conducted a literature review of power system training material, to supplement our engineering education with applied grid operation theory. At the same time, we observed work in the IESO control room intermittently over 31 non-consecutive days. Control room staff included a shift superintendent, two supervisors and four operators together managing electricity generation dispatch, power flows, outages and stability limits.

We conducted critical incident interviews with eight control room operators and supervisors who volunteered to participate (Flanagan, 1954). Participants each had between 3 and 15 years of experience operating the Ontario grid, and many had progressed through control room roles. Participants were asked to describe typical daily work and examples of unusual situations they had faced. They were then asked to recall specific critical incidents where they made either a positive or a negative impact. Each interview lasted approximately 1 h. We collected accounts of 22 critical incidents in total. These interviews informed us of issues with existing tools and procedures, and helped identify priority areas for interface redesign.

We also led six focus group sessions, each with a different crew, comprising 42 operators in total. Each crew first recounted an incident that they had recently encountered as a team. A discussion on operational challenges followed, both those experienced during the incident and those that operators found in day-to-day operations. Finally, participants were asked to fill out a questionnaire on control room ergonomics and tool development. We corroborated findings from field and interview sources with official incident reports and documentation.

The WDA presented in this article was one of the products from this study. We developed it in two stages using standard methods (Naikar et al., 2005). We first developed an initial taxonomy of work domain concepts based on literature review (e.g. EPRI, 2009), and then refined them based on observations and interviews. The WDA was reported to operations staff as a draft for feedback, and an operations manager and an operator independently validated the analysis, providing criticism and suggestions.

6.3 FINDINGS

We developed a high-level WDA of power grid operations by interpreting how to describe domain phenomena using four theoretic concepts:

- *Functional abstraction*: What descriptions of electric grids mediate reasoning between system purposes and physical equipment capabilities?
- *Part–whole decomposition*: How can elements be conceptually 'chunked' in functionally relevant groups to support problem-solving within human short-term memory limitations?
- *Structural means–ends links*: What opportunities for action do system functions provide, and what alternate resources or unintended trade-offs are present?
- *Topographic/causal links*: How can internal system coupling be described at different levels of abstraction and decomposition? How will faults or control actions propagate through the system?

We present the WDA in terms of each concept in turn, after outlining the scope of the analysis.

6.3.1 System Boundary

Due to the exploratory nature of this WDA and limited time and research resources, we selected a small system boundary, which could be expanded for future applications. The geographical scope of the WDA was nominally the IESO-controlled grid, but in principle would be applicable to other (wider) areas such as the North American Eastern Interconnect. Supportive IT was considered out of scope, except as embedded software in grid components (e.g. automated circuit breaker relay communications). Cybersecurity issues were omitted from this WDA. The WDA addressed functions relevant to power grid operation, particularly generation and transmission, although distribution (i.e. radial customer loads) was included in less detail. Electricity supply markets were included as they influenced grid operation, although we omitted many IESO-specific market functional mechanisms. Weather was considered in-scope, since thunderstorms, wind and sun particularly affect power grid operation (Burns et al., 2005). In general, we focused on functions relevant to the timescale of human operator decision-making and action (Memisevic et al., 2007, p. 765).

6.3.2 Levels of Abstraction

Electric grids, being highly engineered technical systems, were well-characterised by the five conventional WDA abstraction levels. We present each level below, starting with the purposes that limit what should be done in grid control, then skipping all the way down to physical objects (POs), since they are concrete and clearly identifiable, and finally working back up to abstract value and priority measure descriptions.

6.3.2.1 Functional Purposes

The three overarching functional purposes (FPs) of power grid operations that we determined are, by priority, the following:

- *Reliability*: Securely operate the power system, respect interconnection reliability standards and prevent damage to generation or transmission equipment.
- *Quality*: Deliver a high-quality electricity service within voltage and frequency standards, with few outages and prevent damage to distribution or customer equipment.
- *Efficiency*: Operate efficient markets, deliver electricity at lowest cost possible and avoid resistive losses.

Grid control staff consider these overriding purposes when assessing risk, although in routine operations regulatory rules and procedures implicitly balance the three purposes. When managing unanticipated events, expert judgement about these purposes is required to make critical decisions – for example, whether to shed customer load (quality) or curtail generation (efficiency) to reduce the risk of stability loss (reliability) or damaging equipment.

6.3.2.2 Physical Objects

At the other end of the abstraction scale, the most concrete manifestation of a power grid is the physical appearance of its material equipment. In critical incident interviews, operators recalled rare, complex, ambiguous physical faults that required PO information to diagnose. This information can be both easy and difficult for the control room team to acquire. Physical appearance is easy to gather information about, because unlike most electrical phenomena, it can be observed by even novice field workers (or cameras). However, it is complicated by the scale of the electric grid. To inspect equipment at any one of hundreds of remote transmission stations can take several hours as a crew mobilises and travels to the site. Power grids can be described at this level using terms listed in Table 6.1 (Figures 6.1 and 6.2).

TABLE 6.1
Physical Objects (POs) That Comprise the Electricity Grid

Element Type	Characteristics
Transmission lines	The physical path of transmission lines across terrain (e.g. Figure 6.1). Relevant information includes line length, support tower locations and crossings above/below other lines (where a tower failure could cause a cross-line short circuit)
Transmission stations	Equipment in transmission stations or switching yards can be described by plan map views and physical appearance (e.g. Figure 6.2)
Generation stations	Physical locations of electricity generators, their appearance and their status (e.g. running, stopped, warming up, on fire)
Power consumers	Physical locations of power consumers, and geographic areas served from different distribution stations
Intertie lines between jurisdictions	Similar to transmission lines, but distinguished since they serve specific functions (imports/exports) and have slightly different purpose trade-offs
Weather	Geographic position of thunderstorms, lightning, clouds, wind and solar flares, all of which affect power grid operations

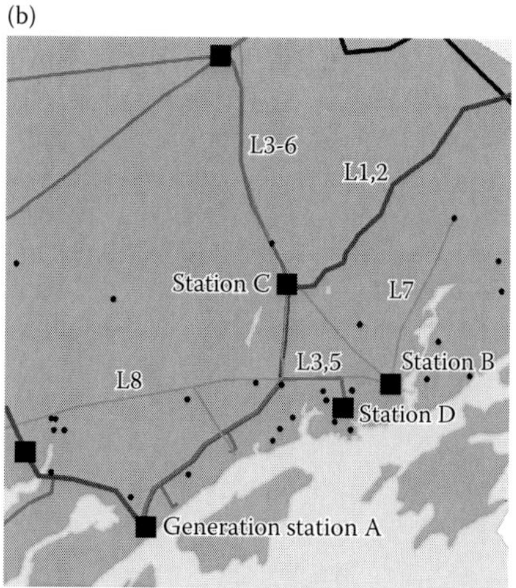

FIGURE 6.1 Satellite map showing the appearance of POs at an Ontario station (a) and their surrounding (forested) environment. (b) An idealised map of the surrounding area showing location of POs. (Adapted from (a) U.S. Geological Survey. 2015. LandsatLook Viewer. http://landsatlook.usgs.gov/; (b) Inergi and Hydro One, n.d. Map04-168.)

FIGURE 6.2 Transformers at an example generation station. Information about appearance of POs can help diagnose equipment failures. (Adapted from Independent Electricity System Operator. n.d. Transformer Fire at Substation.)

Reliability Coordinators such as the IESO are not responsible for monitoring the physical condition of transmission-connected equipment. Rather, equipment owners such as utilities, generators or customers must report degradation or contamination of their assets. Thus, except for weather patterns, control room operators rarely need to consult PO information. Electrical functionality is a much more useful level of abstraction.

6.3.2.3 Physical Functions

Physical functions (PFn) describe what grid elements *are capable of doing*, regardless of their physical appearance. Equipment PFn are well defined and mostly standardised as mature AC grid technology and we found them the default perspective in control room operation. The most common way to describe the power grid at this level of abstraction is through power line schematic diagram notation (e.g. Figure 6.3). The PFn we analysed are listed in Table 6.2.

Schematic circuit diagrams are visually structured according to PFn, showing equipment connected to station buses and the electrical connectivity linking them (topological links, discussed later). A PFn level of abstraction supports diagnosing paths that electric current can follow in present or hypothetical future grid configurations. However, schematic diagrams lay out transmission lines based on visual clarity rather than actual physical arrangement, which obscures implications of some PO failures (e.g. nearby conductors coming into contact due to a failed transmission tower insulator).

6.3.2.4 Purpose-Related Functions

This level of abstraction describes how PFn can be used to achieve system purposes. We found electric power flows the most useful terminology to describe the use of

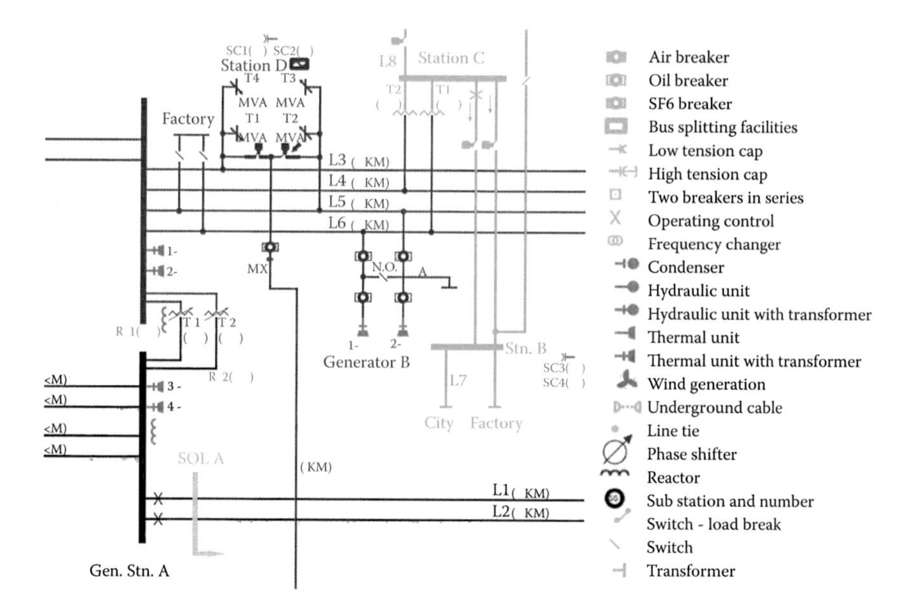

FIGURE 6.3 Schematic (PFn) diagram of the Ontario power grid around the same area as Figure 6.1. Lines between functional elements are topological links. (Adapted from Independent Electricity System Operator. 2012. IESO-Controlled Grid Total System – Revision 16.)

generators and transmission lines (Figure 6.4). Power flows can also capture the functionality of controllable capacitors and reactors through the notions of real and reactive components of AC electric power (Figure 6.5). In WDA, energy is usually the most abstract functional system description, but since the physics of AC electricity is very well characterised, power is less cognitively abstract (Naikar et al., 2005). System elements at a purpose-related function (PrF) level may be described as in Table 6.3.

Power flows support reasoning about how to coordinate physical equipment functions. For example, generators can be substituted to supply real power, and reactive power can be supplied by capacitors, generators or under-loaded transmission lines. Power flows also support inference about un-telemetered grid elements through network power balance equations, as implemented in EMS grid state estimation (Larson et al., 1970). PrF representations can summarise equipment functionality and form a basis for wide-area monitoring displays that capture large geographical areas. Finally, to relate how power flows contribute to achieving the FPs of the electric grid, we propose an abstract functional description.

6.3.2.5 Abstract Functions, Values and Priorities

This abstraction level describes the grid in terms of detailed scientific (and economic) understanding, which we do not claim to have fully captured in this preliminary analysis. Information at this level of abstraction is developed offline by planning engineers, for example, through stability simulations, statistical analyses and precise

TABLE 6.2
Physical Functions (PFn) of the Electricity Grid

Element Type	Function
Transmission network	Conductive paths between equipment, in terms of voltage, current-carrying and characteristic impedance
Transformers (including auto-transformers)	Magnetically coupled paths between transmission networks of different voltages, sometimes adjustable via tap change
Capacitive compensators (static and dynamic)	Increase electric capacitance of a transmission network, subject to voltage limits and control capability
Reactive compensators	Add reactance to a transmission network, as above
Circuit breakers	Split the transmission network; includes equipment necessary to enable automatic circuit protection (relay communications)
Switches	Split the transmission network, but with functional limits on switching under load and cycle time
DC–DC couplers	Power transmission interfaces between separate AC electric transmission networks, usually across jurisdictions
Phase shifters	Shift voltage phase angle to control 'real' AC power flow (a more abstract function described below)
Dispatchable generators	Generate electricity and support voltage, with power output controlled through market signals
Non-dispatchable generators	Generate electricity and support voltage, but power cannot be controlled through market signals (e.g. intermittent solar/wind)
Dispatchable loads	Consume electricity, but functionally responsive to market signals and control orders (e.g. interruptible equipment)
Non-dispatchable loads	Consume electricity and do *not* respond to market signals (e.g. price-regulated residential and small commercial consumers), but can be 'shed' by control orders
Weather (wind, sunlight, lightning, temperature, humidity)	Affect the performance of other functions such as renewable generators, transmission network or air-conditioning loads

technical models. Judgement criteria at this level of abstraction are specified by regulatory agencies, or developed by stakeholder groups and policymakers. The IESO follows criteria developed by the North American Electric Reliability Corporation (NERC), which regulates Reliability Coordinators across North America. Some elements of the system from an abstract function (AF) perspective are listed in Table 6.4.

Since information at this level of abstraction is complex and time-consuming to develop, it is extensively pre-analysed and summarised for the control room as system control manuals or embedded in operating limits. In unanticipated contingency situations, system simulation tools in the EMS control automation are intended to support reasoning at this abstraction level. Expert operators and superintendents must make judgements at this level of abstraction to determine how PFn such as breakers or generators should be controlled to mitigate risks to reliability, quality and efficiency. To help the control room crew make the best decisions about preserving quality or efficiency under threats to reliability, future control room tools should aim to support accurate judgements at this level of abstraction.

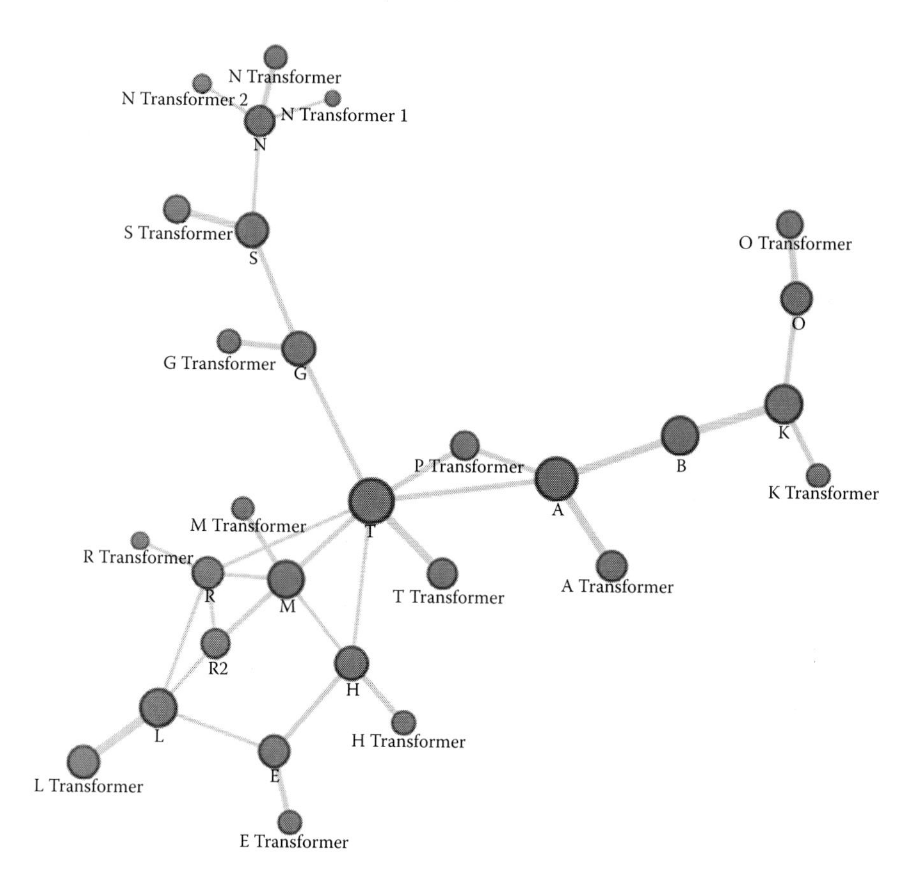

FIGURE 6.4 Node map of potential power flows (PrF) within the highest-voltage (500 kV) Ontario functional transmission network. Power flow functionality is enabled by transmission lines and determined by balance of real and reactive generation and consumption.

6.3.3 Means–Ends Links

The abstraction hierarchy diagram in Figure 6.6 describes means–ends links between elements at an overview level of aggregation (discussed in Section 6.3.4). Means–ends links represent both intended and trade-off effects of equipment on more abstract system properties, and conversely the alternate means at operators' disposal to achieve system purposes. For example, reasoning 'bottom-up' with italics referencing Figure 6.6:

- A PO in a *transmission station* is a means to achieve the PFn of a *capacitor.*
- This *capacitor* is a means to achieve a *reactive power source* at a particular region in the transmission grid.
- *Reactive power sources* are a means to *maintaining dynamic stability* and *robustness* against voltage collapse, when combined with sufficient *transmission.* However, oversupply of reactive power can damage equipment through over-voltage.
- *Dynamic stability* and robustness are a means to *reliability.*

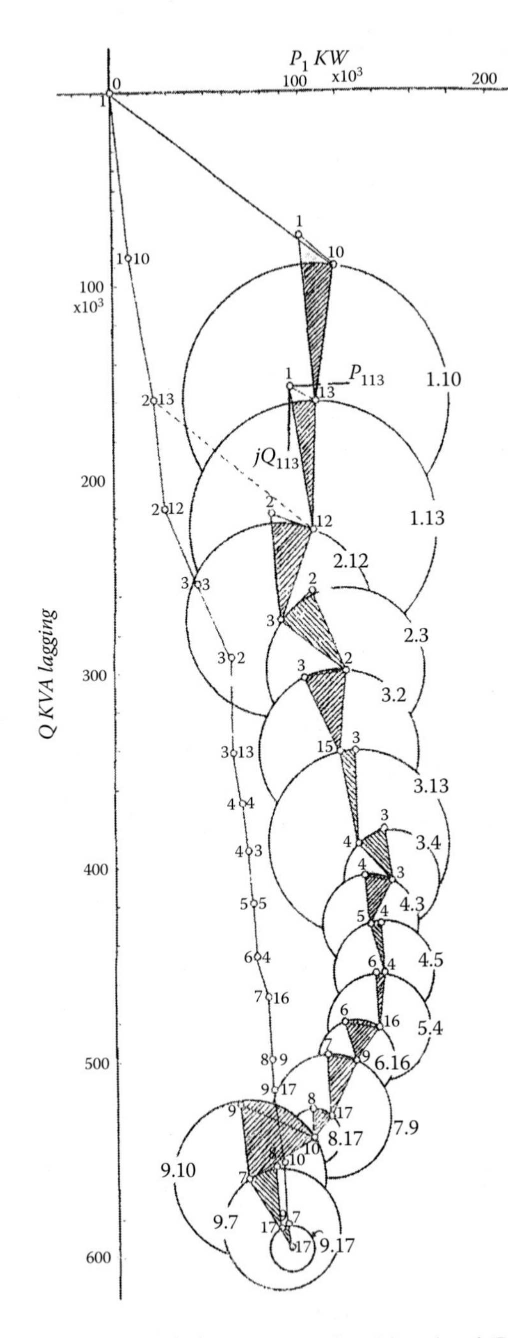

FIGURE 6.5 AC power transmission purpose-related functional (PrF) relationship between real (kW) and reactive (kVA) power flows for receiving ends of 17 transmission and generation stations in Hokkaido District. (Adapted from Ogushi, K. 1950. *Memoirs of the Faculty of Engineering, Hokkaido University*, 8(3-1), 104–115. http://hdl.handle.net/2115/37758.)

TABLE 6.3
Purpose-Related Functions (PrF) of the Electricity Grid

Purpose	Function
Real power transmission	Transmitting real (work-performing) power between generators and consumers (via lines, transformers, breakers and switches)
Real power generation	Generating real power; enabled by dispatchable and non-dispatchable generation stations
Real power consumption	Consuming power, both intended customer loads and undesired resistive losses
Reactive power transmission	Transmitting reactive (out-of-phase) AC electric power; this trades off with real power transmission capability
Reactive power sources	Converting real power to reactive; reactive power can be supplied by capacitive support equipment, generation stations or un-loaded transmission lines
Reactive power loads	Absorbing reactive power by loaded power lines, transformers or reactive control equipment; reactive power balance is coupled to voltage
System protection	Protecting equipment from grounding faults or other disturbances, using circuit breakers and dispatchable generators
Operating reserves	On-demand real and reactive power generation that dispatchable generators, dispatchable loads, tie lines or equipment voltage reductions can provide
Import/export flows	Interchangeable with real (and reactive) power transmission but with distinct policies and equipment

Or, conversely, reasoning 'top-down':

- *Stability* is achieved by *real and reactive transmission* not exceeding their flow limits.
- *Power transmission* can be changed according to *transmission network* impedance by controlling a set of *dispatchable generators* (and/or *loads*).
- The functionality of *dispatchable generators* depends on the physical *generation station* condition (is there water behind the hydroelectric reservoir?).

We did not develop the analysis of means–ends links at finer decomposition levels. However, given the well-documented and engineered nature of the power grid, this should be straightforward. Physical components have clearly defined functionality (Figure 6.3), and are connected to the transmission grid in clearly defined locations (Figure 6.1). Grid structure is discussed next in terms of part–whole and topological links.

6.3.4 Part–Whole Decomposition

Electric grids are straightforward (but unwieldy) to describe in terms of extensive component lists of POs and PFn. In principle, grids can be aggregated at a larger

TABLE 6.4
Abstract Functions (AF), Values and Priorities of Electricity Grid Operations

Abstract Function, Value or Priority	Description
Maintain dynamic stability	How reactive power must be coordinated to minimise phase angle instability risk and maintain synchronicity; described as differential equations, phase angle, margin of safety, etc.
Maintain robust system	Sensitivity of the grid to voltage collapse under disturbances; robustness is reduced by higher power flows due to high power consumption or low power generation
Match power supply and demand	How transmission system instantaneously balances power generation and consumption; described by differential equations, like $M_{tot}(d\omega/dt) + D_{tot}\omega = P_{gen} - P_{load}$ (EPRI, 2009)
Match energy bids to asks	Economic dispatch in the energy market in terms of generator bid curves, load ask curves and market fairness principles
Protect equipment	Limits and policies around respecting equipment owners and market participants' equipment capabilities such as voltage limit, and start-up ramp rates
Minimise power losses	Combines resistive loss equation $P = I^2R$ with a distinction between productive matching of supply and demand and unproductive recirculating 'loop' flows; describes benefits of locating real and reactive power generation and consumption nearby each other
Minimise load shedding	This is a metric for quality of service; shedding load means local blackouts, which may be necessary to balance power supply and preserve dynamic stability
Minimise net inadvertent interchange	This is a metric for controlling the power grid to, on average, limit cross-border power flows to only those financially accounted for

scope than we analysed in our system boundary; the Ontario grid is directly interconnected to most of Eastern North America. Because of how electric grids are coupled, we found intermediate part–whole aggregations of elements at PFn and higher abstraction levels can be situation specific. We discuss this more below, after introducing the principles we used to define decomposition at each abstraction level:

1. *Physical Objects*: Physical grid component objects are geographically grouped by the transmission yards they are located in. Lines can be grouped by which transmission towers they share.
2. *Physical Functions*: Grid component functions are similarly well defined, shown as symbols on schematic diagrams (e.g. Figure 6.3). Equipment can be aggregated by similar function within an area (e.g. four generators in one generating station). However, reasoning on whole groupings is sensible only if sufficient robust transmission capability links them together. A capacitor on one side of the grid cannot substitute for one on the other.

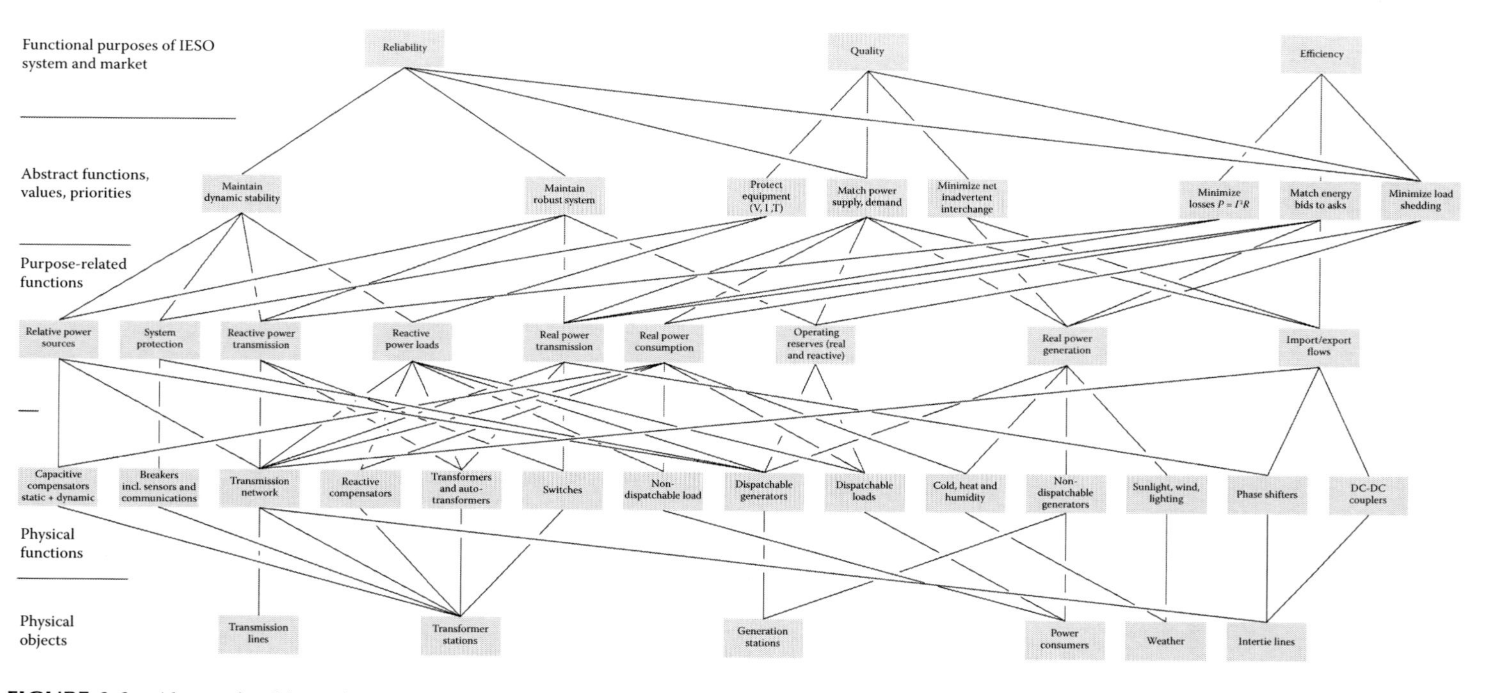

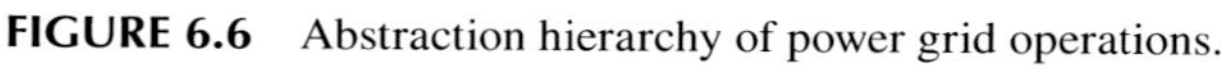
FIGURE 6.6 Abstraction hierarchy of power grid operations.

3. *Purpose-Related Functions*: At its finest decomposition, this level describes power flow along individual lines (Figure 6.3). Power flows can be grouped as pre-analysed System Operating Limits (SOLs) and regional balances as Area Control Error (ACE). However, potentially critical flows and areas can emerge as grid configuration and generation availability changes.
4. *Abstract Functions, Values and Priorities*: The grid at this level is defined in terms of dynamic stability, power-frequency coupling and economic market efficiency. These concepts are very different and decompose along distinct lines.
5. *Functional Purposes*: The parts of 'reliability', 'quality' and 'efficiency' can be broken down according to regulatory requirements (North American Energy Standards Board, 2010; North American Electric Reliabilty Corporation, 2013), particularly requirements that potentially conflict with each other.

Determining correct, useful ways to group abstract electric power grid parts into psychologically relevant wholes is difficult, and presents an opportunity for cognitive support tools. All transmission equipment linked by a conductive path forms a synchronised 'whole'. Coupling within this whole is strengthened by independent and high-capacity conductive paths (topological links). In extreme circumstances, multiple circuit breaker trips may sever parts of the grid from each other. Such 'pockets' with more load than generation will black out. Parts with more generation than load will form 'islands' each out of phase with others. Anticipating and mitigating risks of fragmentation is a crucial role of grid operators and requires reasoning about stability-relevant part–whole decomposition in the context of topological structure.

To anticipate risky situations, engineers pre-analyse grids to find equipment aggregations with weakly coupled boundaries between them, and associated aggregated power flows (SOLs). These pre-analysed groupings are used to conceptually simplify routine grid operations and inform preventive control actions (such as reducing transmission overload by dispatching generation or load within a 'pocket' group). But pre-analysed wholes risk misleading if they do not account for grid topology changes due to outages or contingencies. For real-time support, automated EMS security analysis tools monitor equipment outages, transmission saturation or other potential weakening of coupling, but only at component levels. Operators must interpret these automated analyses in situation-specific context to aggregate grid elements into groups cognitively relevant to problem-solving.

6.3.5 Topographic/Causal Links by Abstraction and Decomposition Level

Topographic links practically define the function of an electric grid and are essential to determining useful intermediate part–whole decompositions at higher abstraction levels. Links between elements mean different things at different levels of abstraction. For example,

1. *Physical Objects*: Links at this level are the physical wires connecting each piece of equipment, or conversely the physical insulation preventing electric faults. For example, the physical route a power line takes across the map of

Ontario (Figure 6.1), or the physical shadow (or lightning) a cloud projects onto equipment physically below it.

2. *Physical Functions*: Links at this level are conductive electrical paths, as shown in single-line diagrams (Figure 6.3). If breakers (or melted-through transmission lines) separate topology, parts of the grid will be islanded from each other. Electrical node-breaker engineering models (Hargreaves et al., 2012) are an example of representations of this type.
3. *Purpose-Related Functions*: Links between elements at this level are in terms of power flows (Figure 6.4), which distribute across every parallel conductive path depending on voltage, phase, line impedance and topology. A computational breaker model (in EMS automation) can derive or infer power flows at this level of abstraction.
4. *Abstract Functions, Values and Priorities*: At this level, links can be considered cause-and-effect relationships or more formally potential synchronous coupling between elements (Motter et al., 2013). These links generally weaken as power flows (PrF) increase, because saturating transmission capacity (PFn) leaves less reactive reserve for robustly maintaining dynamic stability. In principle, links at this level would describe how far a voltage collapse or stability loss will cascade before stopping.
5. *Functional Purposes*: Less relevant or applicable, as purposes are all somewhat interrelated. However, for example, unreliability (a blackout) directly causes inefficiency and low quality but not necessarily the other way around.

The state of topographic links is essential information for grid operation. Automated algorithms to dispatch generators and monitor transmission security depend on an EMS model of present circuit breaker configuration and resulting grid topology. Operators must trace circuit topology to predict the effects of control actions or switching operations. For example, loop current effects mean that operators need overviews of topographic links between a breaker and all adjacent buses and lines in order to make safe operating choices.

Operators also recognise stereotypical topological link configurations associated with risk. For example, an unusual property of AC transmission grids is that two links between grid elements is the most risky configuration. Single links are simple to control, since if the link is severed, the outcome is predictable. Multiple links simplify operation; if one link in a triply redundant connection fails, the remaining two will only be 50% more loaded (assuming identical links). By contrast, link pairs are the least predictable, since the failure of one can lead to 100% more load in a parallel line, or even more in a circuitous alternate path. Whether remaining lines can withstand power flows and maintain synchronous coupling must be carefully considered by engineers and operators.

Human abilities to reason about complex topographic links are limited, hence the need for simulation tools and automated control systems. Finding how to communicate implications of complex topology 'at a glance' is a research opportunity for wide-area monitoring.

6.4 DISCUSSION AND CONCLUSION

This WDA is the first reported in the literature of the power grid operation domain and has practical and methodological implications. Distinctions made between levels of means–ends abstraction distinctions in this WDA are consistent with design concepts in development elsewhere.

6.4.1 Practical Implications

6.4.1.1 Grid Visualisation Methods

Much contemporary research and development are being done on electric grid control, under the tagline 'smart grid'. Some implications of the WDA framework are already anticipated by on-going visualisation research for wide-area monitoring. For example, projects funded by the U.S. Department of Energy have developed visualisation methods for dynamic map (PO), schematic (PFn) and power flow (PrF) views (Power Info, 2014), all generated from existing topological model databases. Others have developed multi-abstraction displays that dynamically visualise power flow (PrF) in the context of grid impedance (PFn) with the intent of informing judgements about dynamic stability and robustness (AF) (Wong et al., 2009). However, while these visualisations dynamically represent topological links at multiple abstraction levels, few visualisations have integrated means–ends structures (Gosink et al., 2014) or supported functionally valid part–whole aggregation (though some aggregate for visual clarity [Wong et al., 2009]). Designing for higher levels of aggregation is essential in order to support accurate mental models of increasingly large and complex modern power grids (Hoffmann et al., 2011).

Means–ends links suggest that visualisations should support navigation and reasoning between abstraction levels, for example, to visualise the stability implications of manipulating equipment, or conversely help problem-solve about which equipment to manipulate to resolve stability risks. Navigation between grid representations at different abstraction levels can be disorienting and may require innovative animated visualisations (e.g. between Figures 6.1 and 6.3). Part–whole links suggest that grid visualisations ought to be able to idealise across the component level used in EMS breaker models. Useful summaries may be achieved through configural visualisation techniques that support perceptual grouping (Wong et al., 2009), or by algorithms that can dynamically group equipment in purpose-relevant ways that respect system functionality (Gosink et al., 2014). Part–whole groupings (e.g. multiple transformers into one) must carefully consider grid topology to be sure they do not obscure implications of component failure (e.g. one transformer may not be as robustly connected as the others).

6.4.1.2 Information Architecture

Furthermore, this WDA emphasised that physical manifestations of weather have important means–ends implications for grid operation and should be incorporated in abstract grid visualisations, particularly as renewable (wind and solar) generation proliferates. Whereas a single hydropower plant is mainly concerned with precipitation patterns (Sanderson et al., 2004), relevant phenomena for the whole power grid extend to lightning, wind, solar irradiation and solar flares. Representing the means–ends implications of weather will require combining meteorological data, forecasting models, renewable telemetry and energy market data.

This WDA has implications for anticipating such information interchange needs between disparate databases in grid control systems. Current arrangements often prevent data exchange between systems due to the use of proprietary data models and software (Wu et al., 2005). This separation requires the human operator to form the link between systems, increasing cognitive workload and creating a potential source of error (Obradovich, 2011). Information security, network load, reliability and maintenance requirements mean that data source integrations must be planned in database design and may not be feasible to implement in-between major software revisions. Extending this WDA in a formative design process may help anticipate data integration needs to support human or automated problem-solving in (smart) grid operation.

6.4.2 WDA Methodological Implications

This analysis encountered two notable complications in applying WDA methodology: modelling power system dynamics and properly representing couplings between topology and part–whole aggregation. Mapping grid operation concepts to PO and PFn abstraction levels was straightforward, but we found it unusual to include power/energy in the next most abstract PrF level (a.k.a. generalised function) (Sanderson et al., 2004). In canonical WDA (Naikar et al., 2005), energy balances and flows are part of the most abstract functional representations. However, in electric systems, power is defined by a single equation ($P = VI$), readily measured by instruments, insulated and well behaved, and can therefore be reasoned about in fairly simple algebraic terms (e.g. $Flow\ 1 + Flow\ 2 = Flow\ 3$). We conclude that power is less abstract in electrical systems. The most abstract representation that we found of AC power grids is as a dynamical system in terms of differential equations (Kundur et al., 2004) and probabilistic robustness against failure. We are not aware of WDAs that have explored how to incorporate dynamical stability as a work domain function.

The second issue this analysis encountered was the relationship between electric grid topology and part–whole groupings. While POs are clearly grouped geographically, other concepts even as concrete as PFn have a tension between part–whole groupings according to function (e.g. generators part of generation fleet, capacitors part of reactive compensation resources) and according to topological proximity (capacitors part of one transmission station's resources). Groupings are a matter of degree rather than distinction – PFn in a well-connected region can substitute for each other, but the weaker the transmission connections, the greater the side effects of functional substitution.

This issue is complicated at more abstract levels by the changing topology of the power grid as load varies, equipment is maintained or unexpected failures occur. Situations vary by how grid parts are coupled through power flow (PrF) or synchronicity (AF). Coupling is particularly important to millisecond-scale dynamic stability (AF), where in certain configurations an overloaded (PrF) part of the grid that would normally collapse into a local blackout could instead trigger cascading failures across a greater whole, as in the North American blackout in 2003 that affected 50 million people (US–Canada Power System Outage Task Force, 2004). Meaningful part–whole groupings may not be entirely pre-analysable and seem to form and shift with grid state. This may require WDA methods extensions to specify dynamic part–whole aggregation rules.

6.4.3 Future Research Areas and Study Limitations

This work leaves many opportunities for future research into cognitive work in power grid operation. While our field studies were extensive, they were limited by being only at one Reliability Coordinator, the IESO. Each Reliability Coordinator has different grid topology, energy market mechanisms and automation structure. However, all address the same AC grid functions and regulations, and exchange expertise with each other. Our field studies relied on retrospective critical events that we did not directly observe. However, we used a well-established knowledge elicitation technique, and corroborated findings across individual interviews, focus group sessions and literature review of incident reports.

Future work could expand the power grid WDA to finer part–whole detail, particularly of purposes and AFs. A Control Task or Strategy Analysis of grid operations would be an opportunity to model how workers interact with the extensive grid control automation, which we did not analyse except as integrated component functionality (e.g. circuit breaker relaying). To apply WDA implications to design, development work might focus on how to computationally or analogically integrate existing functional models such as geographic information systems, breaker models of grid topology, state estimators of power flow and millisecond-level grid stability simulators.

6.4.4 Conclusion

This study found WDA effective at capturing power grid system functional structure, and identified relationships that canonical visualisations do not clearly distinguish (e.g. the non-linear value of topological links). Some WDA concepts are consistent with contemporary information visualisations being developed in 'smart grid' research, while others such as part–whole decomposition are promising candidates for future information integration research. We have applied the WDA to develop grid control room interface requirements; visual style guides; design principles for wide-area monitoring; and concepts for communicating and applying system operating limits.

ACKNOWLEDGEMENTS

This research was funded by the Independent Electricity System Operator (IESO) of Ontario and MITACS.

REFERENCES

Burns, C., Bryant, D. and Chalmers, B. 2005. Boundary, purpose, and values in work-domain models: Models of naval command and control. *IEEE Transactions on Systems, Man, and Cybernetics – Part A: Systems and Humans*, 35(5), 603–616.

EPRI. 2009. *EPRI power systems dynamics tutorial*. Technical Report 1016042. Palo Alto, CA: EPRI. http://www.epri.com/abstracts/Pages/ProductAbstract.aspx?ProductId=000000000001016042.

Flanagan, J. C. 1954. The critical incident technique. *Psychological Bulletin*, 51(4), 327.

Gosink, L., Burtner, E., Obradovich, J. and Dowson, S. 2014. Multi-domain situational awareness for infrastructure monitoring. US Patent App. 14/167,807.

Hargreaves, N., Taylor, G. and Carter, A. 2012. Information standards to support application and enterprise interoperability for the smart grid. *2012 IEEE Power and Energy Society General Meeting*. San Diego, CA, IEEE, 22–26 July 2012, pp. 1–6. http://www.pes-gm.org/2012/

Hoffmann, R., Capitanescu, F., Promel, F., Krost, G. and Wehenkel, L. 2011. Advanced visualisation securing awareness of the overall status for operational monitoring of the European interconnected grid. *IET Conference on Reliability of Transmission and Distribution Networks (RTDN 2011)*. London, UK, IET, 22–24 November 2011, pp. 62–67.

Independent Electricity System Operator. 2012. IESO-Controlled Grid Total System – Revision 16.

Independent Electricity System Operator. n.d. Transformer Fire at Substation.

Inergi and Hydro One. n.d. Map04-168.

Kröger, W. 2008. Critical infrastructures at risk: A need for a new conceptual approach and extended analytical tools. *Reliability Engineering & System Safety*, 93(12), 1781–1787.

Kundur, P., Paserba, J., Ajjarapu, V., Andersson, G., Bose, A., Canizares, C., Hatziargyriou, N. et al. 2004. Definition and classification of power system stability IEEE/CIGRE joint task force on stability terms and definitions. *IEEE Transactions on Power Systems*, 19(3), 1387–1401.

Larson, R. E., Tinney, W. F. and Peschon, J. 1970. State estimation in power systems, Part I: Theory and feasibility. *IEEE Transactions on Power Apparatus and Systems*, PAS-89(3), 345–352.

Memisevic, R., Sanderson, P. M., Wong, W. B. L., Choudhury, S. and Li, X. 2007. Investigating human-system interaction with an integrated hydropower and market system simulator. *IEEE Transactions on Power Systems*, 22(2), 762–769.

Motter, A. E., Myers, S. A., Anghel, M. and Nishikawa, T. 2013. Spontaneous synchrony in power-grid networks. *Nature Physics*, 9(3), 191–197.

Naikar, N., Hopcroft, R. and Moylan, A. 2005. Work domain analysis theoretical concepts and methodology. Technical Report DSTO-TR-1665. Defence Science and Technology Organisation (Australia). Air Operations Division.

North American Electric Reliabilty Corporation. 2013. NERC. http://www.nerc.com. Accessed 30 August 2015.

North American Energy Standards Board. 2010. NAESB. https://www.naesb.org. Accessed 30 August 2015.

Obradovich, J. H. 2011. Understanding cognitive and collaborative work: Observations in an electric transmission operations control center. *Proceedings of the Human Factors and Ergonomics Society Annual Meeting*, 55(1), 247–251.

Ogushi, K. 1950. Power circle diagram of interconnected electric power transmission system. *Memoirs of the Faculty of Engineering, Hokkaido University*, 8(3-1), 104–115. http://hdl.handle.net/2115/37758.

Power Info. 2014. Grid-IE – the Next-Generation Power Grid Visualization Tool. http://www.powerinfo.us/gridIE.html. Accessed 31 August 2015.

Sanderson, P., Memisevic, R. and Wong, W. 2004. Analysing cognitive work of hydroelectricity generation in a dynamic deregulated market. *Proceedings of the Human Factors and Ergonomics Society 48th Annual Meeting*. New Orleans, LA, Vol. 48, pp. 484–488.

Schneiders, C., Vanzetta, J. and Verstege, J. F. 2012. Enhancement of situation awareness in wide area transmission systems for electricity and visualization of the global system state. *2012 3rd IEEE PES International Conference and Exhibition on Innovative Smart Grid Technologies (ISGT Europe)*. Washington, DC, 16–20 January 2012, pp. 1–9.

US–Canada Power System Outage Task Force. 2004. *Final report on the August 14, 2003 blackout in the United States and Canada: Causes and recommendations.* Technical Report, U.S. Department of Energy.

U.S. Geological Survey. 2015. LandsatLook Viewer. http://landsatlook.usgs.gov/

Wong, P. C., Schneider, K., Mackey, P., Foote, H., Chin, G., Guttromson, R. and Thomas, J. 2009. A novel visualization technique for electric power grid analytics. *IEEE Transactions on Visualization and Computer Graphics*, 15(3), 410–423.

Wu, F. F., Moslehi, K. and Bose, A. 2005. Power system control centers: Past, present, and future. *Proceedings of the IEEE*, 93(11), 1890–1908.

7 How a Submarine Returns to Periscope Depth

Neville A. Stanton and Kevin Bessell

CONTENTS

7.1 INTRODUCTION

The discipline of ergonomics has always been concerned with the theory, methodology and analysis of systems (Waterson and Eason, 2009; Dul et al., 2012) in both simple (Walker et al., 2009a) and complex (Walker et al., 2009b). Such analyses have been previously undertaken in aviation (de Carvalho et al., 2009; Harris and Stanton, 2010), healthcare (Carayon and Buckle, 2010; Jun et al., 2010), military (Walker et al., 2009b; Rafferty et al., 2010), road and rail (Stanton and Salmon, 2011) and sports (Salmon et al., 2010) domains. This chapter is concerned with the analysis of a complex sociotechnical system (Walker et al., 2010). Specifically, the aim of the chapter is to help develop an understanding and explicit representations of how a submarine returns to periscope depth. The complexity of this system is in the problem of returning to the surface safely, while being dependent on passive sonar as a means of detecting surface vessels and working as a team comprising personnel, the sound room, control room and ships control, as well as contributions from the rest of the ship. To analyse this system required the application of a method that would assist in identifying key features of the work and clarify the constraints. Cognitive work analysis (CWA) was developed to analyse complex sociotechnical systems such as those found in nuclear power generation (Rasmussen, 1986). This development came from the realisation that an in-depth understanding of the interrelations of social systems and technical systems was required to fully appreciate how constraints act upon the working of system functions (such as the communications and activities in and between the sound room and control room on a nuclear submarine). These systems are made up of numerous interacting parts, both human and non-human, operating in dynamic, ambiguous and often safety-critical domains. The complexity embodied in these systems presents significant challenges for modelling and analysis, and most methods are not well designed to capture the complexity of the interrelations and analyse the layers of interconnection within sociotechnical systems (Rasmussen, 1986; Vicente, 1999; Jenkins et al., 2009). The semi-structured framework presented within CWA helps to guide the analyst through considerations of the various levels of constraints acting on systems and the effects that can have upon the way in which work can be carried out. The attraction of CWA is the flexibility of the approach and the range of domains to which it has been applied (Durugbo, 2012). It is, perhaps, the epitome of a systems ergonomics method.

It is generally accepted that CWA is divided into five main phases, each focussing on different constraint sets and presenting different perspectives on the system; the phase names are listed on the left-hand side in Figure 7.1. In the centre of the figure is an indication of the types of constraint analysis undertaken for these phases. The different forms of representation are listed on the right-hand side of Figure 7.1.

The CWA process has been criticised for being complex and time consuming. In order to address these concerns, and in an attempt to provide some level of guidance and expedite the documentation process, a software tool has been developed (Jenkins et al., 2009). The software tool allows data to be passed between these phases, expediting the documentation process, and facilitating updates and changes. Thus, the tool provides structure to the analysis process and markedly expedites the documentation and presentation of the analysis results. There is the added benefit that it has standardised and proceduralised the process, although we do accept that there may be variations for those not using the tool.

7.2 VISITS, OBSERVATIONS AND INTERVIEWS

The project began with a kick-off meeting on the 1st November 2011 at HMS Collingwood. Following the meeting, the researchers met with ex-sound room operators and controllers at Dstl Portsdown West. An exploratory visit was then arranged to Devonport to observe the Talisman trainer control room and meet with the training staff on the 14th and 15th November 2011. The CWA was started at the meeting with the training staff. The second visit to Devonport was undertaken on

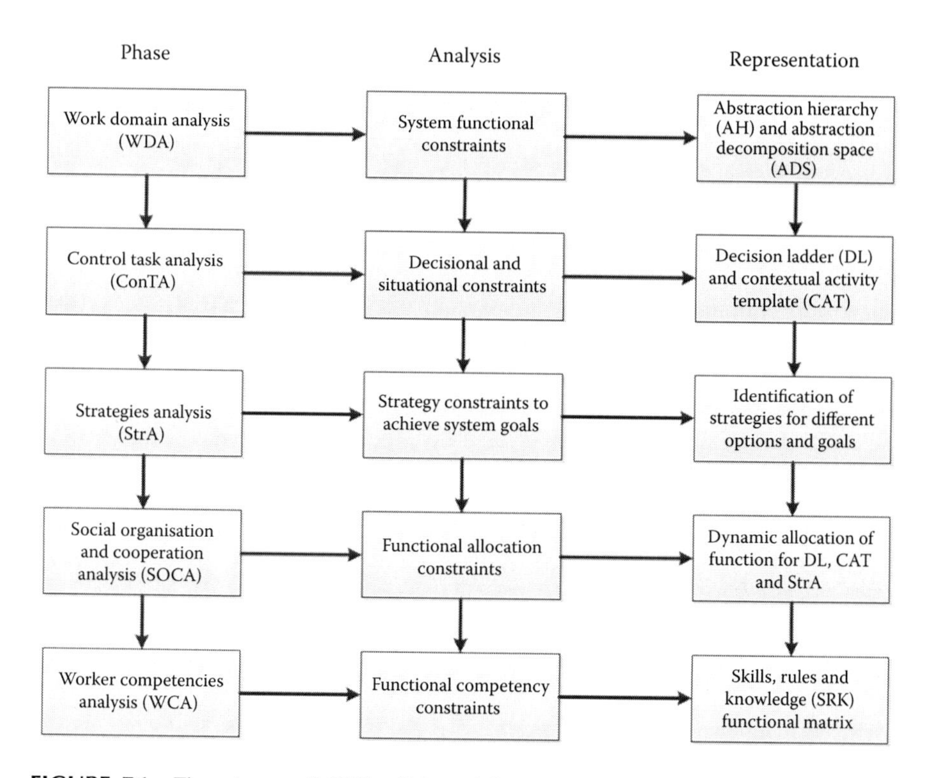

FIGURE 7.1 Five phases of CWA. (Adapted from Vicente, K. J. 1999. *Cognitive Work Analysis: Toward Safe, Productive, and Healthy Computer-Based Work*. Mahwah, NJ: Lawrence Erlbaum Associates.)

the 21st to 24th November 2011. The analysts observed a petty officers' (POs) training course within the Talisman trainer. Interviews were conducted with numerous training staff where the CWA was developed further. The analysts also observed a return to periscope depth (RTPD) undertaken by the training staff. The audio communications and ambient control room communications were recorded for transcription later. The third visit to Devonport was undertaken on the 9th to 11th January 2012. Observation of recertification training for the crew of HMS Tireless was undertaken, recording the audio communications and ambient control room communications for the RTPD tasks. The CWA analysis was further developed with two POs, one of who had recently returned from sea. Between the 6th and 8th February 2012, the analysts visited Faslane, Glasgow to conduct observations in Veracity. This was to ensure that the CWA representations could be generalised to Vanguard class submarines. Review and validation of the current analysis was undertaken with two members of the training staff. This was followed by a visit to the astute trainer. Again, a review and validation of the current analysis was undertaken with a member of the training staff. Finally, on the 28th February 2012, the analysts presented the CWA representations to the Maritime Warfare Centre. In addition to Dstl scientists, the attendees included a number of ex-submariners who were able to verify the analysis.

7.3 CONTROL ROOM LAYOUT IN TALISMAN COMMAND TEAM TRAINER AT HMS DRAKE

The control room and sound room layout in the Talisman Command Team Trainer is generic (Locke, 2012). It is similar, but not identical, to the layout on most of the Trafalgar-class submarines. In the trainer, there is a separate sound room, but this is not always the case, as on some submarines it is in an extended part of the control room. Despite this, there is enough commonality for it to serve as a procedural trainer for training command teams. As can be seen in Figure 7.2, there is a strong communication link between the officer of the watch (OOW), chief petty officer for tactical systems (CHOPS(TS), sometimes called the OpsO) and the sound room controller (SRC, sometimes called the sonar controller or CHOPS(S)). The layout of the control room and sound room is fixed by the equipment layout, but it is likely to have been optimised over the decades to the current design. The photograph in Figure 7.3 shows a view down the control room with the periscope on the right-hand side.

As explained in Chapter 1, CWA comprises five main phases: work domain analysis (WDA), control task analysis (ConTA), strategies analysis (StrA), social organisation and cooperation analysis (SOCA) and worker competencies analysis (WCA). Each of these phases was applied to the system activity of returning the submarine to

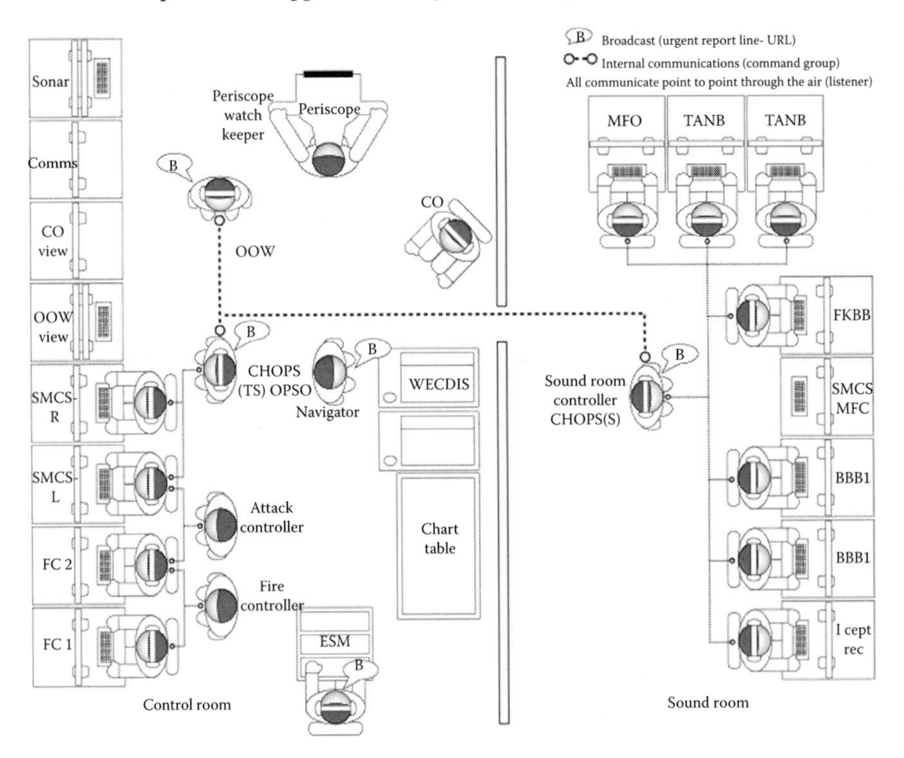

FIGURE 7.2 Control room and sound room layout. (Adapted from Locke, J. 2012. *Situation Awareness to Support Submarine Command Team Tactical Decision Making*. Fareham, UK: Dstl.)

FIGURE 7.3 View down the control room with periscope on right-hand side.

periscope depth. It was assumed that the submarine started out below 30 m (the safe depth below the hull of any ship) before the RTPD operation began. Each of the five phases within the analysis is presented in the following subsections.

7.4 WORK DOMAIN ANALYSIS

WDA identifies the functional constraints on the activity of the system. As such, WDA provides the foundation for all of the subsequent phases. The diagram, known as an abstraction hierarchy (AH), models the system at a number of levels of abstraction; at the highest level, the overall functional purpose of the system is considered, and at the lowest level, the individual components within the system are described. Generally, five levels of abstraction are used: functional purposes (the purposes of the work system and the external constraints on its operation), values and priority measures (the criteria that the work system uses for measuring its progress towards the functional purposes), purpose-related functions (the general functions of the work system that are necessary for achieving the functional purposes), object-related processes (the functional capabilities and limitations of physical objects in the work system that enable the purpose-related functions) and physical objects (the physical objects in the work system that afford the object-related processes).

7.4.1 The Functional Purposes

The AH was created in a top-down and a bottom-up fashion. First, the overall functional purposes of the submarine control room were specified. This is a description of the reasons for the existence of the system. These purposes are independent of time; they exist for as long as the system exists. In this case, it is to 'provide the tactical picture' and 'support the command aims'. These purposes were represented as two nodes at the top of the WDA in Figure 7.4.

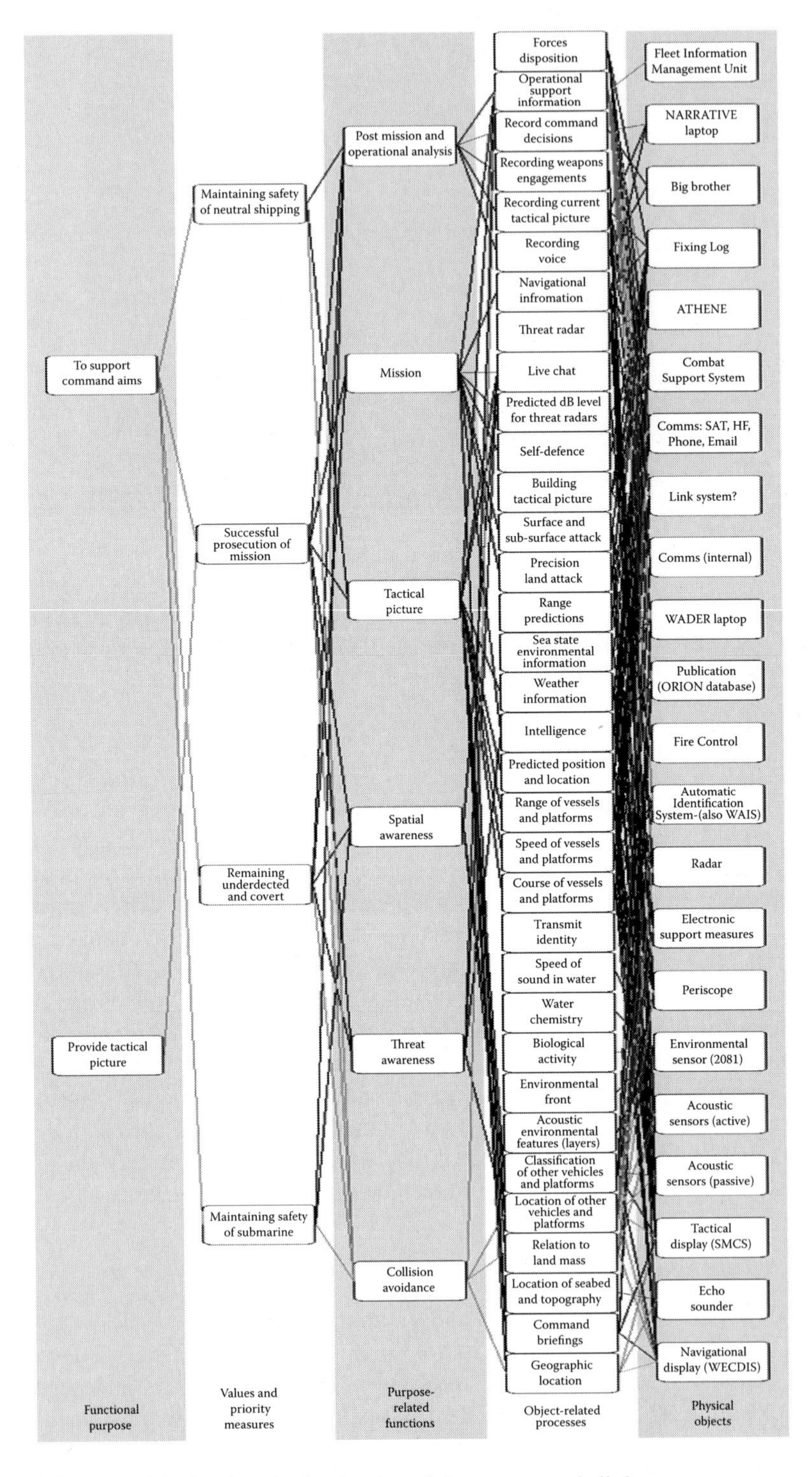

FIGURE 7.4 WDA showing the five levels and the means–ends links.

7.4.2 The Values and Priority Measures

The next step in the development was to identify the values and priority measures. Further constraints on the functional purposes are more explicitly listed. These are measures for determining how well the system is achieving its functional purposes. In this case, they were identified as 'maintaining safety of submarine', 'remaining undetected and covert', 'successful prosecution of mission' and 'maintaining safety of neutral shipping'. The way that the system is configured to meet these needs is likely to be heavily contextually dependent.

7.4.3 The Purpose-Related Functions

In the middle of the AH, the purpose-related functions are listed. These functions have the ability to influence one or more of the values and priority measures. They link the purpose-independent processes with the object-independent functions. They are listed as 'collision avoidance', 'threat awareness', 'spatial awareness', 'tactical picture', 'mission' and 'post-mission operational analysis'.

7.4.4 The Object-Related Processes

The second level from the bottom of the AH, the object-related processes, captures the processes that are conducted by the physical objects in order to perform purpose-related functions. Most importantly, they capture the affordances of the physical objects independently of their purpose. For example, the geographic location is afforded by the combination of the navigation display, echo sounder and tactical display. A complete list of the object-related processes can be found at the level above the base level of Figure 7.5 along with the links that indicate to which object they relate. To aid readability, this list comprises the following, in order from left to right of the figure: 'geographic location', 'command briefings', 'location of seabed and topography', 'relation to land mass', 'location of other vehicles and platforms',

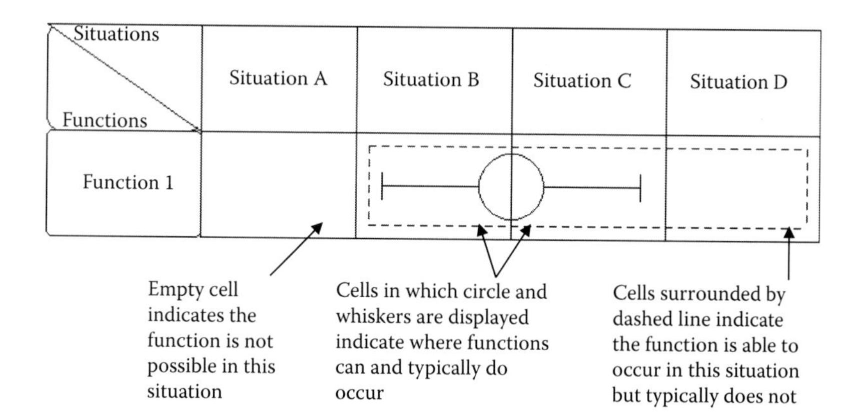

FIGURE 7.5 Layout for the CAT. (Adapted from Stanton, N. A. and McIlroy, R. C. 2012. *Theoretical Issues in Ergonomics Science*, 13(2), 146–168.)

'classification of other vehicles and platforms', 'acoustic environmental features', 'environmental front', 'biological activity', 'water chemistry', 'speed of sound in water', 'transmit identity', 'course of vessels and platforms', 'speed of vessels and platforms', 'range of vessels and platforms', 'predicted position and location', 'intelligence', 'weather information', 'sea state environmental information', 'range predictions', 'precision land attack', 'surface and subsurface attack', 'building tactical picture', 'self-defence', 'predicted dB level for threat radars', 'live chat', 'threat radar', 'navigational information', 'recording voice', 'recording current tactical picture', 'recording weapons engagements', 'recording command decisions', 'operational support information' and 'forces disposition'.

7.4.5 The Physical Objects

At the lowest level of the hierarchy, the physical objects within the system are listed. The boundaries of this analysis have limited this list to the systems of direct relevance to the operation of the control room, rather than every single object. The boundaries of the analysis indicate the levels of fidelity applied here. In an attempt to keep the analysis manageable, the boundary has omitted individual system elements. It would be possible to decompose many of the listed objects into their component parts and describe their affordances more concisely; however, this has also been classified as outside the remit of this initial evaluation. A list of the physical objects that have been analysed can be found at the base of Figure 7.4; this list comprises the following: the 'navigation display (WECDIS)', 'echo sounder', 'tactical display (SMCS)', 'acoustic sensors (passive)', 'acoustic sensors (active)', 'environmental sensor (2081)', 'periscope', 'electronic support measures', 'radar', 'automatic identification system', 'fire control', 'publications (ORION database)', 'WADER laptop', 'communications (internal)', 'link system', 'communications (SAT, HF, Phone, Email)' and the 'combat support system'.

The use of means–ends links and the utility of the AH can be described with an example from Figure 7.4, the node 'collision avoidance' in the purpose-related functions level. Following the links out of the top of this node answers the question 'why is this needed?', in this case to 'maintain safety of submarine', 'remaining undetected and covert', 'successful prosecution of mission' and 'maintaining safety of neutral shipping'. Following the links down from the 'collision avoidance' node, it is possible to answer the question 'how can this be achieved?', in this case by 'geographic location', 'relation to land mass', 'location of other vehicles and platforms' and 'operational support information'.

The hierarchy does not prescribe a particular arrangement for providing this functionality; rather, it lists all of the components that can affect it. In this case, there is redundancy in the system.

7.5 CONTROL TASK ANALYSIS

The second phase of the CWA framework, ConTA, allows the constraints associated with recurring classes of situations to be identified. The contextual activity template (CAT) is a representation of a system in terms of both work situations and work functions. Work situations are those that can be decomposed based on recurring schedules or specific locations. Typically, the work situations are shown along the horizontal axis, and the work functions (from the AH of the WDA) are shown along the vertical axis of the CAT, as illustrated in Figure 7.5.

The circles indicate the work functions with the bars showing the extent of the table in which the activity typically occurs. The dotted boxes around each circle indicate all of the work situations in which a work function can occur (as opposed to must occur), thus capturing the constraints of the system. Figure 7.6 shows the CAT for the object-related processes taken from the AH (see Figure 7.4). The situations were derived from interviews with SMEs and observational studies as outlined in

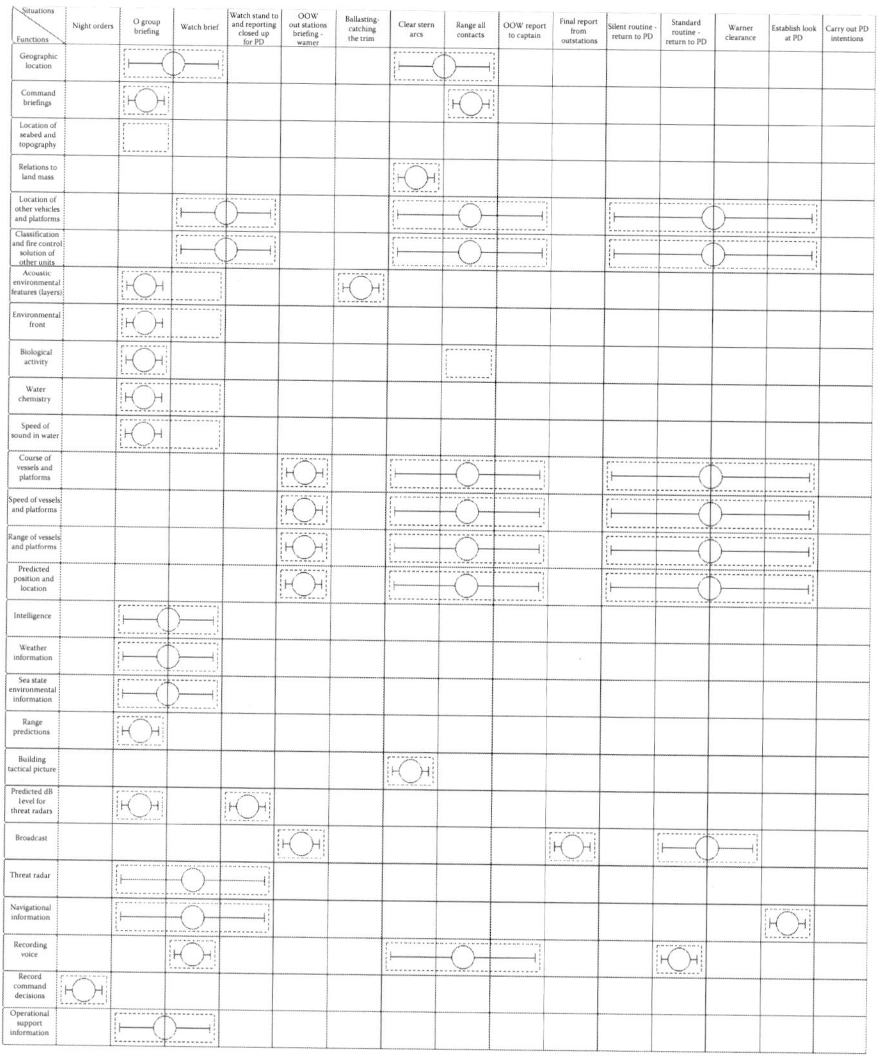

Situations / Functions	Night orders	O group briefing	Watch brief	Watch stand to and reporting closed up for PD	OOW out stations briefing - warner	Ballasting-catching the trim	Clear stern arcs	Range all contacts	OOW report to captain	Final report from outstations	Silent routine - return to PD	Standard routine - return to PD	Warner clearance	Establish look at PD	Carry out PD intentions
Geographic location															
Command briefings															
Location of seabed and topography															
Relations to land mass															
Location of other vehicles and platforms															
Classification and fire control solution of other units															
Acoustic environmental features (layers)															
Environmental front															
Biological activity															
Water chemistry															
Speed of sound in water															
Course of vessels and platforms															
Speed of vessels and platforms															
Range of vessels and platforms															
Predicted position and location															
Intelligence															
Weather information															
Sea state environmental information															
Range predictions															
Building tactical picture															
Predicted dB level for threat radars															
Broadcast															
Threat radar															
Navigational information															
Recording voice															
Record command decisions															
Operational support information															

FIGURE 7.6 CAT for object-related processes.

the introduction to this report. The situations follow the main phases of returning the submarine to periscope depth, as follows:

O group briefing
Watch brief
Watch stand to and reporting closed up for periscope depth (PD)
OOW outstations briefing
Ballasting – catching the trim
Clear stern arcs
Range all contacts
OOW report to captain
Final report all outstations
Silent routine or standard routine
Warner clearance
Establish look at PD
Carry out PD intentions

It is surprising to see so many empty cells in Figure 7.6; perhaps, this is due to the proceduralised nature of the task, but this should be questioned in any future work. One of the most salient features is that, in this system, some function constraints are situationally influenced (where a limited set of functions are applicable to the situation, e.g. ballasting and ranging contacts) whereas others are less so (where most of the functions are applicable to the situation, e.g. O group briefing and watch brief).

7.5.1 Activity Analysis in Decision-Making Terms

The decision ladder, shown in Figure 7.7, has two different types of nodes: the rectangular boxes represent information-processing activities and the circles represent states-of-knowledge that result from information-processing activities. The decision ladder can be used to describe both levels of expertise and novelty of the decision processes. Novice users are expected to follow the decision ladder in a linear fashion, whereas expert users are expected to link the two halves by shortcuts. The left side of the decision ladder represents the observation and information gathering activities to identify the system state, whereas the right side represents the planning and execution of tasks and procedures to achieve a target system state. In between identifying the system state and target state are the options selection activities to meet the desired goal(s). In experts or proceduralised activities, observing information and diagnosing the current system state immediately signals a procedure to execute. This means that rule-based shortcuts can be shown in the centre of the ladder. On the other hand, effortful, knowledge-based goal evaluation may be required to determine the procedure to execute; this is represented in the top of the ladder. There are two types of shortcuts that can be applied to the ladder; 'shunts' connect an information-processing activity to a state of knowledge (box to circle) and 'leaps' connect two states of knowledge (circle to circle); this is where one state of knowledge can be directly related to another without any further information processing. It is not possible to link straight from a box to a box as this misses out the resultant knowledge state.

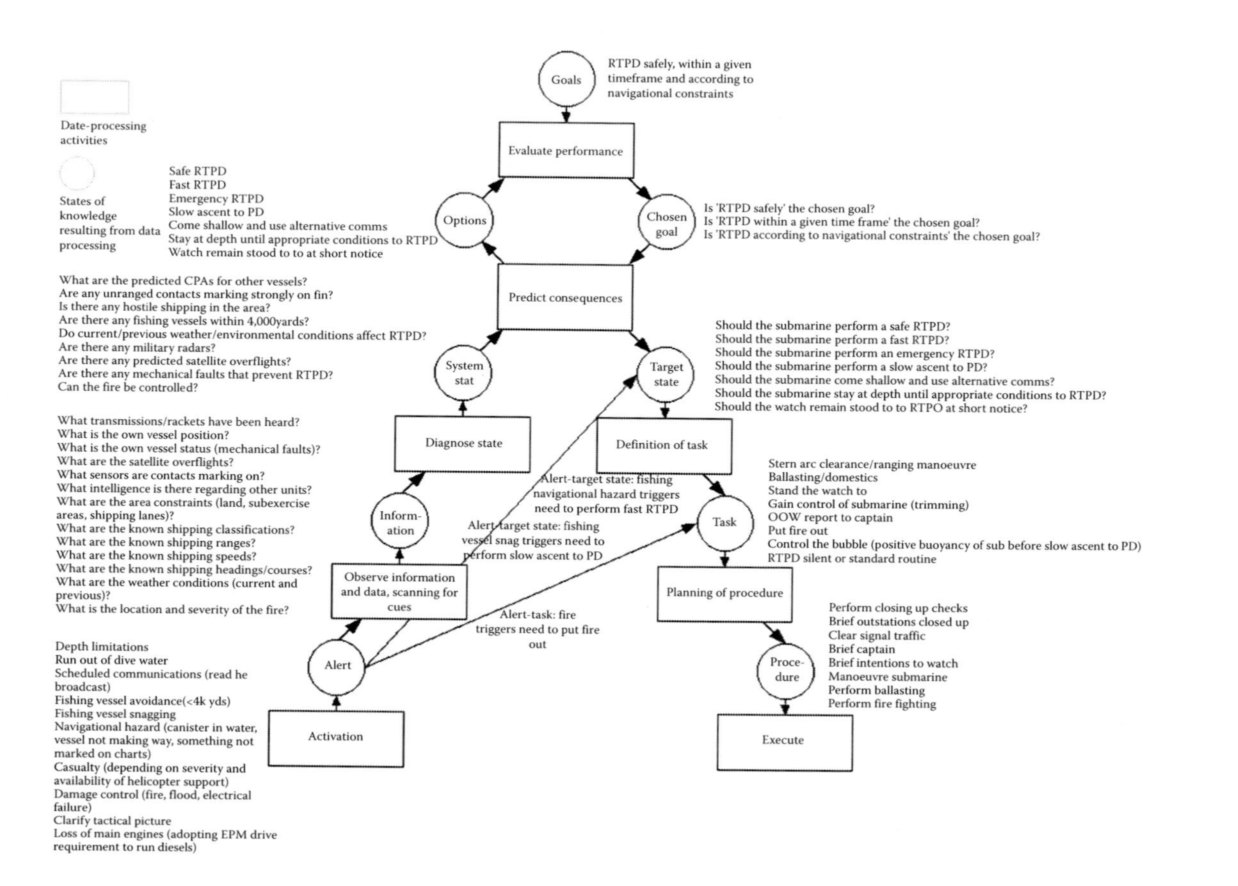

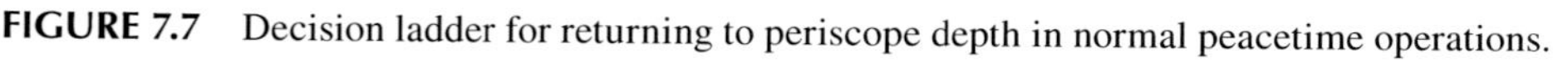

FIGURE 7.7 Decision ladder for returning to periscope depth in normal peacetime operations.

In order to constrain the analysis, the experts recommended that the ladder be completed for a specific scenario, with the submarine acting autonomously, conducting standard operations, under general peacetime conditions, in a home waters environment. As can be seen from Figure 7.7, even under these conditions, a multitude of factors influence the decisions involved in determining how to proceed when returning to periscope depth.

Starting with the goals, the experts identified three constraints acting on the overall goal of returning to periscope depth: to return safely, within a given timeframe and according to navigational constraints. Given that these constraints can be in conflict (the need to return safely may mean that the timeframe cannot be met, for example), this then leads to three possible choices of goal.

A number of alerts, or reasons why the submarine would need to RTPD, were elicited, ranging from the need to make a scheduled communication to a casualty or fire onboard. The information that might subsequently need to be gathered included known shipping classifications, ranges, speeds and so forth, as well as the own vessel position and status in terms of any mechanical faults. The system state elements then reflect what can be ascertained from the available information. So, for example, combining information about own vessel with shipping ranges, headings, etc. enables calculation of the closest points of approach (CPAs) for other vessels.

Options identified were of two types: either different ways in which an RTPD could be conducted (safe, fast, emergency or a slow ascent) or, interestingly, alternatives to returning to periscope depth, namely, coming shallow to use alternative communications, staying at depth, or leaving the watch stood to ready to RTPD at short notice. These options also represented the available target states, selected on the basis of the chosen goal (RTPD safely, within a given timeframe or according to navigational constraints). Tasks listed covered the range of actions that might be necessary to achieve the possible target states, such as standing the watch to and ranging all contacts, along with the procedures for carrying out these tasks (e.g. manoeuvring the submarine). Owing to time constraints, it was not possible to cover the full range of tasks and procedures, and a reference to other phases of the analysis shows that there is more that could be added here. Indeed, further development and refinement of the decision ladder would be the first step in any follow-on work that aimed to take advantage of this particular output.

Finally, a number of 'leaps' were identified by the experts – shortcuts connecting two states of knowledge. One leap connecting the 'alert' and 'task' states of knowledge represented the fact that as soon as a fire is detected, it must be put out. Two leaps connected the 'alert' and 'target state' states of knowledge. First, encountering a navigational hazard, for example, something not marked on the charts, might trigger the need to perform a fast RTPD. Second, snagging a fishing vessel automatically leads to the target state of performing a slow ascent to PD, in order to minimise the risk to those onboard the snagged vessel.

These leaps highlight the fact that the journey around the ladder does not necessarily involve travelling up one side and then down the other in a linear fashion. They also show how the decision-making process can be iterative. In the case of detecting a fire, the leap to formulating the task of putting out the fire would immediately be followed by a return to the left-hand side of the ladder in order to assess the severity of the fire and determine the correct course of subsequent action. This is also true when some of the target states relating to doing something other than returning to periscope depth are selected. If the decision is made to stay at depth, for example, then the situation will continue to be monitored until the system states are such that it becomes possible to RTPD.

Once the decision ladder had been completed, it was then possible to examine how the various elements related to each other. The decision ladder in Figure 7.7 shows the complete range of information, system states, tasks and so forth that can be involved in an RTPD; that is, it is prototypical. It can be useful to determine the relationships between each of these to understand which items of information contribute to which system states, etc. This can be done for both 'legs' of the decision ladder (Jenkins et al., 2011). It should be noted that, in line with the formative nature of the CWA approach, the fact that elements are related does not mean that they 'do' influence each other, but rather that they 'could'. First, the information, system states and options can be related to each other, as shown in Figure 7.8, with the black cells indicating a relationship.

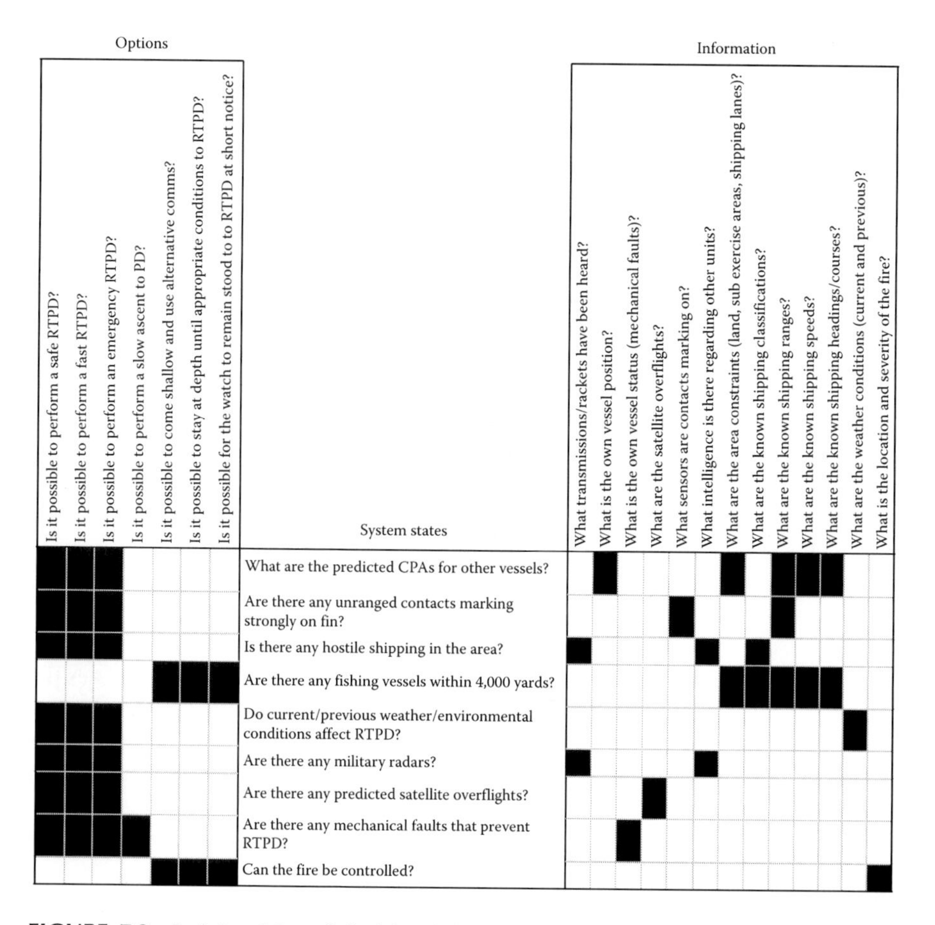

FIGURE 7.8 Left-hand leg of decision ladder linking information to system states and options.

Relating the information and system state elements shows what information could be necessary to diagnose each system state. So, for example, determining whether there is any hostile shipping in the area requires some or all of the transmissions or rackets that might have been heard; any intelligence regarding other units; and currently known shipping classifications. Whether there are any mechanical faults that would prevent RTPD, on the other hand, can be ascertained simply from information about the own vessel status.

The linking of the system states and options indicates how different system states might constrain the available options. It would appear that there is a basic split between states that constrain whether or not it is possible to RTPD, and states that constrain whether it is possible to choose some alternative. Where these constraints are in conflict, the chosen goal may come into play and influence the decision about the most appropriate target system state. For example, when a fishing vessel is detected within 4,000 yards, during peacetime and within home waters, a submarine is required to immediately RTPD (i.e. within a given timeframe). If, however, at the same time an unranged contact is marking strongly on fin (the contact is being detected on a particular sonar array, indicating that it may be extremely close), it could be unsafe to RTPD. In this case, there is a choice to be made between returning to periscope depth and staying deep; the decision-maker may choose 'RTPD safely' as the goal, leading to a decision to stay deep until the contact has been ranged. Alternatively, if 'RTPD within a given timeframe' is chosen as the goal, the submarine may RTPD while accepting the risk of the unranged contact.

Figure 7.9 relates the chosen goal, target states and tasks for the right-hand leg of the decision ladder.

It is apparent that, as expected, the chosen goal has an influencing factor on the way in which the submarine will attempt to RTPD. For example, if safety is the overriding constraint, the RTPD will be neither fast nor emergency, meaning that the standard set of procedures will be followed. Conversely, the submarine will only ever perform a slow ascent to PD or come shallow as a result of some safety consideration (a fishing vessel snag or adverse weather conditions being the likely causes, respectively).

The linking of the target states and tasks provides a similar picture to that presented by the StrA in which the tasks required to RTPD vary depending on the way in which it is to be achieved. Unlike the StrA, however, some further detail is apparent for those situations where the target state is to do something other than RTPD, such as come shallow or stay at depth.

Linking these elements in this way is useful because it provides an insight into how different pieces of information are required to determine the system states that inform option selection, and how different goals lead to various target states and their associated tasks. Once these relationships are understood they can be used for a variety of purposes such as informing interface and simulator design, or decision-making training.

7.6 STRATEGIES ANALYSIS

The StrA phase presents the alternative pathways from one system state to another. The strategy adopted under a particular situation may vary significantly. The term

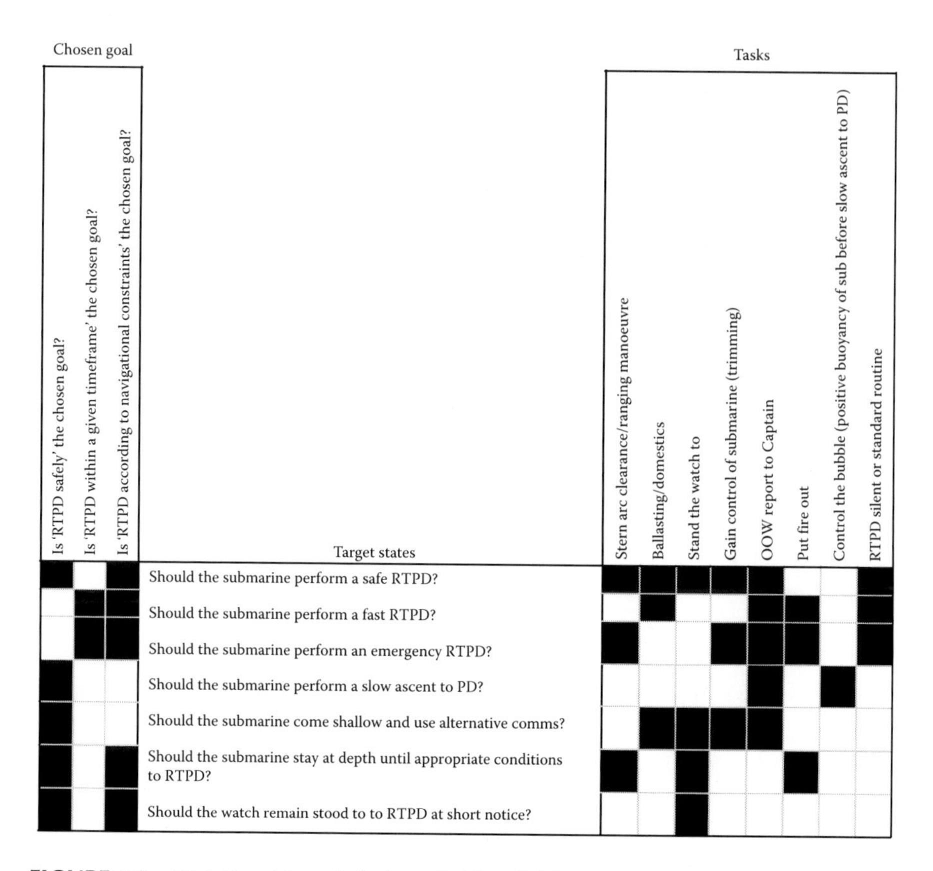

FIGURE 7.9 Right-hand leg of decision ladder, linking chosen goal to target states and tasks.

'agent' is used to signify that the activity could be performed by either a human or automated technology. Different agents may perform tasks in different ways, and the same agent (either human or non-human) might perform the same task in a variety of different ways. The strategy that the agent chooses to select will be dependent on a huge number of variables, including, among others, their experience, training, workload and familiarity with the current situation.

The CAT or decision ladder can be used as a basis for eliciting the StrA required. The matrix of functions and situations illustrates the possible activity within the domain – to perform a complete analysis a StrA should be created for each of the cells where activities can take place. The differences to the strategies available, affected by the situational constraints, can then be examined. Alternatively, the pathways through the decision ladder can show how different strategies can lead to the same end state, as is the case in this example. Within some specialist environments, the way activities are conducted is often mandated by standard operating procedures and emergency operating procedures. For this reason, systems are often configured to limit the available strategies open to an agent. This has both positive and negative attributes. Procedures make activity

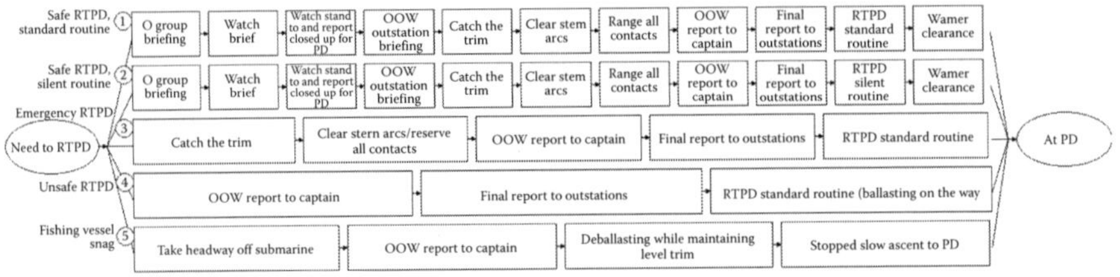

FIGURE 7.10 Five strategies for returning to periscope depth.

predictable although it may limit flexibility. While the procedures for returning the submarine to periscope depth are well defined, there are a number of circumstances in which they might vary. This is reflected in the StrA shown in Figure 7.10, which was constructed based on some of the options identified in the earlier decision ladder (Figure 7.7).

In the next phase of the analysis, roles of personnel are allocated to the functions across the situation. This phase is called SOCA.

7.7 SOCIAL ORGANISATION AND COOPERATION ANALYSIS

SOCA addresses the constraints imposed by allocation of specific agent roles to functions in given situations. The objective is to determine how the social and technical factors in a sociotechnical system can work together in a way that enhances the performance of the system as a whole. SOCA is concerned with identifying the set of possibilities for work allocation, distribution and social organisation. SOCA explicitly aims to support flexibility and adaptation in organisations developing designs that are tailored to the requirements of the various situations. In this way, SOCA supports the idea of dynamic allocation of function, such that function allocation can transfer between agents as the situation changes. Flexible organisational structures are superior to rigid ones because they can adapt to local situations. Rather than defining a single or best organisational structure, SOCA is concerned with identifying the criteria that may shape or govern how work might be allocated across agents. Such criteria might include the following:

- Agent competencies
- Access to information or means for action
- Level of coordination
- Workload
- Safety and reliability
- Availability

The first stage of the process is to define the key agent roles in the system. The role reflects the work at any given time rather than a particular person or machine. It was observed in the control room studies that roles may change depending upon who is available, which shows that this idea has already been embraced. A list of the key roles and their related coding can be seen in Figure 7.11.

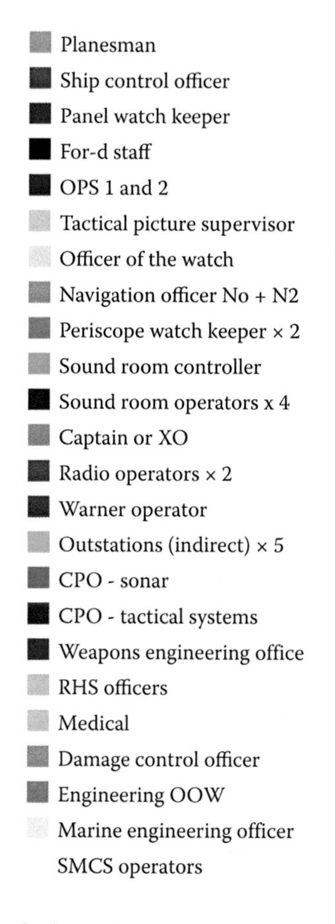

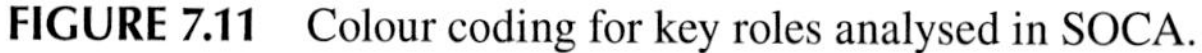

FIGURE 7.11 Colour coding for key roles analysed in SOCA.

As Figure 7.12 shows, many of the functions are performed either exclusively by one role in certain situations or by multiple roles in others. For example, the function 'classification and fire control solution of other units' is performed exclusively by the OOW in the situations 'watch brief' and 'watch stand to and reporting closed up for PD', whereas it is performed by a number of different roles in the situations 'clear stern arcs' and 'range all contacts'. Specifically, those roles are Ops 1 and 2, tactical picture supervisor (TPS; another name for the OpsO), OOW, SRC, sound room operators and right-hand side (RHS) officers.

Turning to the decision ladder, SOCA is used to determine which roles are involved in information-processing activities, and the resultant states of knowledge, at each stage of the decision-making process. It can be seen in Figure 7.13 that the captain or executive officer (XO) is solely responsible for evaluating the various options available against the overall goal, and then deciding against which constraint the goal will be achieved, leading to a desired target state. Conversely, a large number of varied roles can contribute towards alerting to the need to RTPD in the first place and then gathering the required information to diagnose

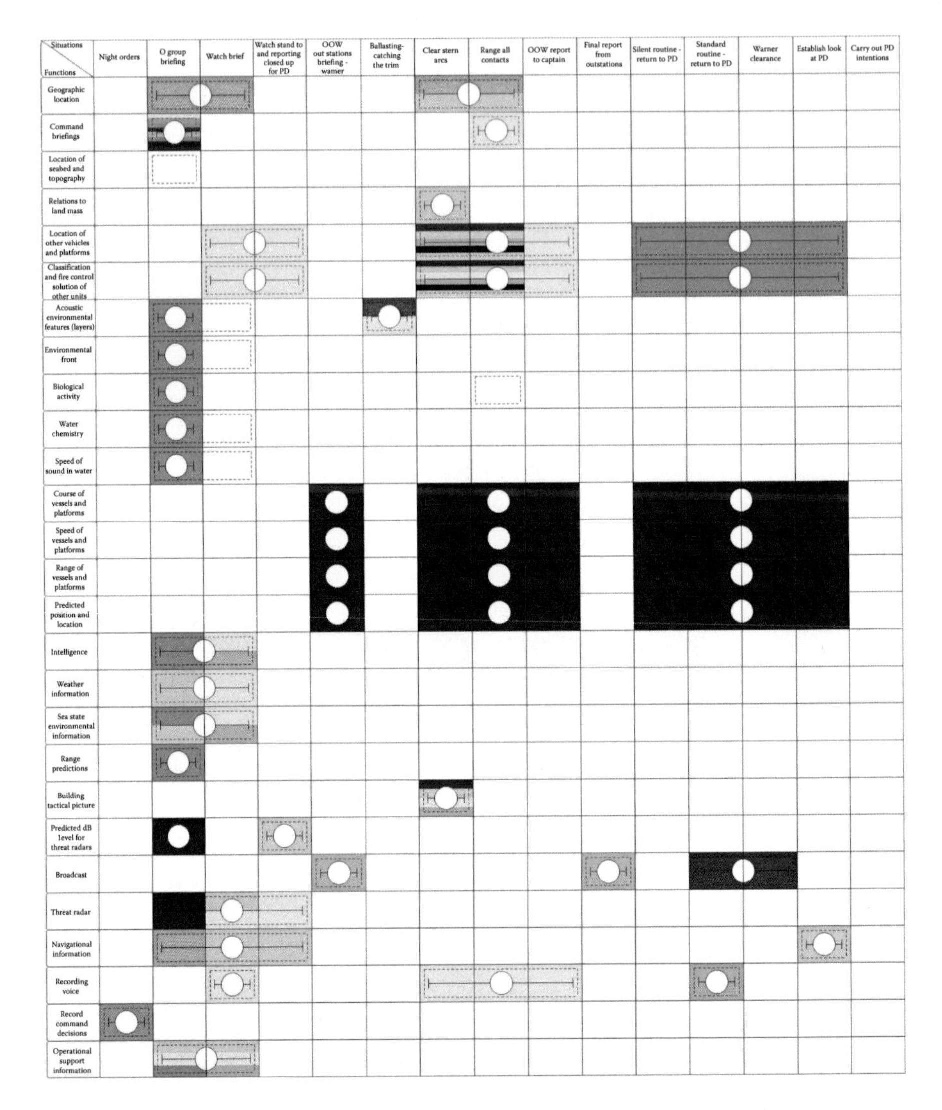

FIGURE 7.12 SOCA-CAT for returning to periscope depth in the control room.

the current system state. Similarly, many different roles are involved in executing the tasks and procedures resulting from the target state determined by the captain or XO.

One noticeable difference in comparing the two legs of the decision ladder is that the planesman, ship control officer and For-d Staff are involved only in responding to the target state, as might be expected given that their roles relate primarily to the control of the submarine. The TPS, OOW and SRC, on the other hand, are involved at all stages of the process (with the exception of those reserved for the captain or XO, as described previously).

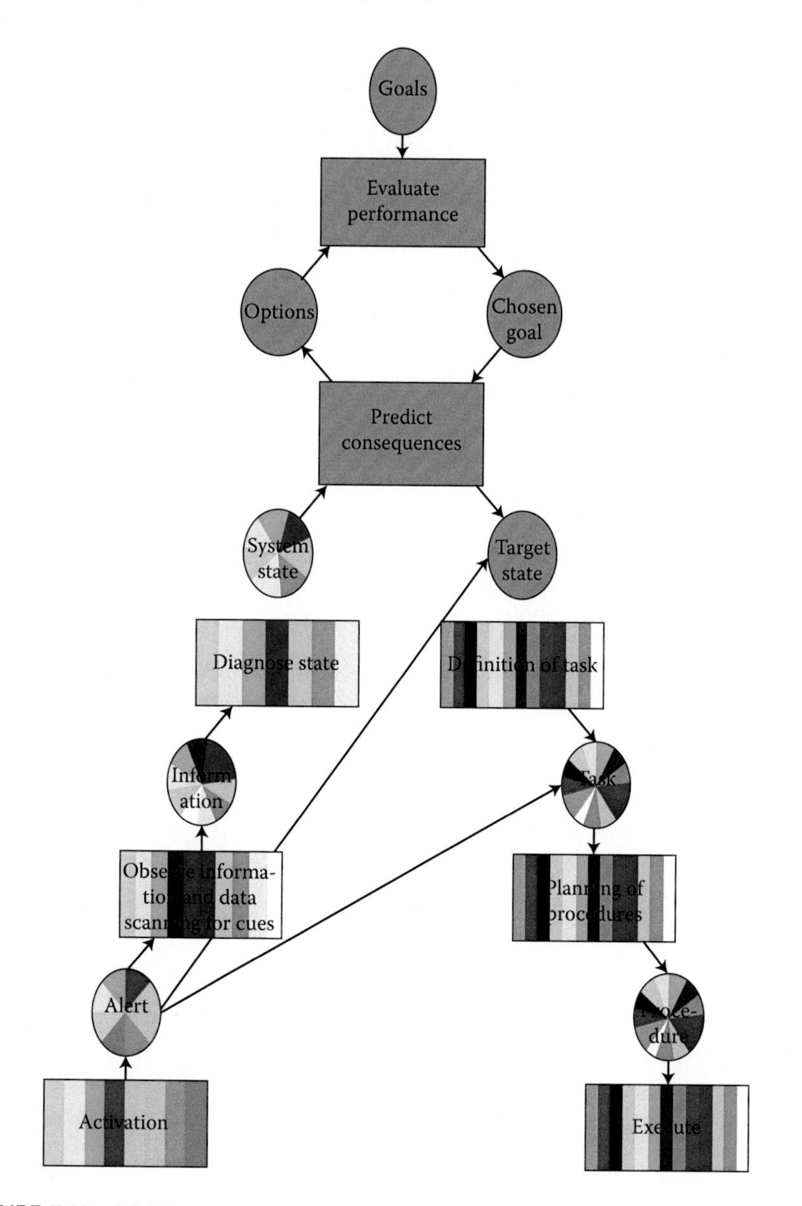

FIGURE 7.13 SOCA decision ladder (SOCA-DL) for returning to periscope depth in the control room.

Finally, SOCA can be applied to the StrA in order to determine which roles are involved in the activities that make up each strategy, and this is shown in Figure 7.14. What is striking about this representation is that the roles for each activity do not vary across strategies. That is to say that the strategy being enacted does not affect who is involved in each activity, but rather the activities that occur and their

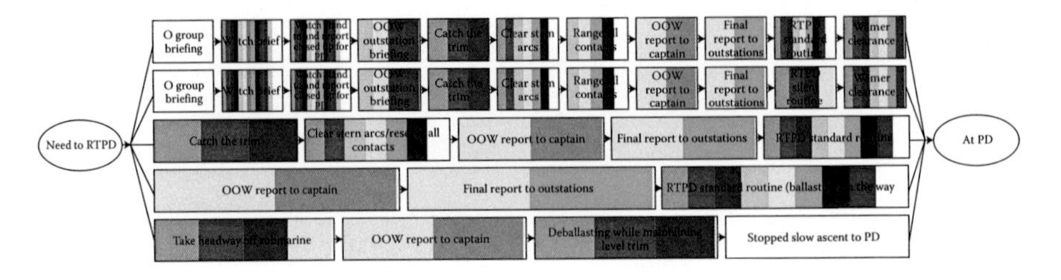

FIGURE 7.14 SOCA-StrA for returning to periscope depth in the control room.

sequencing. This implies that each role is well defined in terms of the activities for which they are responsible.

7.8 WORKER COMPETENCIES ANALYSIS

In the final phase of the CWA framework, WCA addresses the constraints of agent skill within different functions. WCA investigates the behaviour required by the humans and automation within the system required to undertake the functions. Typically, this behaviour is modelled using the skills–rules–knowledge (SRK) taxonomy. Traditionally, the WCA has been based on the information-processing steps in the decision ladder, but a modification uses the functions from the CAT to be consistent with the functional analysis produced by CWA (McIlroy and Stanton, 2011).

Skill-based behaviour (SBB) consists of automatic actions in response to environmental cues and events. Very little or no conscious effort is required to perform an action. Rule-based behaviour (RBB) is characterised by a set of stored rules and procedures. These rules guide behaviour without the actor having to consider individual system goals and are acquired either through experience or learned from supervisors, former operators and instructional materials. Unlike SBB, the process requires conscious cognitive processing. When more advanced reasoning is required, the actor must use knowledge-based behaviour (KBB). This behaviour is slow and effortful; the actor is required to apply conscious attention, carefully considering the functional principles that govern the system. This type of behaviour is most common in novel, unanticipated events. In the current analysis, the inventory was informed by the object-related processes level of the AH as this offered consistency with the formative nature of CWA and would allow WCA to be considered much earlier in the design life cycle. The SRK inventory presented in Table 7.1 has four columns; working from left to right, these are the following:

- Functions taken from the object-related processes level of the AH
- The 'SBBs' that apply to those functions
- The 'RBBs' that apply to those functions
- The 'KBBs' that apply to those functions

The WCA can be used to inform allocation of function. By describing the behaviour required from an agent to fulfil a function, it is possible to ascertain which

TABLE 7.1
Worker Competencies Analysis

	Skill-Based Behaviour	Rule-Based Behaviour	Knowledge-Based Behaviour
Definition	Observed behaviour of experts	The IF-THEN rules that allow people or machines to identify the conditions that determine if a task should be done	First principles that novices have to acquire to understand the task
Geographic location	Check plan, determine future position	Display navigational plan – distance against speed of advance. Human: Follow rule book	Theory of navigation and use of WECDIS
Command briefing	Briefing Commanding Officer on all information he requires to safe RTPD	Follow rule book – O group: Ops, CSS, WE, marine engineers, CHOPS, etc.	Individual competencies in each of the specialisms: Ops, CSS, WE, marine engineers, CHOPS, etc.
Location of seabed and topography	Advise command on dangers associated with proximity of the seabed	Apply knowledge and experience to establish location of seabed and topography	Understand the nature of the seabed, electronic charts and echo sounders
Relation to land mass	Advise command on all aspects of submarine operations and safety with relation to land mass	Rules are determined by documentation with regards to operating close or far from land	Understand the differences between operating the Submarine close or far from land
Classification and fire control solution of other units	Supervise the command team, providing direction and guidance to ensure pertinent information is gathered and disseminated	Use classification information to determine parameters and apply values regarding target speed, range, course and depth	target motion analysis (TMA) taught during first career course
Acoustic environmental features (layers)	Use of gained knowledge to advise command on how to best evade threat, either above or below the layer and where best to place sensor to aid detection	Rules based on assimilation of layer information; gained from organic environmental sensors and third-party information	Taught basic oceanography, from initial training and throughout career

(*Continued*)

TABLE 7.1 ***(Continued)***
Worker Competencies Analysis

	Skill-Based Behaviour	Rule-Based Behaviour	Knowledge-Based Behaviour
Environmental front	Use of gained knowledge to advise command on how to best evade threat and operate the submarine tactically	Rules based on assimilation of environmental information; gained from organic environmental sensors and third-party information	Taught basic oceanography, from initial training, throughout career
Biological activity	Plotting of gained information in order to advise command on best areas of high ambient noise that will provide best evasion route; carry out environmental impact assessment, if sonic mammals exist	If biological activity exists, determine type, density and location; pass information to Tactical Data Handling System (TDHS)	Taught how to best identify different types of biological activity, by their sound
Water chemistry	Use of knowledge to make tactical decisions	Rules based on assimilation of water chemistry information; gained from organic environmental sensors	First principles gained from On Job Training (OJT)
Speed of sound in water		Refer to layers information	
Intelligence	Full understanding of the overall intelligence and geo-political situation with regards to strategic decisions	Correlation of intelligence information with organic sensors in order to make tactical decisions	No formal training, however information disseminated by briefings
Weather information	Plotting and briefing weather information; providing command with courses of action	Gather weather information at mandated intervals	Depending on rank, training delivered at various stages of career
Sea state environmental information	Take overall responsibility to establish safety of personnel and equipment	React to conditions dictated by the environment	No training other than an understanding gained from experience

(Continued)

TABLE 7.1 (*Continued*)
Worker Competencies Analysis

	Skill-Based Behaviour	Rule-Based Behaviour	Knowledge-Based Behaviour
Range predictions	Use information gathered to determine range prediction	If prevailing factors dictate, fire expendable sensor to ascertain true environmental conditions in water column	Taught basic understanding at initial training, progressing to a more comprehensive level with promotion
Building tactical picture	Use of experience to consider all aspects of the tactical picture, that is, shipping lanes, fishing vessel (FV) grounds and all sensor clues	Tactical information passed to TDHS; operator conducts TMA to provide overall air, surface and subsurface picture	Formal training given as a core component of all career courses
Understanding and applying dB level for threat radars	Have a full understanding and advise command on all risks which involve counter detection	If the dB level or a parametric value reaches a threshold condition, inform command immediately	Principles of radar taught on career courses
Threat radar	Mitigate against risk of threat by applying threshold conditions	Operate equipment within boundaries of standard operating procedures (SOPs); inform command if threat radar exists	Theory of radar principles and understanding of specific threat
Navigational information	Take overall responsibility for submarine safety	Rules governed by fleet documentation; mandated by the International Maritime Organisation (IMO)	Navigation equipment (WECDIS) and chart-work principles taught in training
Recording voice	Quality control of recorded information for dispatch as records	If events dictate, ensure vital information is recorded	No training other than OJT
Record command decisions	Information recorded in all narratives is quality controlled for submission for records	If a tactical comment or decision is made, it is written into the narrative	Narrative training given

functions are more suited to automation and which are more suited to human workers (as well as ensuring people end up with integrated roles). The descriptions of the skills, rules and knowledge required to perform this function can also inform interface design such that both novice and expert behaviour are supported. An expert will generally rely on SBB and RBB to guide actions. To support novice use, KBB needs to be supported. An important point to make here is that these details should not be on display unless the user requests them to be; in the majority of situations, an expert would consider this information to be unnecessary. The continual presentation of superfluous information would hinder rather than help the expert user. Another advantage brought by the SRK is to enable the analyst to draw attention to skill gaps. Comparing information about the agents currently in the system with the description of the competencies required to perform system functions will allow the researcher to highlight which, if any, competencies required by the system are currently absent. If a function cannot adequately be performed by the agents in the current system, then extra training, recruitment or the procurement of automation would be necessary. This training, recruitment or automation procurement could be guided by the descriptions contained within the SRK inventory.

7.9 CONCLUSION

CWA as a sociotechnical method acknowledges that systems are inherently complex and multiple perspectives on the problem are required to more fully appreciate the relationships between the social and technical aspects of the system. The sociotechnical approaches accept that systems are intertwined and analyse the whole system, rather than component parts. In summary, this analysis of activities involved in returning to periscope depth in the control room using CWA has shown how six sets of constraints may be mapped out in terms of functions, situations, decisions, strategies, roles and competencies. This provides a basis for comparison of future configuration of the system. The eight outputs (namely, AH, CAT, DL, StrA, SOCA-CAT, SOCA-DL, SOCA-StrA and WCA) offer a graphical representation of the system constraints.

CWA has described the control room in terms of system functions and explored the way in which those functions were used in situations, decisions, strategies, role allocations and required competencies. The nature of these functional constraints needs to be brought into question when considering improvements and design of future systems. For example, new systems may change the nature of the functions being performed, change the type of decisions and informational cues (to make decisions simpler and more transparent, and use prompts for tasks to reduce ambiguity and memory load), change function allocation (such as semi-automating or fully automating some functions and/or changing role allocations) and change competency requirements (to simplify training and work demands). In summary, CWA was able to characterise the activities in the control room of the Trafalgar-class submarine by revealing the inherent complexities of the work domain and relations between functions and activities. Thus, it is considered capable of being applied to a complex sociotechnical 'system of systems'. CWA offers complementary descriptions of the requirements for the RTPD task using the current command system.

RTPD is much proceduralised due to the inherent complexity and its safety-critical nature. This means that many of the functions are unique to particular situations. Nevertheless, experts' performance differences were uncovered in the decision analysis, enabling the shortcut 'shunts' and 'leaps' across the ladder. The analyses also show that the activities for RTPD vary according to the situation. Future analysis should attempt to characterise future command systems so that the multiple perspectives can be compared and 'so-what' questions can be asked as ideas for design of the social and technical aspects of the system co-evolve. By exploring the system through different perspectives, it is possible to explore which constraints are hard (i.e. cannot be removed or changed) and which are soft (i.e. may be changed through training, procedures and work design) to understand how the system may be changed and improved (Stanton et al., 2013). Insights from CWA are contained within the individual representations as well as taking the analysis as a whole. CWA presents multiple perspectives on the activities in the system, which is a necessary requirement for sociotechnical analysis. More than any other approach in ergonomics, it presents a joined-up set of representations that all develop from a functional description of the system.

ACKNOWLEDGEMENT

This work from the Human Factors Integration Defence Technology Centre (HFI-DTC) was part funded by the Human Capability Domain of the UK Ministry of Defence.

The CWA software tool is available free-of-charge as shareware from the HFI-DTC (www.hfidtc.com). The software requires the '.net' (dot-net) framework to be installed on the host computer before it can be run. Contact n.stanton@soton.ac.uk for details of how to obtain a copy.

REFERENCES

Carayon, P. and Buckle, P. 2010. Editorial for special issue on applied ergonomics on patient safety. *Applied Ergonomics*, 41(5), 643–644.

de Carvalho, P. V. R., Gomes, J. O., Huber, G. J. and Vidal, M. C. 2009. Normal people working in normal organizations with normal equipment: System safety and cognition in a mid-air collision. *Applied Ergonomics*, 40(3), 325–340.

Dul, J., Bruder, R., Buckle, P., Carayon, P., Falzon, P., Marras, W. S., Wilson, J. R. and van der Doelen, B. 2012. A strategy for human factors/ergonomics: Developing the discipline and profession. *Ergonomics*, 55(4), 377–395.

Durugbo, C. 2012. Work domain analysis for enhancing collaborative systems: A study of the management of microsystems design. *Ergonomics*, 55(6), 603–620.

Harris, D. and Stanton, N. A. 2010. Aviation as a system of systems. *Ergonomics*, 53(2), 145–148.

Jenkins, D. P., Stanton, N. A., Salmon, P. M. and Walker, G. H. 2011. A formative approach to developing synthetic environment fidelity requirements for decision-making training. *Applied Ergonomics*, 42(5), 757–769.

Jenkins, D. P., Stanton, N. A., Walker, G. H. and Salmon, P. M. 2009. *Cognitive Work Analysis: Coping with Complexity*. Aldershot, UK: Ashgate.

Jun, G. T., Ward, J. and Clarkson, P. J. 2010. Systems modelling approaches to the design of safe healthcare delivery: Ease of use and usefulness perceived by healthcare workers. *Ergonomics*, 53(7), 829–847.

Locke, J. 2012. *Situation Awareness to Support Submarine Command Team Tactical Decision Making*. Fareham, UK: Dstl.

McIlroy, R. C. and Stanton, N. A. 2011. Getting past first base: Going all the way with cognitive work analysis. *Applied Ergonomics*, 42(2), 358–370.

Rafferty, L. A., Stanton, N. A. and Walker, G. H. 2010. The famous five factors in teamwork: A case study of fratricide. *Ergonomics*, 53(10), 1187–1204.

Rasmussen, J. 1986. *Information Processing and Human-Machine Interaction: An Approach to Cognitive Engineering*. New York: North-Holland.

Salmon, P. M., Williamson, A., Lenné, M., Mitsopoulos-Rubens, E. and Rudin-Brown, C. M. 2010. Systems based accident analysis in the led outdoor activity domain: Application and evaluation of a risk management framework. *Ergonomics*, 53(8), 927–939.

Stanton, N. A. and McIlroy, R. C. 2012. Designing mission communication planning: The role of rich pictures and cognitive work analysis. *Theoretical Issues in Ergonomics Science*, 13(2), 146–168.

Stanton, N. A., McIlroy, R. C., Harvey, C., Blainey, S., Hickford, A., Preston, J. M. and Ryan, B. 2013. Following the cognitive work analysis train of thought: Exploring the constraints of modal shift to rail transport. *Ergonomics*, 56(3), 522–540.

Stanton, N. A. and Salmon, P. M. 2011. Planes, trains and automobiles: Contemporary ergonomics research in transportation safety. *Applied Ergonomics*, 42(4), 529–532.

Vicente, K. J. 1999. *Cognitive Work Analysis: Toward Safe, Productive, and Healthy Computer-Based Work*. Mahwah, NJ: Lawrence Erlbaum Associates.

Walker, G. H., Stanton, N. A., Jenkins, D. P. and Salmon, P. M. 2009a. From telephones to iPhones: Applying systems thinking to networked, interoperable products. *Applied Ergonomics*, 40(2), 206–215.

Walker, G. H., Stanton, N. A., Salmon, P. M., Jenkins, D. P. and Rafferty, L. A. 2010. Translating concepts of complexity to the field of ergonomics. *Ergonomics*, 53(10), 1175–1186.

Walker, G. H., Stanton, N. A., Salmon, P. M., Jenkins, D., Stewart, R. and Wells, L. 2009b. Using an integrated methods approach to analyse the emergent properties of military command and control. *Applied Ergonomics*, 40(4), 636–647.

Waterson, P. and Eason, K. 2009. '1966 and all that': Trends and developments in UK ergonomics during the 1960s. *Ergonomics*, 52 (11), 1323–1341.

8 Cognitive Work Analysis
Lens on Work

Catherine M. Burns

CONTENTS

8.1 INTRODUCTION

Cognitive work analysis (CWA) is a five-phase analytical approach to understanding work in complex environments (Chapter 1). CWA emerged from the work of Jens Rasmussen in the 1970s and 1980s (Rasmussen, 1983, 1985, 1990) and coalesced into an analytical approach through the collaboration of Rasmussen and Vicente (Rasmussen and Vicente, 1989; Vicente, 1999). CWA originated in work geared towards understanding the complexities of its time, which in this were the problems in nuclear power, most notably, Three Mile Island. These problems clearly identified that human work had become complex beyond the ability of teams of operators to understand such situations. Furthermore, these accidents showed that decision support for such complex work was lagging far behind what was needed. A disruptive shift in the approaches of human factors engineering and human–computer interaction was badly needed. Human factors as a field had evolved the extremely successful concept of 'user-centred design' that, although creating a needed re-orientation in the technology of the time in terms of considering the user perspective, was limited when applied to complex work situations. This limitation resulted from the need to elicit requirements from user interviews. While often a very good strategy, in large and complex systems users were often unable to articulate the full complexity of relationships in these systems and, as a result, this approach was unable to elicit full requirements. The advantage of cognitive work analysis was that it presented an approach to knowledge elicitation that remained user centric, but also emphasised the need for the designer to explore the functional relationships that were built into the system when it was designed. In this way, cognitive work analysis added the designer's intent for functionality, to the user-centric concern, to work towards safe and correct requirements that would help operators to make better decisions and operate the system correctly.

CWA entered the human factors literature with the early studies of the DURESS II simulator. This micro-world had been designed to capture some of the complexities that could be seen in larger process control situation at a scale that could be operated by students with a reasonable amount of training. The first full CWA was conducted on this micro-world, an ecological interface was designed, and the results showed clearly that CWA could be used to reveal information that could help people operate better in fault scenarios, compared with typical designs at the time (Vicente et al., 1995; Christoffersen et al., 1996; Pawlak and Vicente, 1996). These early successes confirmed that CWA was certainly an approach of potential in human factors.

Despite the potential of CWA, there were still vocal detractors and opponents of the approach. Some felt it was too hard to learn, too engineering oriented and the leap from the analysis to design seemed like a magical leap of faith or design talent. As can be typical of new ideas, CWA was more of an 'approach' than a method; examples were few and learning to use it was challenging. At best, some regarded it as a limited approach for those human factors practitioners with an engineering orientation working in complex industrial contexts.

For those of us working with CWA, we saw something different. We saw an approach that looked at the world of human work from multiple lenses, several of which had been barely explored. We saw an approach that allowed designers to ask insightful questions that added to the user-centred perspective and could help in developing a more robust design. We also saw questions of how to support work move from process control to new domains like military command and control, automated domains, domains with human components, and work with teams. While the CWA structure still worked in these new contexts, these examples encouraged people to look more flexibly at what they could do with the CWA approach.

8.2 BUILDING BETTER MODELS

Working and developing CWA further seemed to face some key challenges. First, CWA needed to be used on real-world systems, which are messier and more difficult to define than laboratory simulations. In exploring real-world systems, the boundaries of analysis, and the dangers of bounding an analysis needed to be considered. As well, it quickly became apparent that workers satisfy far more than just physical constraints and often negotiate the physical capabilities with an equally real set of intentional and value-driven constraints. Finally, working with people, either in teams or in larger groups, was an important dimension of real cognitive work that required explicit treatment to be understood well in the framework of CWA.

8.2.1 Testing the Boundary

The first step was to take CWA to problems in the real world. In a constructed micro-world, like DURESS, the edges of the system are well defined. For the most part, the same is true for an industrial plant; there are limited flows in and out of the building boundary and, with the exception of monitoring intake parameters or emissions, the

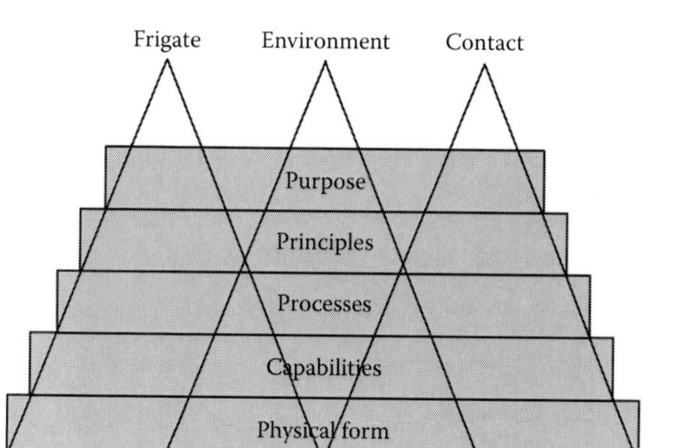

FIGURE 8.1 Abstraction hierarchies for an open system. (Reprinted with permission from Burns, C. M., Bryant, D. B. and Chalmers, B. A. Boundary, purpose and values in work domain models: Models of naval command and control. *IEEE Transactions on Systems, Man and Cybernetics Part A*, 35(5), 603–616 © 2005 IEEE.)

boundary is relatively well defined. However, when we started exploring CWA in the context of naval command and control, boundary issues became very salient. When you are controlling a ship, moving around the world, and interacting with others outside of the hull of the ship, what is the system of analysis? Clearly, the ship itself could be treated as a contained power plant, and managing the ship is an important function. However, to model the command and control domain properly, the ship must be modelled in the context of its military function – in this case, determining and controlling naval contacts. This consideration resulted in a work domain model that contained three interlinked abstraction hierarchies: one for the operation of the ship; one for the environment the ship navigates; and one for the contact the ship must monitor or control (Figure 8.1).

These models with multiple interacting domains have been useful in many other situations. Hajdukiewicz et al. used them to model the anaesthesiology, where the domains were the patient and the anaesthesiologist (Hajdukiewicz et al., 1998). Euerby and Burns modelled communities of practice, where one domain was the community itself with its growth and sustainability, and the other domain was the problem the community was trying to solve (Euerby and Burns, 2012, 2014). Recently, these models were also used to model medication reconciliation with domains of the patient, family physician and hospital (Grindrod et al., in press). Other examples are a gambler versus the casino (Burns and Proulx, 2002).

8.2.2 It Is Not All Physical

In some of our early explorations of CWA, it became apparent that operators were often thinking of more than just physical constraints when making decisions. In fixing a problem, or taking an action, you would elicit additional concerns like 'well we couldn't just do that', or 'this specifies what we can do' and the source of the concern would be a rule of engagement, a particular value, or a social organisational

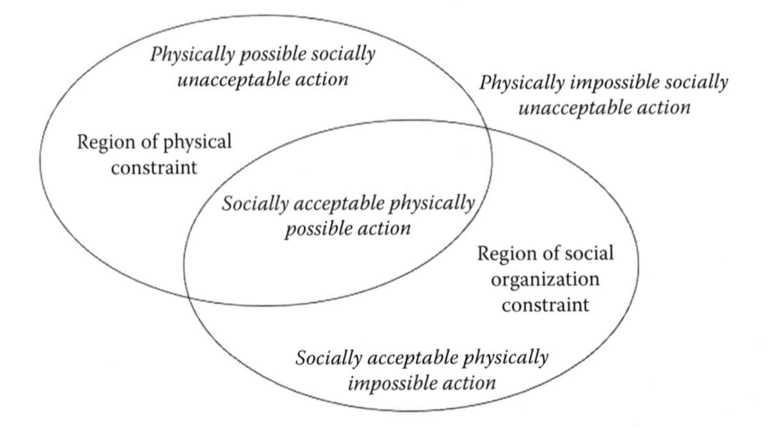

FIGURE 8.2 Showing how intentional constraints limit the space of likely action. (Reprinted with permission from Burns, C. M., Bryant, D. B. and Chalmers, B. A. Boundary, purpose and values in work domain models: Models of naval command and control. *IEEE Transactions on Systems, Man and Cybernetics Part A*, 35(5), 603–616 © 2005 IEEE.)

constraint. In determining appropriate action, these 'intentional' or 'social' constraints could be as constraining as the physical capabilities of the system (Figure 8.2). Early examples where these non-physical constraints were explored are intentional constraints in emergency ambulance dispatching (Hajdukiewicz et al., 1999) and social–organisational constraints in Burns et al. (2005). These can be handled in the abstraction hierarchy (AH) of CWA by incorporation in the values and priority measures (or abstract function) level of the AH (e.g. Naikar, 2013), by building an intentional domain model (Hajdukiewicz et al., 1999), or by integration through the AH (Burns et al., 2005). Since that time it has become common practice to include these constraints in a CWA. Examples of intentional constraints in an abstraction hierarchy include fun:cost ratio (Burns and Proulx, 2002), balancing institutional priorities (Burns et al., 2008), and valuing a healthy life (Rezai and Burns, 2014).

8.2.3 What Do We Do about People?

In CWA, the intent is to understand the work system. The role of people within the work system has tended to be confusing. CWA considers people explicitly in the worker competencies phase and the social organisational phase; however, the role of people in the other phases has been less clear. Within the work domain analysis, people have been modelled in three ways:

1. As components within the work system, with certain capabilities, and the ability to be monitored and controlled
2. As the work system itself, in healthcare applications where the patient is a core part of the domain of work
3. As patient and user, when the patient must take action on themselves, for example, self-treatment

These views of people in work have specific implications in CWA models. When people are modelled as components in the work system, they are distinct from the operator, who must supervise and control the work system. In this situation, people are modelled at the components level and their skills and capabilities contribute to various functions in the work system. A good example of this situation occurs in military command and control, where various personnel have various skills and roles, and may also have various conditions such as fatigued, injured or rested. The commander must understand the capabilities and conditions of his or her team to manage the work domain effectively.

In contrast, it is common in healthcare applications to need to model the patient as a work domain (Hajdukiewicz et al., 1998 was one of the earlier examples). In work domain models of patients, it has been common to adjust the abstraction hierarchy labelling to reflect physiological processes, balances and anatomy. An added challenge in models of patients or biological systems can be the consideration of embedded control (Miller, 2004). In engineered systems, the control systems and automation can often be considered with some independence from the core functions of the system. However, in biological systems, the control systems are often multifunctional, creating complex and tight interactions that cannot be easily considered independently.

More recently, we have had to consider people as the work domain, a system component and the operator, in cases of self-treatment or self-monitoring of health (e.g. Rezai and Burns, 2014; Grindrod et al., in press). An example of this is medication adherence, where the medication has a functional role in the maintenance of the patient's health, but the patient is also responsible for the choice to take the medication. The patient's cognitive capabilities, psychological assessment of the medication or its side effects and understanding of the role of the medication in meeting his or her life purposes play a role in how well this work domain operates. As an example, a patient experiencing cognitive impairment (patient as component) may not take his or her medication (patient as operator), with negative consequences for his or her health (patient as work domain). A work domain analysis built for medication reconciliation (Grindrod et al., in press) is shown in Figure 8.3. Within the patient, not only have their physical processes been modelled, but also cognitive processes have been modelled as well.

8.2.4 What Do We Do about Many People, Teams and Communities?

Knowing that people play an integral role in cognitive work, understanding work in the context of groups, teams and communities was important. The social organisational analysis of CWA provides a structure for this, but when examining teams in more detail, we could gain a richer view of the team by using CWA in new ways. In particular, a small modification of the decision ladder (shown in Figure 8.4) into a wheel allowed intra-team interactions to be shown more clearly (Ashoori and Burns, 2013). Each wheel represents a team unit, and the exchanges between teams could be modelled. The decision wheel shows which teams or team members become engaged in knowledge-based decision-making (at the top of the wheel) and which ones work at skill- and rule-based levels. Synchronous and asynchronous activities

	Clinic or Hospital	Patient	Patient Systems (e.g. Cardiovascular with hypertension)
Purposes	Maintain and improve client's health	Maintain and improve patient health	Maintain blood pressure in normal range
Principles Priorities and Balances	Patient flow Treatment principles Information flow Resource flow	Balance recovery and restore function	Principles of fluid mechanics, cardiovascular function
Processes	Patient movements from home to FHT to other care locations Information flow (sources patient, other care sources) sinks (patient, caregivers, pharmacies) transforms (med rec) Medication flow Treatments (from FHT and from other care sources)	Healing Disease process Homeostatic processes Psychological and cognitive processes Process of taking medication	Physiological processes (osmosis, circulation, pumping, etc.) Pharmacological processes of the prescribed med (Diuretic, beta-blocker, ACE inhibitor, etc.)
Anatomy and Objects	Patients Caregivers Care institutions Physicians Pharmacists Information	Body as a whole Medications	Circulatory system (heart, blood vessels) Endrocrine system Nervous system Active Ingredients of medication Active ingredients of food (e.g. salt)
Details and Attributes	Location and capabilities of each	Age, weight, gender of patient Cognitive capacity Type dose and concentration of medications	Blood pressure level, heart rate, heart condition, medication type and dose, food type and amount

FIGURE 8.3 Work domain analysis for medication reconciliation.

could be modelled, which showed that a new way of looking at decision was making through the ladder.

Ashoori et al. made similar extensions to the work domain analysis, strategies analysis and worker competency analysis (Ashoori et al., 2014). In particular, by looking at strategies, Ashoori et al. could show that teams would reconfigure under certain conditions (Figure 8.5). For example, an emergency team might have a

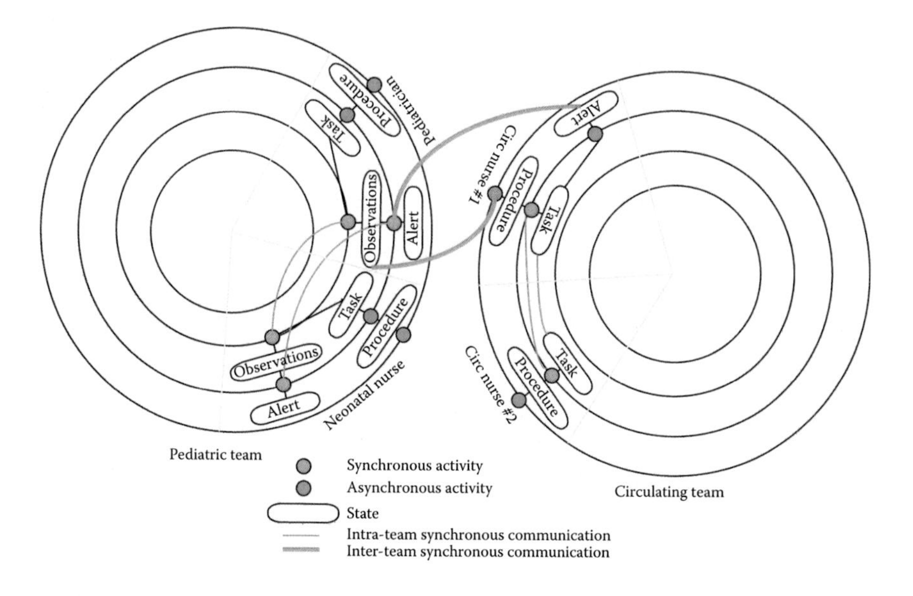

FIGURE 8.4 Example of the decision wheel for nursing. (Adapted from Burns, C. M. 2014. Using team cognitive work analysis as an approach for understanding teamwork in health care. In A. Bisantz, C. M. Burns and T. Fairbanks (Eds.). *Cognitive Engineering in Medicine.* CRC Press, pp. 27–42.)

different structure and roles than a normal operation team. Some of these reconfigurations are to bring different skills and knowledge to the team, and some are to manage workload levels more appropriately. In the same way that operators might change their operational strategies in response to emergencies or higher workload, teams also reconfigure, and this is a critical part of managing and distributing the cognitive work. All of this work built on, and complemented the work of others who looked at team activities (e.g. Naikar et al., 2006; Jenkins et al., 2008a,b; Stanton et al., 2013).

In Ashoori's work, she showed how the foundation of CWA could continue to serve as a lens on each of these new interactions. Euerby and Burns (2012) used the work domain analysis to model communities of practice with the two-model approach discussed earlier. In this approach, one model was for the goals of the community, what they wanted to accomplish. The second model was for the growth and sustainability of the community itself, or what they needed to do to be a community of practice.

8.3 CONCLUSION: LENS ON WORK

In reflection, CWA has provided a very powerful way of looking at the world. In particular, CWA creates different views from functional relationships, decision-making and information processing, strategies, individual capabilities and social organisational factors. The core structures of the abstraction hierarchy and decision ladders are useful in multiple contexts and in many different ways. These structures can be applied multiple times, in different views, with different kinds of connections, and they continue to yield insights because the fundamental

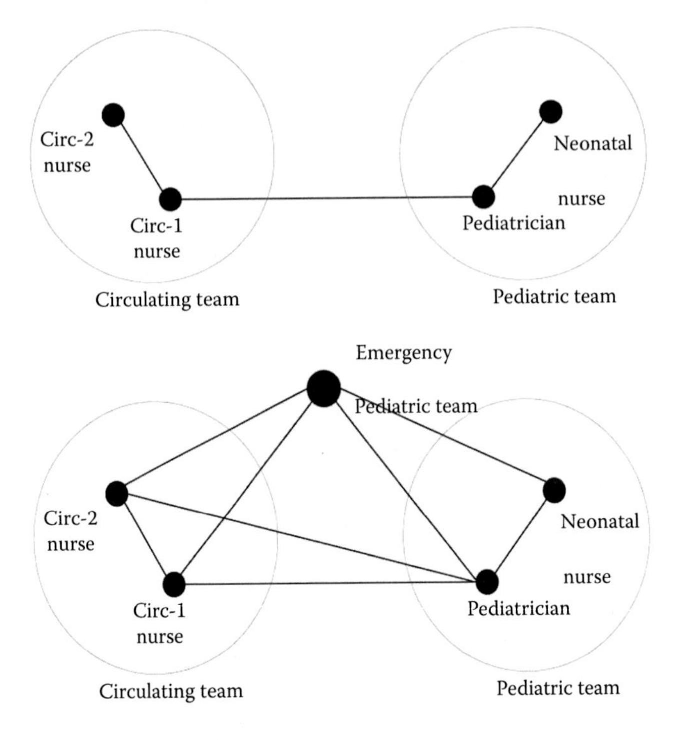

FIGURE 8.5 Different team configurations as a strategy. (Adapted from Ashoori, M. et al. 2014. *Ergonomics,* 57(4), 973–986.)

structures are core to human decision-making, design and cognition. They have been adapted to different domains, different contexts, and different combinations of individuals, teams and communities and continued to be helpful and useful analytical tools.

The second core value from CWA is its ability to teach the designer. I have taught CWA to hundreds of students at this point and watched them try it out for the first time, in a new environment they wanted to explore. Despite their initial struggles, or worries about 'doing it right', or 'what goes at what level', it does not take very long before they are uncovering new insights about the world. By building a CWA, designers need to ask key questions of their domain – what is the purpose? What makes it complex? What are the different strategies that can be used? This makes the analyst a more targeted observer and a more insightful question-asker during interviews. It is not uncommon for a CWA to spark a discussion between the workers participating about some of the roles, complexities and strategies involved. In this way, CWA can provide a very powerful lens on human work.

REFERENCES

Ashoori, M. and Burns, C. M. 2013. Team cognitive work analysis: Structure and tasks. *Journal of Cognitive Engineering and Decision Making*, 7, 123–140.

Ashoori, M., Burns, C. M., Momtahan, K. and d'Entremont, B. 2014. Using team cognitive work analysis to reveal healthcare team interactions in a labour and delivery unit. *Ergonomics*, 57(4), 973–986.

Burns, C. M. 2014. Using team cognitive work analysis as an approach for understanding teamwork in health care. In A. Bisantz, C. M. Burns and T. Fairbanks (Eds.). *Cognitive Engineering in Medicine*. Boca Raton, FL: CRC Press, pp. 27–42.

Burns, C. M., Bryant, D. B. and Chalmers, B. A. 2005. Boundary, purpose and values in work domain models: Models of naval command and control. *IEEE Transactions on Systems, Man and Cybernetics Part A*, 35(5), 603–616.

Burns, C. M., Enomoto, Y. and Momtahan, K. 2008. A cognitive work analysis of cardiac care nurses performing teletriage. In: A. Bisantz and C. M. Burns (Eds.). *Applications of Cognitive Work Analysis*. Mahwah, NJ: Lawrence Erlbaum and Associates, pp. 149–174.

Burns, C. M. and Proulx, P. 2002. Influencing social problems with interface design: A student project. *Ergonomics in Design*, 10(4), 12–16.

Christoffersen, K., Hunter, C. N. and Vicente, K. J. 1996. A longitudinal study of the effects of ecological interface design on skill acquisition. *Human Factors*, 38(3), 523–541.

Euerby, A. and Burns, C. M. 2012. Designing for social engagement in online social networks using communities of practice theory and cognitive work analysis: A case study. *Journal of Cognitive Engineering and Decision Making*, 6, 194–213.

Euerby, A. and Burns, C. M. 2014. Improving social connection through a communities of practice inspired cognitive work analysis approach. *Human Factors*, 56, 361–383.

Grindrod, K., Tran, P. and Burns, C. M. 2015. Medication reconciliation: More than just a check. *2015 International Symposium on Human Factors and Ergonomics in Health Care: Improving the Outcomes*, 4(1), 206–212. doi: 10.1177/2327857915041014.

Hajdukiewicz, J. R., Burns, C. M., Vicente, K. J. and Eggleston, R. G. 1999. Work domain analysis for intentional systems. *Proceedings of the Human Factors and Ergonomics Society Annual Meeting*, 43, 333–337.

Hajdukiewicz, J. R., Doyle, D. J., Milgram, P., Vicente, K. J. and Burns, C. M. 1998. A work domain analysis of patient monitoring in the operating room. *Proceedings of the Human Factors and Ergonomics Society Annual Meeting*, 42(14), 1038–1042.

Jenkins, D. P., Stanton, N. A., Salmon, P. M., Walker, G. H. and Young, M. S. 2008a. Using cognitive work analysis to explore activity allocation within military domains. *Ergonomics*, 51(6), 798–815.

Jenkins, D. P., Stanton, N. A., Walker, G. H., Salmon, P. M. and Young, M. S. 2008b. Applying cognitive work analysis to the design of rapidly reconfigurable interfaces in complex networks. *Theoretical Issues in Ergonomics Science*, 9(4), 273–295.

Miller, A. 2004. A work domain analysis framework for modelling intensive care unit patients. *Cognition Technology and Work*, 6, 207–222.

Naikar, N. 2013. *Work Domain Analysis: Concepts, Guidelines, and Cases*. Boca Raton, FL: CRC Press.

Naikar, N., Moylan, S. and Pearce, B. 2006. Analyzing activity in complex systems with cognitive work analysis: Concepts, guidelines, and case study for control task analysis. *Theoretical Issues in Ergonomics Science*, 7(4), 371–394.

Pawlak, W. S. and Vicente, K. J. 1996. Inducing effective operator control through ecological interface design. *International Journal of Human-Computer Studies*, 44(5), 653–688.

Rasmussen, J. 1983. Skills, rules, knowledge: Signals, signs, and symbols, and other distinctions in human performance models. *IEEE Transactions on Systems, Man and Cybernetics*, 13, 257–266.

Rasmussen, J. 1985. The role of hierarchical knowledge representation in decision making and system management. *IEEE Transactions on Systems, Man and Cybernetics*, 15, 234–243.

Rasmussen, J. 1990. Mental models and the control of action in complex environments. In D. Ackermann and M. J. Tauber (Eds.). *Mental Models and Human-Computer Interaction 1*. North Holland, the Netherlands: Elsevier Science Publishers, pp. 41–46.

Rasmussen, J. and Vicente, K. J. 1989. Coping with human errors through system design: Implications for ecological interface design. *International Journal of Man-Machine Studies*, 31, 517–534.

Rezai, L. S. and Burns, C. M. 2014. Using cognitive work analysis and a persuasive design approach to create effective blood pressure management systems. *Proceedings of the International Symposium on Human Factors and Ergonomics in Health Care*, 3, 36–43.

Stanton, N. A., McIlroy, R. C., Harvey, C., Blainey, S., Hickford, A., Preston, J. M. and Ryan, B. 2013. Following the cognitive work analysis train of thought: Exploring the constraints of modal shift to rail transport. *Ergonomics*, 56(3), 522–540.

Vicente, K. J. 1999. *Cognitive Work Analysis: Toward Safe, Productive, and Healthy Computer-Based Work*. Mahwah, NJ: Erlbaum and Associates.

Vicente, K. J., Christoffersen, K. and Pereklita, A. 1995. Supporting operator problem solving through ecological interface design. *IEEE Transactions on Systems, Man and Cybernetics*, 25(4), 529–545.

9 Exploring the Constraints of Modal Shift to Rail Transport

Neville A. Stanton, Rich C. McIlroy, Catherine Harvey, John M. Preston, Simon Blainey, Adrian Hickford and Brendan Ryan

CONTENTS

9.1 INTRODUCTION

Despite a relatively small number of dissenters, the vast majority within the scientific community and beyond agree that climate change is upon us, and that it is a significant and widely impacting challenge that must be tackled. Many of the world's governments have proposed a wide variety of policies and strategies designed specifically to deal with the issues of climate change, and, in particular, CO_2 emissions.

However, high costs and differing levels of priority given to these measures have inevitably meant that the potential benefits are not always realised. The transport sector is one of the key areas in which change must occur; according to the IPCC (2007), as of 2004 transport was not only responsible for 23% of the world's energy related greenhouse gas emissions, but also had the fastest growing CO_2 emissions rate of any energy sector. Of transport's share of carbon emissions, 90% can be attributed to road transport and in the United Kingdom in 2008 private vehicle use was responsible for 44% of total greenhouse gas emissions across all modes of transport (including air, maritime, rail, road freight, buses, coaches, trams, tubes and taxis; Department of Transport, 2009). It is unsurprising, given these statistics, that there is a current push to reduce individuals' reliance on the car and increase their use of alternative means of transport. This reflects the IPCC's recommendation for *modal shift* from personal car use to public transport as a key climate change mitigation strategy (IPCC, 2007). Unfortunately however, there are a number of constraints preventing wide scale modal shift.

Constraints to modal shift can be defined as the temporal, financial, physical, mental (cognitive) and/or emotional (affective) effort required to use a particular mode of transport for a journey which are greater than the effort required to use an alternative mode (Accent, 2009; Blainey et al., 2009) or undertake an alternative activity which achieves equivalent results. For travel behaviour to change, an individual will need both the motivation to change (desire) and the means to facilitate such change (ability) by overcoming the constraints that exist (Department of Transport, 2006). Human factors addresses the role of the human within a system and therefore offers an important perspective on constraints to modal shift, particularly regarding passengers' perceptions of the disadvantages of certain modes. This chapter focuses particularly on rail and the constraints to modal shift to train travel. In recent years, there has been a growth in rail human factors research, prompted in part by a number of high profile accidents and resulting focus on human error (Wilson and Norris, 2005a) and also the need to transfer passenger miles away from road towards rail (Wilson et al., 2001). Human factors is particularly important for the modern sociotechnical railway system (Wilson et al., 2007) as it is dependent upon 'the synergy between human beings and engineering assets' (Shepherd and Marshall, 2005, p. 719) to ensure that user and organisational needs are successfully met (Pledger et al., 2005; Wilson and Norris, 2005b). By taking a human factors view of the constraints to rail travel, this analysis will provide important insights into how to potentially overcome some of these issues in terms of human factors solutions. A human factors view is likely to offer some solutions to improving the perception of rail travel among passengers, which can be implemented at significantly lower cost and with less disruption than many of the engineering solutions that have been suggested (e.g. the re-opening of lines and the purchase of new, high-capacity rolling stock [Cox et al., 2006]).

The first aim of this chapter is to present the constraints to modal shift towards rail, highlighting human behavioural factors which could limit the use of rail public transport over the use of the private car, and provide suggestions for particular measures which could help promote such modal shift. The second aim of the chapter is to identify, through the use of cognitive work analysis (CWA; Rasmussen et al., 1994; Vicente, 1999; Jenkins et al., 2009), which of the constraints are most constraining to modal shift, and to which group(s) of actors each constraint is most applicable. This will enable policymakers to target specific actor groups with strategies tailored to their characteristics, thus enhancing the likelihood of modal shift.

9.2 KEY CONSTRAINTS IDENTIFIED

The constraints to modal shift identified in this chapter are derived from two main sources. The first, 'integrated transport – perception and reality' (Association of Train Operating Companies [ATOC] and Passenger Focus, 2010), described the perceptions of and constraints to using rail by non-users and infrequent users of rail, with particular emphasis on the station access, egress and interchange, compared with other modes. The second, 'passengers' priorities for improvements in rail services' (MVA Consultancy, 2010), gives the results of a recent survey of passenger priorities carried out by Passenger Focus, to identify which attributes of rail services passengers would most like to see improved.

9.2.1 Cost/Value for Money

On a trip-by-trip basis, rail travel can be more expensive than car travel (Derek Halden Consultancy, 2003; Eriksson et al., 2008), and it is suggested that cost is the most important factor deterring rail non-users from modal shift (Accent, 2009). For car travel, expenditure tends to be viewed as a necessity (University of Oxford Transport Studies Unit [UOTSO], 1995), while expenditure on rail travel for many trips is seen as a luxury. Existing rail passengers perceive value for money to be poor (MVA Consultancy, 2010), and the current fare structure is viewed as being complex and confusing (Passenger Focus, 2009).

Although cost is not strictly a human factors issue, the perception of cost by passengers can be influenced by the accessibility of cheap rail fares to both current and non-users. For example, constraint, it may be possible to stimulate some modal shift by increasing publicity of cheap rail fares among non-users and increasing the ease with which these can be purchased. This is because infrequent rail users are often unsure how to obtain the best value rail fares, and if a decision is made to shift mode they may in some cases find that rail travel is cheaper than expected (Accent, 2009).

9.2.2 Punctuality and Reliability

Reliability describes how closely actual journey times relate to advertised or expected schedules, and is one of the most important factors affecting individual travel decisions (Derek Halden Consultancy, 2003). A distinction is also sometimes made between punctuality (whether or not a service arrives on time) and reliability (whether or not a service runs at all), but the two terms are not entirely mutually exclusive (Bates et al., 2001), since if a train is delayed so that it runs later than the following service on the same route then as far as passengers are concerned it might as well have been cancelled. Car travel tends to be associated with 'control' and flexibility over arrival times (Bates et al., 2001), whereas rail is often viewed as being unpredictable or unreliable. Punctuality is highly valued by travellers (Bates et al., 2001; MVA Consultancy, 2010), with an excess of delays and cancellations identified as an important constraint by 42% of respondents (ATOC and Passenger Focus, 2010). From a human factors perspective, unpredictability in journey times also contributes to higher levels of stress for commuters (Cox et al., 2006).

One hundred per cent reliability will never be achieved, as the knock-on effects of increased journey times and reduced network capacity outweigh reliability benefits; therefore, complementary measures may be needed to promote modal shift. For example, the MobiHarz project in Germany concluded that there was a need to restructure the public transport offered to suit the requirements of visitors and occasional users, with backup options (such as taxis) provided so that passengers were confident they would not be stranded short of their destination (Hoenniger, 2003). This is evidence of a human factors approach, in which the restructuring of the system was based around the needs of users, with passenger choice an important element in the proposed solution.

9.2.3 Frequency of Trains

If the frequency of trains over a particular route is low, or the timetabled departure and arrival times do not correspond with the needs of potential users, this can act as a constraint to rail use (Eriksson et al., 2008). Poor service frequencies increase wait times, which are a major disadvantage of using public transport and tend to incur a much greater disbenefit per minute than travel time, particularly for shorter trips, where the wait time can form a major component of the total travel time.

Inconvenient timetabling of services, particularly the operating hours of rail services, has been shown to be a constraint to rail use: examples include services failing to run sufficiently late for Glasgow revellers (Derek Halden Consultancy, 2003), the need for more evening and Sunday services in the West Midlands (Passenger Focus, 2006) and demand for more late evening services in the South Central area, particularly at weekends (Passenger Focus, 2008). To alleviate these types of issues, rail operators should be encouraged to engage in passenger feedback exercises in order to establish when users and non-users actually want trains to be provided.

9.2.4 Comfort/Cleanliness

There is a continuing tendency for improvements in the comfort and facilities offered by private cars to occur at a faster rate than those offered by trains, and the perception of rail travel as comparatively uncomfortable (Kogi, 1979) may act as a constraint to mode shift. It has been suggested that the provision of comfortable trains with sufficient seats is the improvement second most likely (after fare reduction) to encourage rail use among infrequent and non-rail users (ATOC and Passenger Focus, 2010). There is likely to be a relationship between comfort and crowding levels; while trains may be viewed as being comfortable when sufficient seats are available, things are likely to change when they are operating at crush capacity. It is therefore likely that any increase in discomfort caused by overcrowding will deter passengers away from rail, toward the car (Howarth et al., 2011). Thomas et al. (2005) found that although the actual risk to passenger safety (i.e. injury or fatality) caused by high passenger density on the train or platform was only 0.1%, the perceived risk was much higher, at over 5%. This suggests that while interventions to reduce overcrowding would have a positive impact, it is also passengers' perceptions of this issue that need to be addressed before constraints to rail travel can be overcome (Cox et al., 2006). Perception of comfort is linked to 'social forces', which 'reflect interpersonal psychological relations that attract or repel the pedestrian from their surroundings' (Howarth et al., 2011) and it is these forces that need to be fully understood in order to reduce some of the constraints to rail travel.

9.2.5 Travel Time

The difference in the time taken to travel from an origin to a destination by different modes will obviously play a major role in mode choice decisions (Lyons et al., 2007). Long door-to-door journey times have been cited as being a significant constraint to rail use (Hine and Scott, 2000; ATOC and Passenger Focus, 2010), particularly in comparison to car journeys. A reduction in train journey times is not easily achievable and there is also the compounding effect of travel to and from the station, which further increases overall journey time. On the other hand, it seems unlikely that road or air journey times will reduce significantly in the future, therefore increasing the comparative appeal of rail. Alternative, user-centred solutions have also been suggested, including highlighting the 'positive utility' of travel time (Lyons et al., 2007). Advertising the positive ways in which passengers could spend their travel time could reduce the extent to which this is seen as a constraint to rail travel and increase the appeal relative to other modes, such as road transport, in which it is more difficult (or impossible for the driver) to use a laptop or read a book during a journey. Human factors can contribute to facilitating the use of these technologies in order to increase the positive value derived from journeys.

9.2.6 Interchange/Station Facilities

The need to interchange between rail services when making some journeys can be a constraint to modal shift. There are a number of disbenefits associated with the interchange: these include variable disbenefits, related to the amount of wait and transfer time required; and fixed disbenefits, related to the inconvenience and risks involved (Wardman and Hine, 2000). The quality of the interchange environment and the availability (or lack) of suitable waiting facilities may also be an issue (Wardman and Hine, 2000), as may concern about unnecessary walking and worries about personal safety (Wardman et al., 2001). If it is necessary to transfer between platforms in the course of an interchange, then this may act as a constraint to mobility-impaired passengers (Hine and Scott, 2000). The need to interchange may also lead some travellers to start their journeys early to reduce the risk of missing a connection (ATOC and Passenger Focus, 2010), therefore incurring an increased journey time penalty.

While the most obvious solution to the constraint posed by interchange is to run more direct services, this may lead to increased delays as a result of problems in one area being transmitted further through the network, and in any case may not be a feasible option. In terms of human factors, there is potential to reduce this constraint by improving the interchange environment in order to support passengers to make quick and comfortable transfers, although it is not clear whether this would have a significant effect on mode shift.

9.2.7 Safety and Security/Staffing

Concerns over personal security may deter some potential passengers from using rail (Mackett and Sutcliffe, 2003), particularly during the hours of darkness and at smaller stations. In focus groups of existing rail users, 75% claimed to have fears of safety when waiting on station platforms after dark, with almost as many experiencing safety fears when approaching the station after dark (Cozens et al., 2003). Over half also had concerns about the safety of travelling by train at night. Some studies have identified a lack of CCTV provision as forming a constraint to modal shift (Cozens et al., 2003; Derek Halden Consultancy, 2003), but even CCTV may not make people feel safe at smaller stations where feelings of isolation make security a particularly important issue (MVA Consultancy, 2010).

The presence of staff at the railway station has been identified as an important issue with regard to safety, and could be particularly effective when combined with other measures; for example, staff-monitored CCTV was found to be the most popular initiative among passengers for improving safety at the station entrance according to a survey conducted by Thomas et al. (2005). Staff presence also needs to be made obvious to passengers, so the security system needs to be designed with high levels of transparency. However, the cost of staff provision can be substantial, and substantial levels of mode shift would be necessary to justify this expenditure at smaller stations. Furthermore, staff provision will not be able to address the problem of personal security when travelling to stations, and this is an area where rail travel compares unfavourably to car travel given the door-to-door security offered by the latter mode. It is not clear how significant a constraint such concerns are to mode shift, but if it is significant, then it may prove to be a comparatively difficult issue to address. Wider trends in and perceptions of crime rates and the risk of crime will influence the importance of this constraint in the future.

9.2.8 Station Access

The functioning of collective modes of transport, such as rail, is highly dependent on the level of connectivity with the other modes, which offer transport to and from the railway stations, as well as the levels of passenger accessibility to the individual modes (Napper et al., 2007). Where rail is on average faster for centre-to-centre journeys, it tends to lose out when travellers' origins and destinations are some distance from stations. Rail users tend to allow too much time for station access in the early stages of a modal shift (Accent, 2009), and this may give access time exaggerated importance as a constraint to infrequent and non-rail users. Station egress can be even more of a problem than access, because travellers will tend to be less familiar with their destination than their origin.

The constraints associated with station access are somewhat dependent on the mode of access. For example, the use of the car as a means of accessing rail is dependent upon parking facilities, road signage around the station and the availability of pick-up/drop-off points. The use of public transport (e.g. the bus) to get to a railway station is influenced by many of the same issues as discussed here for rail travel, including cost, frequency and reliability. For those relatively few rail users who cycle to the station, the constraints tend to be related to lack of good cycle routes to the station, and insufficient or non-secure cycle storage facilities at the station, as well as uncertainty associated with on-train cycle carriage facilities. Walking to access rail is affected by the provision of suitable routes and also by the distance to the station.

9.2.9 Journey Planning and Information Provision

Non-regular travellers tend to require much more effort to plan a trip by public transport (perceived as complex) than by car (perceived as simple) (Kenyon and Lyons, 2003), which may act as a constraint to rail use, particularly for car drivers (Hine and Scott, 2000). Even when public transport is faster and cheaper, people may see it as being easier just to get into their car than to check the timetable, find money to pay the fare and walk to the bus stop or railway station (Derek Halden Consultancy, 2003).

Problems relating to rail journey planning are often based on poor past experiences and inaccurate perceptions. Indeed, non-rail users are often surprised by the ease with which good quality information can be obtained at various stages of rail journeys (ATOC and Passenger Focus, 2010). Inadequate information on the available public transport alternatives and inadequate knowledge of how to use them also mean that people are 'locked' into high levels of car use (SUSTRANS, 2002; Derek Halden Consultancy, 2003). Personalised journey planning techniques may be able to overcome such constraints to modal shift; they can be relatively straightforward to implement, since there is no requirement to alter the current transport provision. Information provision during the journey is equally important, as is information about multimodal transport options and 'softer' aspects such as comfort and convenience (Kenyon and Lyons, 2003). Provision of good quality, reliable, up-to-date information can be one of the cheapest ways to change users' perceptions of transport and therefore achieve modal shift (SUSTRANS, 2002), particularly by targeting those groups whose circumstances yield a higher use of public transport. Publicity does however need to be considered carefully, since typical public transport marketing tends to be more successful in encouraging existing customers to make more journeys than in attracting new customers out of their cars (Blainey et al., 2009).

9.2.10 Ticketing

Studies have shown that some people view the rail ticketing system as being too complex for occasional travellers to understand (Derek Halden Consultancy, 2003). While the rail industry recently attempted to simplify the ticketing structure (with the introduction of 'anytime', 'offpeak' and 'advance' tickets), in practice there still seems to be a great deal of confusion over differences in ticket validity. This issue, together with inaccurate perceptions of the cost of rail travel, may form a significant constraint to modal shift.

Greater provision of intermodal zonal ticketing and of carnets for frequent but irregular travellers could lead to modal shift from car to public transport (Derek Halden Consultancy, 2003). Comprehensive intermodal ticketing might encourage public transport use, as shown by the continuing success of the London Travelcard scheme, although it should be emphasised that integrated ticketing on its own is unlikely to achieve modal shift – integrated and good quality services and information are also required. The importance of ticketing as a constraint to mode shift is unlikely to change in the future, although if booking offices are replaced by ticket machines and internet ticket sales then the lack of a human guide to the vagaries of the ticketing system may make simplicity more important. As self-service ticketing becomes more common, the potential for human factors to aid in the design process will increase.

9.2.11 Constraints Analysis

The previous sections described 10 constraints to modal shift towards rail travel and presented the human factors issues that are relevant to each constraint. The following section describes a CWA which was applied in order to examine these constraints in a systematic way by highlighting the links between the purposes, functions, processes and objects within the rail system from a human factors perspective. CWA is suited to the analysis of large, sociotechnical systems, of which the rail transport system is an example. It was anticipated that the constraints could be linked to the relationships between purposes, functions, situations, journey types and actor groups in the analysis. So rather than just identifying constraints per se, this analysis offers a contextual interpretation of those constraints.

9.3 COGNITIVE WORK ANALYSIS

CWA (Rasmussen et al., 1994; Vicente, 1999) is a burgeoning technique in the human factors and ergonomics domain that is ideally suited to the analysis of large, complex, sociotechnical systems. It has been shown to be useful in a wide range of applications including designing cognitive artefacts (Jenkins et al., 2010a), understanding accidents (Jenkins et al., 2010b), designing large-scale military systems (Bisantz et al., 2003), evaluating design concepts (Naikar and Sanderson, 2001) and understanding the constraints imposed on train drivers (Jansson et al., 2006). The CWA framework comprises five distinct phases, each of which considers the system under analysis from different perspectives. All five phases need not, however, be completed; each phase may be used in isolation, or a number of phases may be used in conjunction, depending on the requirements of the analysis. It is a formative analysis, focussing on the boundaries of a system such that rather than modelling how the system *should* perform (normative modelling) or *does* perform (descriptive modelling), it focuses on how the system *could* perform given its constraints. This chapter describes work in the first, second and fourth stages of CWA. Work in the third and fifth phases of CWA (strategies analysis and worker competencies analysis; see Rasmussen et al., 1990, 1994; Jenkins et al., 2009) is not used; hence, it is not described here. The rationale for focussing on these phases was to capture the constraints of systems functions, situations and journey types (as these were likely to reveal the biggest constraints to modal shift).

The first phase, work domain analysis, considers the functions, purposes and physical objects of which the system, and the environment in which that system is situated, consists (Rasmussen, 1985; Vicente, 1999). The abstraction hierarchy (AH) describes the system based on five different levels of abstraction, ranging from its physical objects at the bottom (the physical components of the system) of the hierarchy, up to the overriding functional purpose at the top (the system's reason for existence). The hierarchy is characterised by the means–ends links describing the relationships within the system (Rasmussen et al., 1990); these linkages are made through the use of a 'how–what–why' triad. Take any node in the hierarchy to answer the question 'what'. All connected nodes on the level immediately below that node can be taken to answer the question of 'how' that function is to be achieved or fulfilled. Considering the connected nodes on the level immediately above will answer the question of 'why' that particular function is required (see Figure 9.1). It is worth noting that there are one-to-many and many-to-one mappings within these triads, as can be seen in Figures 9.4 to 9.6.

The second stage of CWA, control task analysis, addresses system activities in terms of the situation in which they are to be performed. These situations can be defined temporally, spatially or a combination of the two. This phase can be modelled using the contextual activity template (CAT), initially proposed by Naikar et al. (2006). CAT has the advantage of modelling functions by situations to distinguish between hard constraints (e.g. those that require structural changes in the environment or technology to remove) and soft constraints (e.g. those that require changes in attitudes and behaviour to remove) in systems. The horizontal axis is typically populated using nodes taken from either the object-related processes or the purpose-related functions level of the AH. The matrix is then populated according to the situations in which each function *can* be carried out, and in which it is *typically* carried out (see Figure 9.2). Again, this output describes the constraints placed on activity in terms of where each function can and cannot be undertaken. At this stage, the analysis is entirely independent of the actor or group to which each function is attributable.

The fourth phase of CWA is social organisation and cooperation analysis (SOCA). In this stage of the analysis, attention is paid to the actors in the system; it looks at the constraints imposed by social and organisational structures and specific actor roles or definitions. The SOCA phase builds on the previously completed analyses, using differential shading of the diagrams to indicate where each actor or group can perform given tasks and activities. The shading can be applied to any

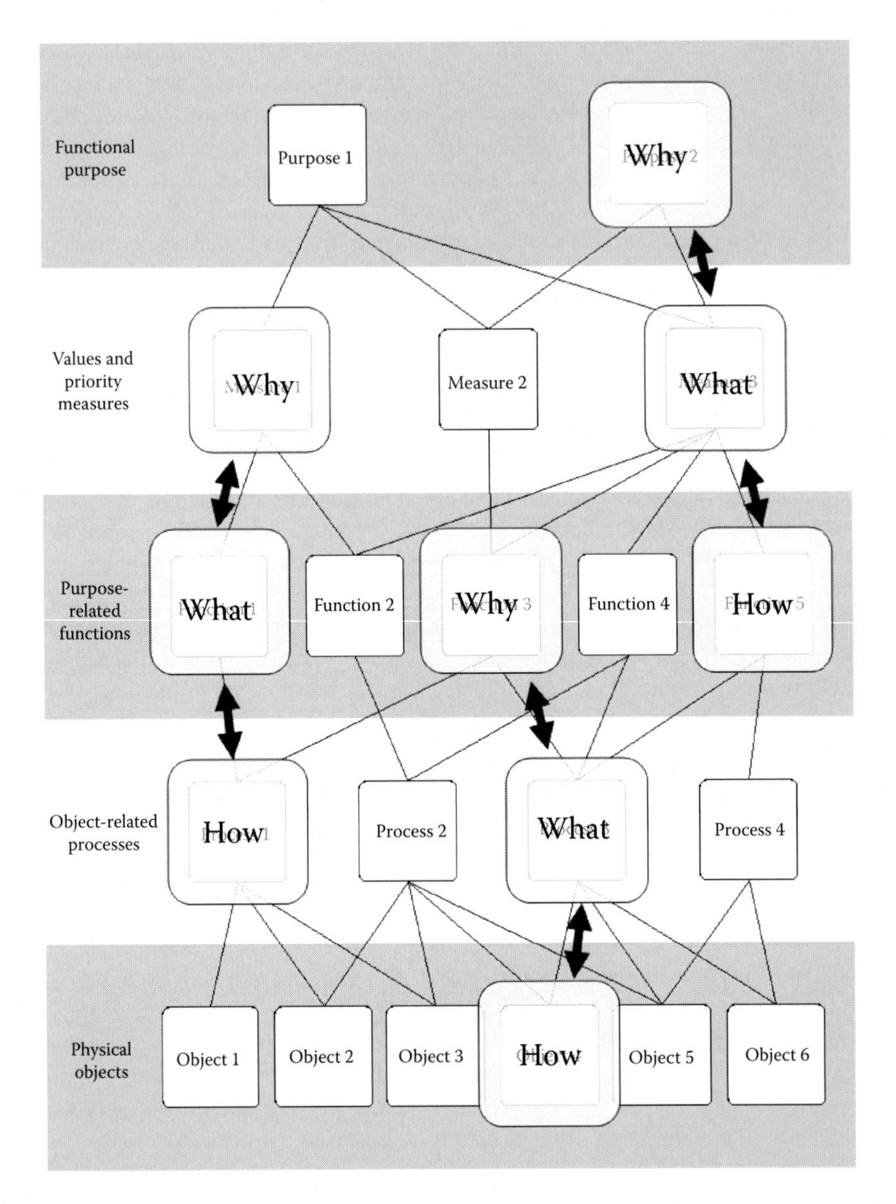

FIGURE 9.1 AH structure, with means–ends links overlaid. (Adapted from McIlroy, R. C. and Stanton, N. A. 2012. *Theoretical Issues in Ergonomics Science*, 13, 450–471.)

of the diagrams constructed thus far; hence, there are a number of different SOCA views. An important point to make here is that the shading is used to represent where actors are *able* to carry out functions, rather than where they typically do or should. As with the other stages of CWA, it is a constraints-based analysis; its aim is to identify possibilities given system boundaries, rather than current practices or standard procedures.

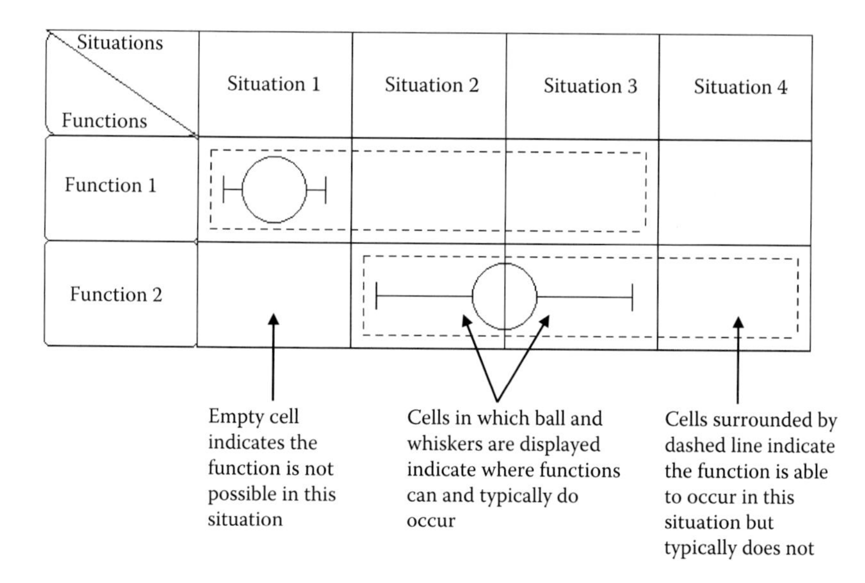

FIGURE 9.2 Explanatory figure of the CAT.

9.3.1 CWA Results

9.3.1.1 Abstraction Hierarchy

The constraints identified above were used to inform the construction of an AH, which consists of five levels of abstraction:

- *Functional purposes* describe the purposes of a system and any external constraints that affect its operation.
- *Values and priority measures* are the criteria that determine how the system progresses towards its functional purposes.
- *Purpose-related functions* describe the general functions necessary for a system to achieve its functional purposes.
- *Object-related processes* refer to the functional capabilities and limitations of objects within a system that affect the functional purposes.
- *Physical objects* are the objects within a system, to which the object-related processes refer.

This analysis focuses predominately on the values and priorities measures, which are the metrics by which system performance is judged (Rasmussen, 1974). As aforementioned, constraints to modal shift describe some of the reasons why an individual chooses to use road transport over rail; this implies that measures taken to reduce or remove those constraints would increase the likelihood of an individual opting for the train. They are, in effect, the methods by which passengers judge the performance of the system. Reducing the constraints, or to put it another way, improving the service on that particular aspect of the system, will increase the system's performance, at least from the perspective of the passenger. As this report focuses on the rail system from the perspective of the passenger, only minimal adaptation of the

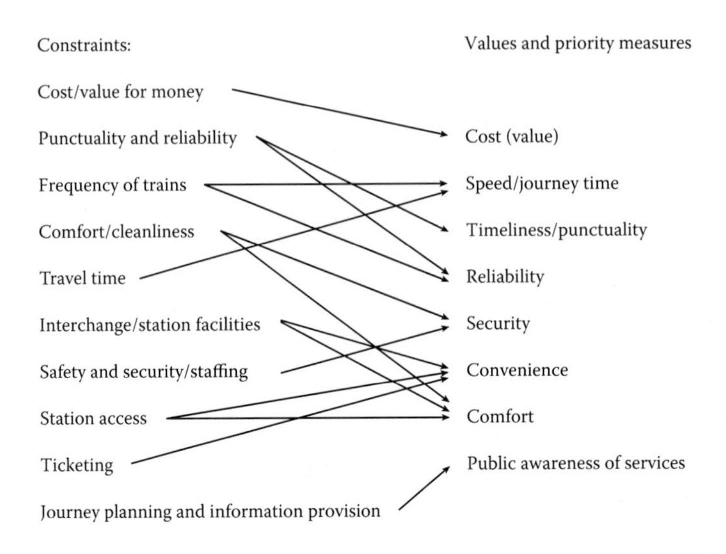

FIGURE 9.3 Translation from constraints into value and priority measures.

constraints was required to populate the values and priority measures. The translation from the constraints to rail travel into values and priority measures in the AH is illustrated in Figure 9.3.

Cost/value for money (Section 9.2.1) needed no change to become a value and priority measure. The punctuality/reliability of services constraint (Section 9.2.2) was split into two nodes, namely, timeliness/punctuality and reliability, to reflect the differences between these concepts. The reliability node also encompasses train frequency (Section 9.2.3), as reliability issues and train frequency are interrelated; for example, if a train runs every 2 min (as in the London Underground), reliability of one train will provide less of a constraint than if that train only comes once an hour because with a more frequent service the waiting time for the next train will be acceptably low. The speed/journey time node in the AH primarily reflects travel time (Section 9.2.5), though train frequency (Section 9.2.3) also has a part to play, as the frequency with which trains run will often affect the length of the journey through waiting times for journeys which have not been planned in advance ('turn up and wait' journeys). Comfort/cleanliness (Section 9.2.4) is mainly accounted for in the comfort value and priority measure, though it is also partly covered by security, as this encompasses crime deterrence, which includes the deterrence of vandalism and littering. The security node in the AH accounts entirely for the safety and security/staffing constraint (Section 9.2.7). Interchange/station facilities (Section 9.2.6) and station access (Section 9.2.8) are covered in the comfort and convenience nodes. Convenience also covers part of the issues surrounding the ticketing complexities constraint (Section 9.2.10), as incomprehensible ticketing systems will necessarily be an inconvenience. Finally, the constraint relating to journey planning and information provision (Section 9.2.9) is reflected in the public awareness of services node in the AH. The value and priority measure was edited as such to reflect the importance of passengers' *knowledge* of available information; it is not enough to have that information available, and the public must be aware of that information to make use of it. The node therefore takes into account issues such as advertising.

A review of rail documentation along with a number of consultations with experienced researchers in the rail domain provided the information required to complete the remainder of the AH. As the purpose of this chapter is to identify constraints to passengers moving to rail, the analysis boundaries (for one must decide what lies within the scope of the analysis) were set in terms of the system as experienced by the passenger. As such, system components such as signals, gantries, engines and power lines were not included as physical objects. The functional purpose of the system also reflects this passenger perspective; it is to provide *safe*, *efficient* and comfortable transport of passengers and their belongings.

The purpose of this analysis was to examine the constraints to rail travel from a human factors perspective. The AH was created using the constraints to rail travel, along with the functional purpose of rail travel, as a starting point to identify purpose-related functions, object-related processes and physical objects. Three of the purpose-related functions were identified as most strongly related to human factors: information availability, passenger protection and cater for task needs. These functions have potential to be affected by human factors interventions and were therefore explored in more detail, as shown in the excerpts from the AH in Figure 9.4

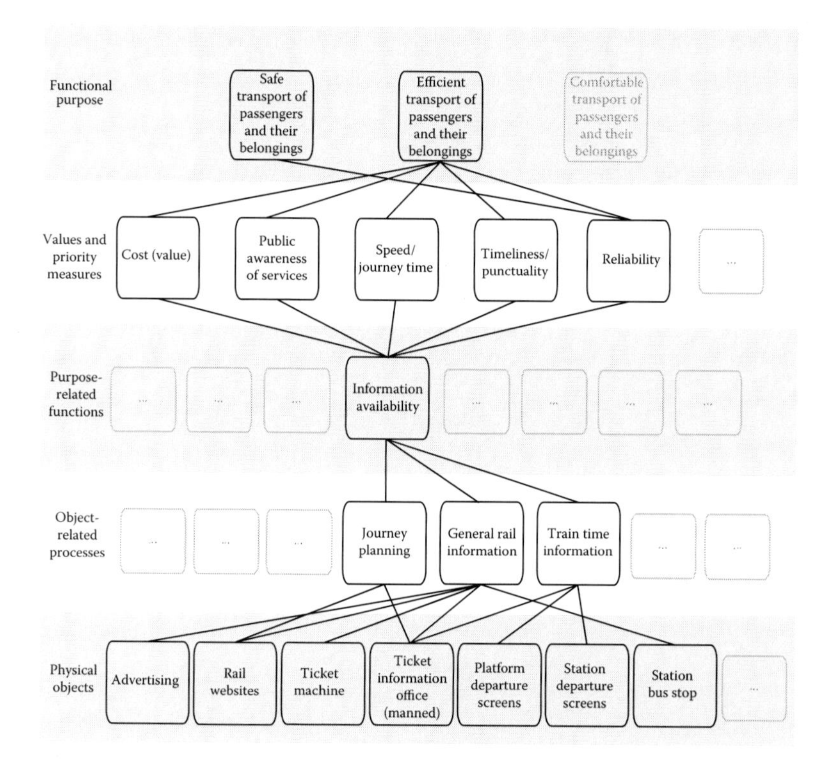

FIGURE 9.4 Excerpt from the AH showing links to and from the information availability function.

(information availability), Figure 9.5 (passenger protection) and Figure 9.6 (cater for task needs).

Taking Figure 9.4 as an example, the AH shows that information availability links to processes of journey planning, obtaining general rail information and obtaining train time information. This shows that these processes are required for the function of information availability. The processes are also linked to a number of physical objects in the lowest level of the hierarchy: advertising, rail websites, ticket machines, manned ticket information office, platform departure screens, station departure screens and information at the station bus stops. It is via these physical objects that information is made available to passengers. This analysis shows that there is potential for human factors interventions to be targeted at particular processes and objects in order to improve information availability. This in turn could contribute to a reduction of the constraints to which information availability is linked, namely, cost (value), public awareness of services, speed/journey time, timeliness/punctuality and reliability. When linked to the value and priority measures in the AH, the three human factors-related functions (information availability, passenger protection and cater for task needs) cover all of the constraints to rail travel

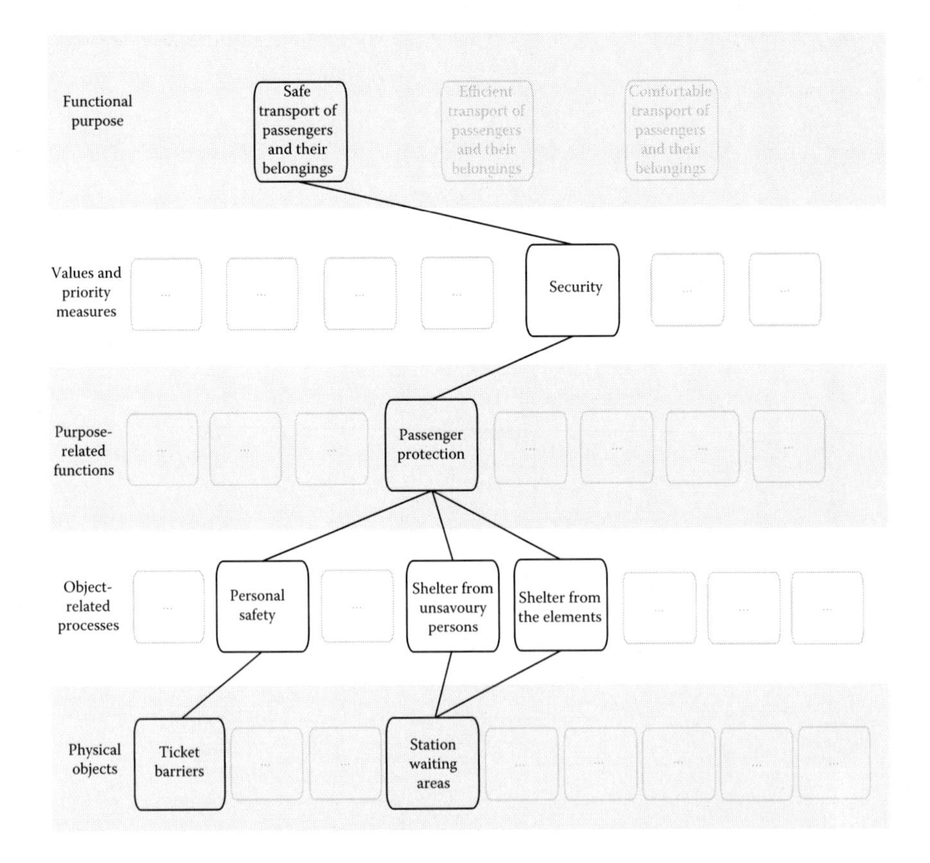

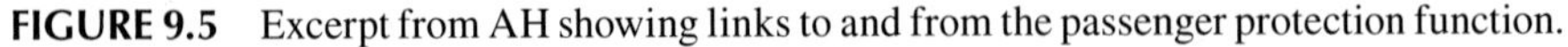

FIGURE 9.5 Excerpt from AH showing links to and from the passenger protection function.

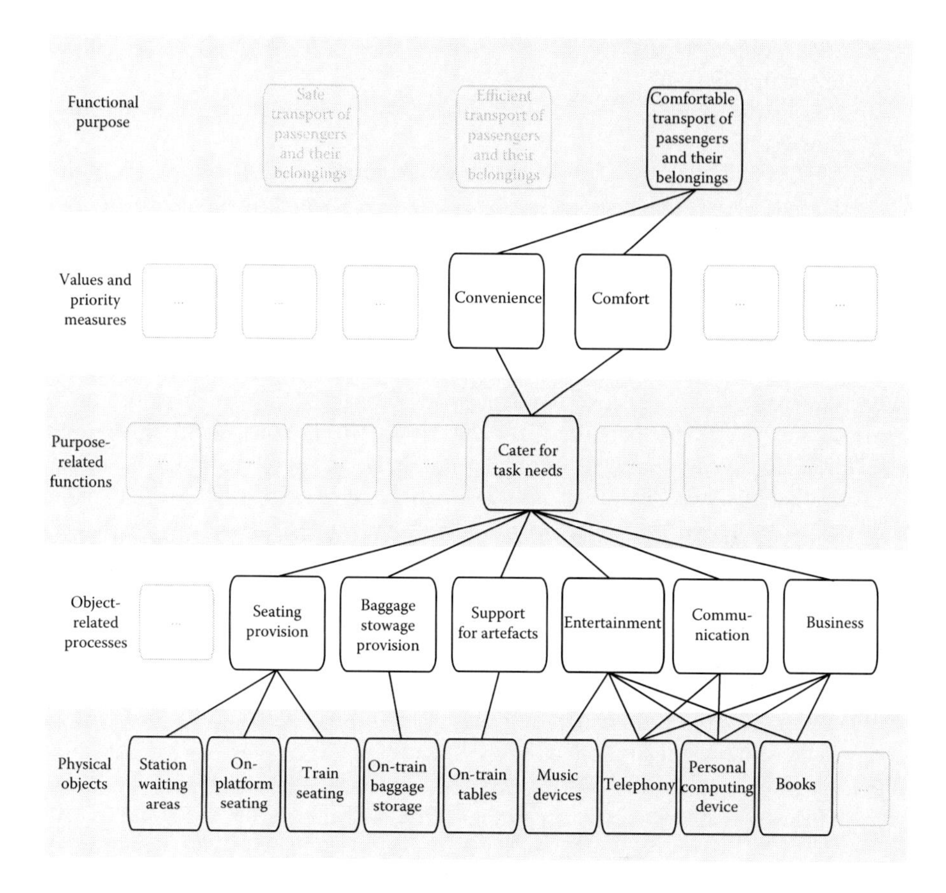

FIGURE 9.6 Excerpt from the AH showing links to and from the cater for task needs function.

identified in this study. This shows that human factors can have some impact on all constraints, and therefore on improving the safe, efficient and comfortable transport of passengers and their belongings.

The object-related processes level of the AH is explored further using control task analysis. In this study, the three human factors-related functions identified by the AH are linked to a number of object-related processes:

- Journey planning
- General rail information
- Train time information
- Personal safety
- Shelter from unsavoury persons
- Shelter from the elements
- Seating provision
- Baggage stowage provision

- Support for artefacts
- Entertainment
- Communication
- Business

In the following sections, these processes are investigated further in terms of the needs of different passenger types and journey types at various stages of the journey.

9.3.1.2 Control Task Analysis

As described above, this stage of the analysis involves the construction of CATs (Naikar et al., 2006). Six temporally and spatially separated situations, representing the various stages of a passenger's journey, were identified and are displayed on the horizontal axis of Figure 9.4. The vertical axis contains the functions defined in the AH. For this analysis, functions were taken from the object-related processes level.

The situations identified are origin/destination, en-route to transport, at station, on train, rail interchange and en-route to destination. Origin and destination are included under one heading as the destination for one journey will often be the origin for the next; their roles are interchangeable, thus functions should apply equally to both. En-route to transport and en-route to destination cover the journey to and from the train station (both access and egress). These have been separated as different variables will apply; for example, an individual driving to the station and parking will not have his or her car available to him or her for the station to final destination journey. Rail interchange is defined as the stations at which a passenger must get off one train and board another, without leaving said station. It is distinct from 'at station' in that it is neither the start nor the end of the journey; rather, it is a mid-journey stopover.

It can be seen from the diagram that 'origin/destination' supports all functions, other than 'check tickets', 'inter-platform movement', and 'inter-station transport'. The function of checking tickets refers to the act of having tickets checked either by a machine or by a member of staff, before, after or during the journey. This function can therefore happen only when in rail-owned situations such as stations and trains (as opposed to en-route or at the home/workplace). Moving between platforms can necessarily happen only at stations (be they the start or end point, or the interchange); moving between stations also has these constraints (for this analysis, moving between stations was taken as an act of travelling by train rather than by other mode).

In the situation 'en-route to transport', there are a number of functions that could occur but typically do not (dashed box only). In a number of instances, this reflects the possibilities brought about by advances in technology; for example, general rail information, journey planning, savings discounts and business (defined as any business other than telephone communication) would be possible through the use of the internet, a widely available resource on mobile telephones. This type of use does not, however, represent the norm; these activities are most often carried out on a computer when at home (or work) or, with the exception of business, arranged at the station.

While the majority of activities can take place at the station, 'baggage stowage provision' cannot; this function refers to the provision of specific areas in which bags may be stored temporarily. Since the vast majority of U.K. train stations no longer have lockers, this is no longer possible. 'Support for artefacts' and 'business' can happen at stations, but typically do not. Generally, stations do not provide tables on which to place artefacts (e.g. laptops, books, writing pads); this in turn hinders the performance of 'business' (again, it is important to note that the 'business' function excludes that carried out by telephone). Importantly, however, 'business' can be, and typically is, conducted while onboard the train (Figure 9.7).

9.3.1.3 Social Organisation and Cooperation Analysis

For this phase of the analysis, two CATs were shaded with respect to different actor groups. In one of the diagrams, differential shading is used to distinguish between different types of passenger, with the other representing different journey types. The choice of which journey types to use was based on the *Passenger Demand Forecasting Handbook* (ATOC, 2009) and were defined as commuters – work, commuters – education, business, leisure – shopping, leisure – visit friends/family, and leisure – holidays. The coded CAT is presented in Figure 9.8.

The choice of passenger type was based on discussions with rail industry experts and the types were defined as old age pensioner (OAP), family, youth/student, disabled and able-bodied adult. The separation of passenger groups intended to identify groups of individuals with distinct needs and requirements, though it is noted that some overlap is likely to be present (e.g. OAPs and disabled passengers are likely to have some similar requirements in terms of mobility). The coded CAT is presented in Figure 9.9.

As Figures 9.8 and 9.9 show, there are some hard constraints (i.e. where the function is not currently possible in a situation, as indicated by the empty cells) and some soft constraints (i.e. where the function is possible but not currently supported, as indicated by the dotted lines without the ball-and-whiskers). For example, advanced purchase of tickets while on the train (the cell function 'advance purchase' by the situation 'on train' in Figures 9.8 and 9.9) shows that it is currently not possible to purchase advanced travel tickets while on the train, even though this would offer greater convenience for the passenger and revenue for the train operating company.

By interrogating each of the hard and soft constraints in turn, each can be addressed as a potential constraint to modal shift in a systematic manner. Addressing each of the potential constraints in terms of journey type and passenger type enables a more comprehensive analysis than addressing either alone. An example is provided in Table 9.1, which shows how each of the function × situational constraints can be examined in turn (consideration of journey type and passenger type reveals which group the strategy would most likely help). As table five shows, on the face of it, five strategies remove most of the potential constraints. Removal of these constraints might indeed promote modal shift in the short term, provided that they were physically, socially, technically and economically viable.

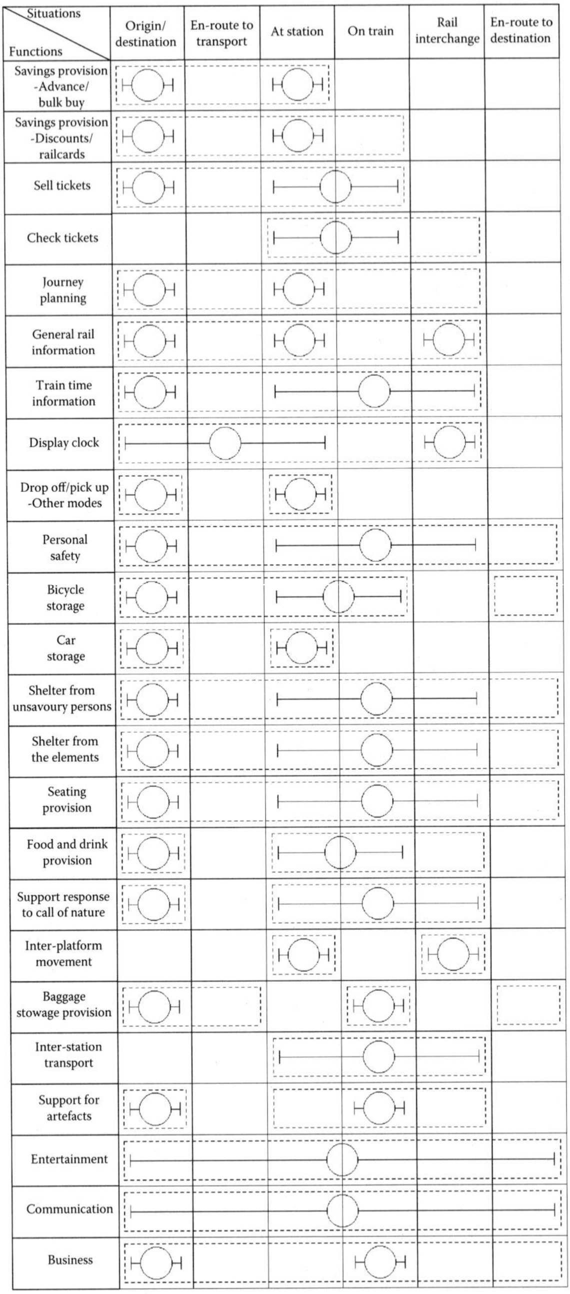

Situations / Functions	Origin/ destination	En-route to transport	At station	On train	Rail interchange	En-route to destination
Savings provision -Advance/ bulk buy						
Savings provision -Discounts/ railcards						
Sell tickets						
Check tickets						
Journey planning						
General rail information						
Train time information						
Display clock						
Drop off/pick up -Other modes						
Personal safety						
Bicycle storage						
Car storage						
Shelter from unsavoury persons						
Shelter from the elements						
Seating provision						
Food and drink provision						
Support response to call of nature						
Inter-platform movement						
Baggage stowage provision						
Inter-station transport						
Support for artefacts						
Entertainment						
Communication						
Business						

FIGURE 9.7 CAT for the rail system.

FIGURE 9.8 CAT coded by journey type.

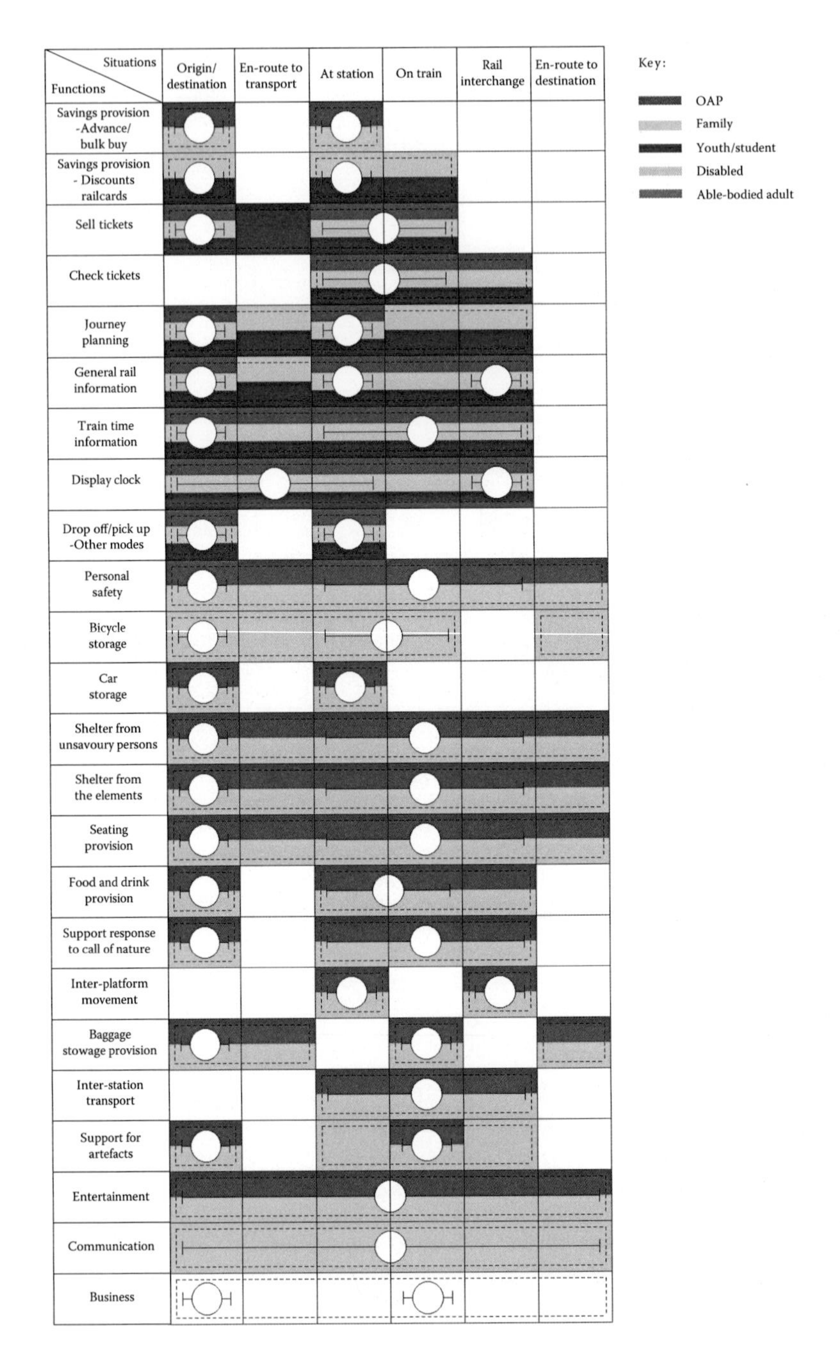

FIGURE 9.9 CAT coded by passenger type.

TABLE 9.1
Analysis of Hard and Soft Constraints from SOCA-CAT

Functions	Situational Constraints	Type	Constraint Removal Strategy
Journey planning	En-route	Soft	Dynamic adaptable journey planner that enables traveller to make and change travel plans from smartphone
	On train	Soft	Dynamic adaptable journey planner that enables traveller to make and change travel plans from onboard conductor
General rail information	En-route	Soft	Up-to-the-minute information on train arrival, departures, delays, cancellations and replacement services from smartphone
	On train	Soft	Up-to-the-minute information on train arrival, departures, delays, cancellations and replacement services from in-carriage information screen
Train time information	En-route	Soft	Up-to-the-minute information on train arrival, departures, delays, cancellations and replacement services from in-carriage information screen
Personal safety	En-route	Soft	Door-to-door journey service with connections via local minicab, car-share and on-demand bus companies
Shelter from unsavoury persons	En-route	Soft	Door-to-door journey service with connections via local minicab, car-share and on-demand bus companies
Shelter from elements	En-route	Soft	Door-to-door journey service with connections via local minicab, car-share and on-demand bus companies
Seating provision	En-route	Soft	Door-to-door journey service with connections via local minicab, car-share and on-demand bus companies
Baggage stowage provision	En-route	Soft	Door-to-door journey service with connections via local minicab, car-share and on-demand bus companies
	At station	Hard	Provision of security-screened baggage storage
	At interchange	Hard	Provision of security-screened baggage storage
Support for artefacts	En-route	Hard	Door-to-door journey service with connections via local minicab, car-share and on-demand bus companies
	At station	Soft	Airport-style passenger facilities with information screens, seating, electrical sockets, tables and internet facilities
	At interchange	Soft	Airport-style passenger facilities with information screens, seating, electrical sockets, tables and internet facilities
Entertainment	N/A	N/A	N/A
Communication	N/A	N/A	N/A
Business	En-route	Soft	Door-to-door journey service with connections via local minicab, car-share and on-demand bus companies
	At interchange	Soft	Airport-style passenger facilities with information screens, seating, electrical sockets, tables and internet facilities

9.4 CONCLUSION

Comparing the constraints raised in the literature review to those raised by CWA reveals some interesting findings. Of the 10 constraints raised in the literature, 7 were identified at the 'value and priority measures' level of the AH in CWA. The remaining 3 constraints were identified at the 'purpose-related functions' or 'object-related processes' level of the AH. This may have implications for the way in which the analysis is conducted or the way in which constraints are represented in CWA. Each of the levels within the AH offers an analysis of constraints (or boundaries) that may affect modal shift from the private motor vehicle to rail transport. While the literature review is a useful list of topics and issues, CWA offers a much more structured analysis of the problems. The AH shows explicitly how the issues are interrelated. For example, the issues of 'cost and value' are linked to 'passenger enticement', 'ticketing' and 'information availability', which in turn are linked to 'discounts', 'advance purchase', selling ticket', check tickets', 'journey planning' and 'general rail information' in the AH. Change in any one of these functions is likely to have an effect on the others, in a systemic manner. Potentially, the analyses in CWA could examine this effect, although that has not been undertaken here. Consideration of the functions against situation together with the journey types and passenger types allows a systems analysis of the constraints. For example, grouping functions like safety, shelter, seating, entertainment, communication and business together can encourage consideration of new travel concepts, such as end-to-end journey provision, where a single ticket covers the provision of all of these functions from origin to destination. What CWA offers over traditional approaches is the systematic approach to representation and analysis. So while CWA has not necessarily revealed any new constraints, it has shown explicit relationships between those constraints. Taking a systems view, it is important to understand how addressing one constraint will affect another. One would not wish to impose new constraints or constraints in addressing the question of modal shift from the private motor vehicle to rail transport.

In conclusion, CWA seems to be a semi-structured approach that can help analysts focus on relevant parts of the system and also provide focus for the search for more information. The very large numbers of functions and constraints that have been identified and the likely challenges in collecting information on these (and also dealing with these) provides a very considerable challenge for modal shift. Future work could use this understanding of the constraints to generate visions or interpretations of rail travel with constraints removed (i.e. brought to life), with the intention of evaluating how travellers could be encouraged towards rail travel.

ACKNOWLEDGEMENT

The funding for this research was made available through the RRUK Factor 20 Feasibility Account – EPSRC ref. EP/H024743/1.

REFERENCES

Accent. 2009. *Perception towards Integrated Transport: Literature Review.* London: Passenger Focus.

Association of Train Operating Companies (ATOC). 2009. *Passenger Demand Forecasting Handbook Version 5.* London: ATOC and Passenger Focus.

Association of Train Operating Companies (ATOC) and Passenger Focus. 2010. *Integrated Transport – Perception and Reality.* London: ATOC and Passenger Focus.

Bates, J., Polak, J., Jones, P. and Cook, A. 2001. The valuation of reliability for personal travel. *Transportation Research Part E: Logistics and Transportation Review*, 37, 191–229.

Bisantz, A. M., Roth, E., Brickman, B., Gosbee, L. L., Hettinger, L. and McKinney, J. 2003. Integrating cognitive analyses in a large-scale system design process. *International Journal of Human-Computer Studies*, 58, 177–206.

Blainey, S., Hickford, A. and Preston, J. M. 2009. *Barriers to Modal Shift: Literature Review.* Unpublished report to the Association of Train Operating Companies. Transportation Research Group, University of Southampton.

Cox, T., Houdmont, J. and Griffiths, A. 2006. Rail passenger crowding, stress, health and safety in Britain. *Transportation Research Part A*, 40, 244–258.

Cozens, P., Neale, R., Whitaker, J. and Hillier, D. 2003. Managing crime and the fear of crime at railway stations – A case study in South Wales (UK). *International Journal of Transport Management*, 1, 121–132.

Department for Transport. 2006. *A Review of the Effectiveness of Personalised Journey Planning Techniques.* http://www.dft.gov.uk/pgr/sustainable/travelplans/ptp/personalisedtravelplanningev5774. Accessed 12 October 2010.

Department for Transport. 2009. *Travel Trends* (2009 Edition). London: Stationery Office.

Derek Halden Consultancy. 2003. *Barriers to Modal Shift.* Edinburgh: Scottish Executive Social Research.

Eriksson, L., Friman, M. and Gärling, T. 2008. Stated reasons for reducing work-commute by car. *Transportation Research Part F: Traffic Psychology and Behaviour*, 11, 427–433.

Hine, J. and Scott, J. 2000. Seamless, accessible travel: Users' views of the public transport journey and interchange. *Transport Policy*, 7, 217–226.

Hoenniger, P. 2003. MobiHarz project: Integrated mobility management and services for visitors. *ECOMM Managing Transport Demand to Attain Sustainable Transport Demand and Economic Effectiveness – Why and How?* Karlstad, 21–23 May.

Howarth, H. V. C., Griffin, M. J., Childs, C., Fujiyama, T. and Hodder, S. G. 2011. *Comfortable Sardines: The Balance between Comfort and Capacity.* Southampton, UK: University of Southampton, Institute of Sound and Vibration (ISVR).

Intergovernmental Panel on Climate Change. 2007. *The IPCC Fourth Assessment Report on Climate Change.* Synthesis Report. IPCC: Geneva.

Jansson, A., Olsson, E. and Erlandsson, M. 2006. Bridging the gap between analysis and design: Improving existing driver interfaces with tools from the framework of cognitive work analysis. *Cognition, Technology and Work*, 8, 41–49.

Jenkins, D. P., Salmon, P. M., Stanton, N. A. and Walker, G. H. 2010a. A new approach for designing cognitive artefacts to support disaster management. *Ergonomics*, 53, 617–635.

Jenkins, D. P., Salmon, P. M., Stanton, N. A. and Walker, G. H. 2010b. A systematic approach to accident analysis: A case study of the Stockwell shooting. *Ergonomics*, 53, 1–17.

Jenkins, D. P., Stanton, N. A., Salmon, P. M. and Walker, G. H. 2009. *Cognitive Work Analysis: Coping with Complexity.* Farnham, UK: Ashgate Publishing Limited.

Kenyon, S. and Lyons, G. 2003. The value of integrated multimodal traveller information and its potential contribution to modal change. *Transportation Research Part F: Traffic Psychology and Behaviour*, 6, 1–21.

Kogi, K. 1979. Passenger requirements and ergonomics in public transport. *Ergonomics*, 22, 631–639.

Lyons, G., Jain, J. and Holley, D. 2007. The use of travel time by rail passengers in Great Britain. *Transportation Research Part A*, 41, 107–120.

Mackett, R. and Sutcliffe, E. B. 2003. New urban rail systems: A policy-based technique to make them more successful. *Journal of Transport Geography*, 11, 151–164.

McIlroy, R. C. and Stanton, N. A. 2012. Specifying the requirements for requirements specification: The case for work domain and worker competencies analyses. *Theoretical Issues in Ergonomics Science*, 13, 450–471.

MVA Consultancy. 2010. *Passengers' Priorities for Improvements in Rail Services*. London: Passenger Focus.

Naikar, N., Moylan, A. and Pearce, B. 2006. Analysing activity in complex systems with cognitive work analysis: Concepts, guidelines, and case study for control task analysis. *Theoretical Issues in Ergonomics Science*, 7, 371–394.

Naikar, N. and Sanderson, P. M. 2001. Evaluating design proposals for complex systems with work domain analysis. *Human Factors*, 43, 529–542.

Napper, R., Coxon, S. and Allen, J. 2007. Bridging the divide: Design's role in improving multi-modal transport. *30th Australasian Transport Research Forum*. Melbourne, Australia, Australian Department of Infrastructure and Transport.

Passenger Focus. 2006. *What Passengers Want from the New West Midlands Franchise: An Executive Summary*. London: Passenger Focus.

Passenger Focus. 2008. *What Passengers Want from South Central*. London: Passenger Focus.

Passenger Focus. 2009. *Fares and Ticketing Study*. London: Passenger Focus.

Pledger, S., Horbury, C. and Bourne, A. 2005. Human factors within LUL – History, progress and future. In Wilson, J. R., Norris, B. J., Clarke, T. and Mills, A. (Eds.). *Rail Human Factors: Supporting the Integrated Railway*. Aldershot, UK: Ashgate, pp. 497–507.

Rasmussen, J. 1974. *The Human Data Processor as a System Component: Bits and Pieces of a Model*. Report no. Risø-M-1722. Roskilde, Denmark: Danish Atomic Energy Commission.

Rasmussen, J. 1985. The role of hierarchical knowledge representation in decision making and system management. *IEEE Transactions on Systems, Man and Cybernetics*, 134, 257–266.

Rasmussen, J., Pejtersen, A. and Goodstein, L. P. 1994. *Cognitive Systems Engineering*. New York: Wiley.

Rasmussen, J., Pejtersen, A. and Schmidt, K. 1990. *Taxonomy for Cognitive Work Analysis*. Roskilde, Denmark: Risø National Laboratory.

Shepherd, A. and Marshall, E. 2005. Timeliness and task specification in designing for human factors in railway operations. *Applied Ergonomics*, 36, 719–727.

SUSTRANS. 2002. The Challenge of Changing Travel Behaviour. Information sheet FF36.

Thomas, L. J., Rhind, D. J. A. and Robinson, K. J. 2005. Rail passenger perceptions of risk and safety and priorities for improvement, In Wilson, J. R., Norris, B. J., Clarke, T. and Mills, A. (Eds.). *Rail Human Factors: Supporting the Integrated Railway*. Aldershot, UK: Ashgate, pp. 473–482.

University of Oxford Transport Studies Unit (UOTSU). 1995. Car Dependence. RAC Foundation for motoring and the Environment.

Vicente, K. J. 1999. *Cognitive Work Analysis: Toward Safe, Productive, and Healthy Computer-Based Work*. Mahwah, NJ: Lawrence Erlbaum Associates.

Wardman, M. and Hine, J. 2000. *Costs of Interchange: A Review of the Literature*. Working Paper 546, Institute of Transport Studies, University of Leeds.

Wardman, M., Hine, J. and Stradling, S. 2001. *Interchange and Travel Choice* (Vol. 1). Edinburgh: Scottish Executive Central Research Unit.

Wilson, J. R., Cordiner, L., Nichols, S., Norton, L., Bristol, N., Clarke, T. and Roberts, S. 2001. On the right track: Systematic implementation of ergonomics in railway network control. *Cognition, Technology and Work*, 3, 238–252.

Wilson, J. R., Farrington-Darby, T., Cox, G., Bye, R. and Hockey, G. R. J. 2007. The railway as a socio-technical system: Human factors at the heart of successful rail engineering. *Proceedings of the Institution of Mechanical Engineers, Part F: Journal of Rail and Rapid Transit*, 221, 101–115.

Wilson, J. R. and Norris, B. J. 2005a. Editorial: Special issue on rail human factors. *Applied Ergonomics*, 36, 647–648.

Wilson, J. R. and Norris, B. J. 2005b. Rail human factors: Past, present and future. *Applied Ergonomics*, 36, 649–660.

10 Conducting Cognitive Work Analysis with an Experiment in Mind

Sean W. Kortschot, Cole Wheeler, Aimzhan Zhunussova and Greg A. Jamieson

CONTENTS

10.1 INTRODUCTION

Nuclear power accounts for an estimated 21% of Organisation for Economic Cooperation and Development (OECD) countries' total energy production (World Nuclear Association, 2015). Although nuclear represents a viable option for large-scale carbon-free power production, the potential consequences of disaster are severe, with the radiation from accidents such as Fukushima and Chernobyl still being felt today (Fushiki, 2013). Although nuclear accidents can occur for a variety of reasons, post-accident reports from Chernobyl, the Three-Mile Island accident and Fukushima found that human error either directly contributed to the accident (Meshkati, 1998; Lau et al., 2012) or significantly exacerbated the accident's impact (Suzuki, 2014). Consequently, designers have placed a much greater emphasis on understanding and improving the ways in which the human operators interact with and control nuclear power plants. One of the signature features of this improvement in recent years has been the inclusion of large screen displays (LSDs) in the control room (Myers and Jamieson, 2013).

LSDs are typically wall-mounted displays that enable multiple control room operators to share a common point of reference (Myers and Jamieson, 2013). LSDs are postulated to improve both operator situation awareness (SA), and collaboration and communication within operating teams (Roth et al., 1998). Although these claims seem intuitively logical, an operating experience review by Myers and Jamieson (2013) revealed (1) that there is little empirical evidence substantiating their inclusion in nuclear control rooms and (2) that there has been minimal research directed at evaluating what information content and structure best evoke these postulated benefits. Our work attempts to amend these gaps in the literature by conducting an experiment that will (1) investigate how effective LSDs perform in relation to other display types and (2) investigate different methods for displaying plant parameters on those LSDs.

10.1.1 The Power Plant

We conducted our analysis on a CANDU 6® nuclear reactor. This reactor uses heavy water (D_2O) as the coolant, which allows for the use of natural uranium fuel. The plant operates similarly to any other power plant: heat is generated in the reactor and transferred to coolant, the coolant transfers the heat to the secondary side by converting liquid H_2O into steam, and the steam is then used to drive a turbine to generate electricity. Although the fundamental processes driving nuclear power have not drastically changed in recent years, the control room has and with it, so to have the displays.

10.1.2 Nuclear Control Room Displays

There have been two main transitions in the design of nuclear control room displays. The first transition saw the original panel-based displays, which consisted of analogue meters and dials, being digitised into *mimic* displays, which could be displayed on monitors (Hurlen et al., 2015). Mimic displays map information on the display according to the plant configuration. Although this transition allowed for some reconfiguration of the controls and for task-related information to be displayed, the actual depiction of plant parameters was still rooted in technology that was developed around the physical constraints of the analogue controls. This limitation was partially addressed by the second transition, which saw new advanced displays implementing trend information, configural graphics, computerised procedures and alarm lists (Owre et al., 2002; Burns et al., 2008; Figure 10.1). This additional information afforded operators a better understanding of recent plant behaviour and how the behaviour of one system can affect the behaviour of another.

Although advanced displays address some of the prominent limitations of previous display technology, they are not perfect. The remaining problems revolve around certain tasks being too arduous and mentally taxing and are best explained through Rasmussen's (1983) skills–rules–knowledge (SRK) taxonomy. The SRK taxonomy categorises different types of control room behaviour that an operator engages in across three levels: skill-based behaviour (SBB), rule-based behaviour (RBB) and knowledge-based behaviour (KBB). SBB, the lowest level, is more automatic than the other behaviours and therefore demands the least cognitive resources. RBB is

FIGURE 10.1 Advanced control room from Qinshan power plant. The LSDs can be seen in front of the operators. (Adapted from L-3 MAPPS Inc. 2011. Galerie de photos de centrales [online]. http://www.mapps.l-3com.com/photo_gallery_power_fr.html. Accessed 1 September 2015.)

behaviour conducted in accordance with a predefined set of rules. The highest level, KBB, is much more conscious and intentional and thus demands the most cognitive resources. Each class of behaviour has advantages and disadvantages and it is therefore the task of the control room interface to exploit these by appropriately allocating different tasks to different behaviours. For example, a menial task can be delegated to SBB since it likely will occur more frequently and have less potential for consequences (McIlroy and Stanton, 2015). A complex task, on the other hand, should be delegated to RBB or KBB since that is likely what is required to adequately perform the task. The problem with current nuclear displays is that they place too heavy a demand on KBB, thereby taking up excessive cognitive resources for tasks that may not necessarily require them. This can increase cognitive workload, error rates and omissions (McIlroy and Stanton, 2015).

In an effort to better account for this, Rasmussen (1983) developed ecological interface design (EID), a theoretical approach to interface design that was developed specifically for complex work domains. Ecological interfaces are designed to reflect the constraints of a work environment in a way that is perceptually accessible and psychologically relevant to the operators who use it by displaying functionally related information together in configural graphics (Burns and Hajdukiewicz, 2004). Past studies have supported EID theory and shown that in certain situations ecological displays outperform those used in practice (e.g. Burns et al., 2008). The architecture of the ecological display is typically determined through cognitive work analysis (CWA), which identifies a set of information requirements from both ecological and cognitive perspectives on the system.

This chapter describes our CWA and the novel ecological interface that we designed using its results. We describe the methodology that we used for evaluating how our ecological interface compares to the existing advanced displays when they are placed on an LSD. Furthermore, this case study provides insight into the challenges of conducting CWA in a large, complex domain, as part of a larger project whose end goal is to compare operator performance across the displays in that domain. It presents creative approaches that overcome these challenges by allowing the analysis to remain flexible yet focused, and includes examples of less commonly used CWA methodologies and how they led to a novel ecological interface.

10.2 ANALYSIS

The objective of our study is to evaluate the efficacy of LSDs in nuclear power plant control rooms. Our analytic efforts were directed by this goal and were therefore a means to realising the experiment designed to achieve it. Therefore, before starting our CWA, we developed the foundation of our experimental design and determined key features such as the characteristic of our experimental tasks and participants. Taking this step before beginning our CWA allowed us to ensure that our analysis would be sufficient for meeting the objectives without being superfluous.

10.2.1 Experimental Design

The design of an experiment within the context of a multi-year project can often seem like an afterthought since it is frequently done after the analysis and design. However, this step is integral in eliciting the desired performance metrics from an evaluation. The experiment brings together all of the work that has preceded it and determines the quality and ease of the work that comes after it.

The characteristics of the experimental participants are key determinants for much of the analytical work in a display evaluation study and should therefore be determined early. Our participants will be chemical production and power engineering students rather than expert operators. They will have experience and training with generic process control systems but likely not with many of the systems peculiar to the nuclear power domain. For example, they will have some familiarity with steam generators since they are common in power facilities, whereas a component specific to nuclear such as the reactor will likely be foreign to them. These characteristics are significantly different from other possible operator groups, and therefore establishing them early in the process allowed us to tailor much of the subsequent work to their unique demands.

10.2.2 Scenario Selection

The test scenarios that we will use during experimentation will all involve variations of steady-state monitoring wherein we will trigger a fault in a specific parameter. In order to ensure that the scenarios will allow us to address our experimental questions, we determined that they need to meet two initial requirements. First, each scenario must be conducive to the full breadth of operator behaviours that will allow us address the experimental questions regarding SA and communication. Second, they need to accommodate the experimental constraints, such as the demands of the participants and the restrictions of the simulator. After these high-level requirements refined the list of candidate scenarios, we used three further criteria to ensure that the scenarios will elicit the necessary metrics for achieving our experimental goal.

The first inclusion criterion is that the scenario must fall within normal operating conditions. There are three main reasons that this is important for our study. First, because they represent the majority of operations in the plant, measuring display performance in these conditions is more representative than studying emergency conditions. Second, by studying parameter deviations within normal operating procedures, we are studying the precursory conditions to emergency scenarios, which are critical to preventing emergency scenarios from developing (Mumaw et al., 2000). Finally, asking novice operators to monitor and control a plant in emergency situations would likely exceed their process knowledge and possibly exceed their ability to cope with stressful situations. Performance under these conditions may not reflect the effectiveness of the interface. We therefore aimed to find scenarios that were simple enough for novice operators to understand, but challenging enough to present a need to exercise control on multiple levels.

The second inclusion criterion is that the scenarios need to test the claimed benefits of LSDs: that they enhance SA and that they improve communication and collaboration (Roth et al., 1998). Although the definition of SA can differ depending on the domain in question (Lau et al., 2013), it is generally understood to be as knowledge of overall system behaviour. In an effort to determine if a scenario will call into question SA according to this general definition, we identified scenarios that required a holistic understanding of the relationship between different systems. One way to ensure that the scenarios require this holistic understanding is to have any parameter deviation manifest itself in an area of the plant with several degrees of separation from the actual problem. For example, a leak in the heat transport system (HTS) will reveal itself in the D_2O storage tank, which is several tanks removed from the actual source of the problem. In order to diagnose this problem, an operator would need a sufficient understanding of the overall behaviour of the plant, which requires an understanding of the relationships between subsystems over time, which therefore demands high-level SA. The second claim, which pertains to communication and collaboration, can be more difficult to ensure in an experimental scenario since these behaviours are dependent on the individuals in a team and the dynamic of that team. However, it has been shown that in high task-load and emergency scenarios, operators communicate less with one another (Juhasz and Soos, 2007). Therefore, our decision to avoid emergency scenarios is consistent with the goal of finding scenarios that facilitate communication.

The final criterion was that no control action will be required. This is the most controversial of all of our criteria and is largely based on subject matter expert recommendation. The rationale for this is threefold: (1) operators of a highly automated process take few control actions under steady-state conditions; (2) the addition of control actions will further complicate the scenarios, thereby complicating the analysis, design and training and (3) the majority of control actions are done at the panel. Since the focus of this project is on evaluating LSDs, having participants interact with the control panel would detract from the true purpose of the study. Omitting control actions from the experiment will allow participants to focus on a smaller range of tasks that are more akin to what nuclear power plant operators are doing and form a higher level of proficiency with the system.

In addition to a fault introduction in steady-state monitoring, we will be asking the participants to perform testing on one of the shutdown systems in the plant. This is included mainly to serve as a distractor from the monitoring task.

10.2.3 Breadth and Depth of Analysis

The analysis of any work domain as large and complex as a nuclear power plant requires careful understanding of the necessary scope of analysis or it can quickly become unmanageable. By using the characteristics of our candidate scenarios as guidelines, we were able to constrain the scope of our analysis in two key ways. First, we included only systems that were relevant to the main task of monitoring the plant and the distractor task of testing the shutdown system. Because of the interconnected nature of the power plant, it was difficult to omit any systems from analysis. However, we were able to use the literature (e.g. Davey, 2000) and subject matter experts to determine which specific systems and subsystems operators tend to dedicate greatest effort to, and focus our analysis on these.

Second, we further constrained the scope by considering which systems the participants in the experiment will be most familiar with. As mentioned before, we will be using novice participants who will be unfamiliar with systems specific to nuclear. Since many of these systems are integral to the operations of the plant, we could not simply remove them from our representation of the system. However, we were able to simplify their inner workings while retaining the same inputs and outputs. This prevented any additional systems from being affected while reducing them to a comprehensible level of complexity. For example, although understanding the processes inside a nuclear reactor (e.g. reactivity flux) requires deep domain knowledge, the idea of fuel being inserted, experiencing a reaction, and that reaction producing heat, is fairly straightforward. Although this limits the representativeness of isolated systems, it increases the validity of the overall experiment since it will allow our novice operators to more closely achieve the level of expertise that is required of nuclear power plant operators.

Two non-experimental factors were also relevant to our consideration of our analytic scope. First, since many members of our team were new to the nuclear domain, we needed to use our analytic work to gain a sound understanding of the systems. We therefore decided to analyse the chosen systems at a depth sufficient for our own understanding, which was often deeper than that would be necessary if creating a set of information requirements was the only goal. Second, we analysed the main systems in the plant regardless of their direct relevance to the experiment due to the interdependencies of individual systems. In other words, in order to understand the inputs of one system, we needed to understand the outputs of the system feeding into it.

10.2.4 Work Domain Analysis

We approached our work domain analysis (WDA) with the understanding that it would need to adapt to our evolving understanding of our experimental needs. We created an abstraction hierarchy (AH) for each of the systems that fit within our scope and arranged the majority of these hierarchies at three levels of aggregation (see Figure 10.2). An overall plant hierarchy formed the highest level and provided basic context for each system that was analysed. The second level consisted of the main systems in the plant and allowed us to capture sufficient breadth for our analysis. The third level consisted of relevant subsystems within these systems and allowed us to capture sufficient depth where necessary. By using these nested hierarchies, we were able to capture both the necessary breadth of our work domain and the necessary depth.

While most of the analyses fit well into the traditional AH framework, our analysis of the shutdown system required a more unusual approach. This system was included because of the specific testing task that operators will be required to perform during experimentation. Since we included this system in our analysis because of this specific testing task, we needed to analyse the system from the perspective of that task rather than from the actual functional purpose of the system. We therefore adopted an *object worlds* perspective for this analysis. Object worlds exist in work domains where individual systems are used for multiple purposes. In the case of shutdown system testing, there are three unique purposes, each representing a distinct object world: activation, testing, and maintenance (see Figure 10.3). Each of these purposes represents an individual work domain that is defined with respect to the set of stakeholders interested in that particular purpose (Torenvliet et al., 2008). In our case, the stakeholders remain the same but their relationship with the system changes since they are in the act of testing the shutdown system rather than using it and because

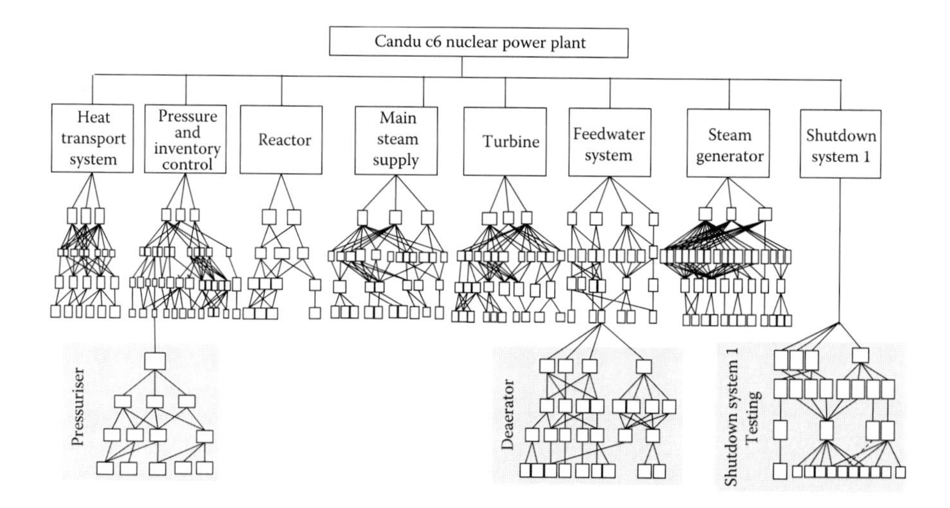

FIGURE 10.2 AH skeleton detailing the different levels of depth that we went into for the systems relevant to our experimental scenarios.

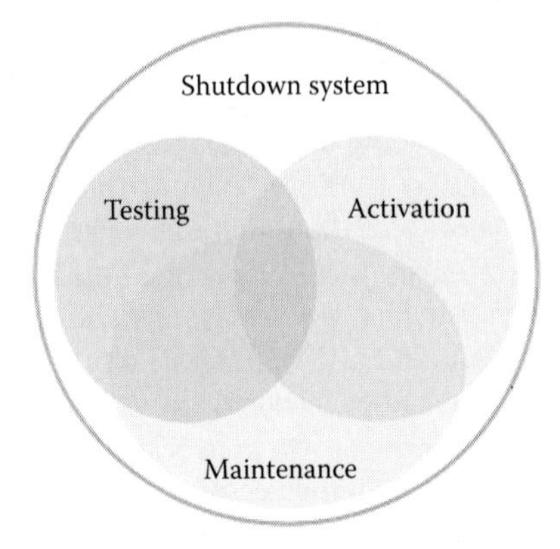

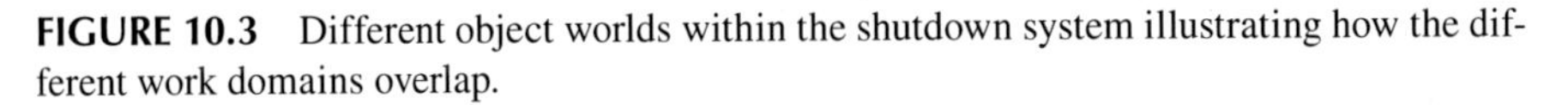

FIGURE 10.3 Different object worlds within the shutdown system illustrating how the different work domains overlap.

they will be using only the shutdown system for this specific task, they are focused only on the components necessary for that task. For example, while the shutdown rods and their release are essential during the activation of the shutdown system, they are not part of testing and were therefore omitted from the AH in favour of the components that are necessary for testing such as circuitry and indicator lights.

The hierarchy for shutdown system testing is also unique in that it employs a different nomenclature for the four hierarchy levels (Torenvliet et al., 2008). This system is based more on procedures and regulations than on natural laws. The terminologies used to describe the different levels of abstraction were therefore altered to reflect this. This is most pronounced at the second level, which adopts the *values and priorities* label instead of *abstract functions*. Instead of describing the physical laws that drive a system, we describe the regulatory nature of the system here (see Figure 10.4).

10.2.5 Control Task Analysis and Strategies Analysis

Once the scenarios had been identified and the relevant systems had been analysed, it was important to consider how the operators will navigate through our experimental tasks. Identifying the underlying structure of different scenarios and tasks can facilitate the development of interfaces capable of supporting these tasks (Vicente and Rasmussen, 1992). In highly procedural tasks, these internal activities that are traditionally depicted in the second phase of CWA, control task analysis (ConTA), are typically prescribed in the form of operating manuals. This limits the variation in control tasks, and therefore greatly limits the useful insights that can come from ConTA. Since tasks in the nuclear domain are so highly regulated, we did not rely as heavily on ConTA as we did on strategies analysis (StrA).

Although StrA typically depicts strategies used to perform a control action (Kilgore et al., 2008), we instead used it to depict diagnostic strategies since no control actions will be required in our experiment. Therefore, our information flow maps (IFMs) illustrate the different paths an operator can take to identify the root

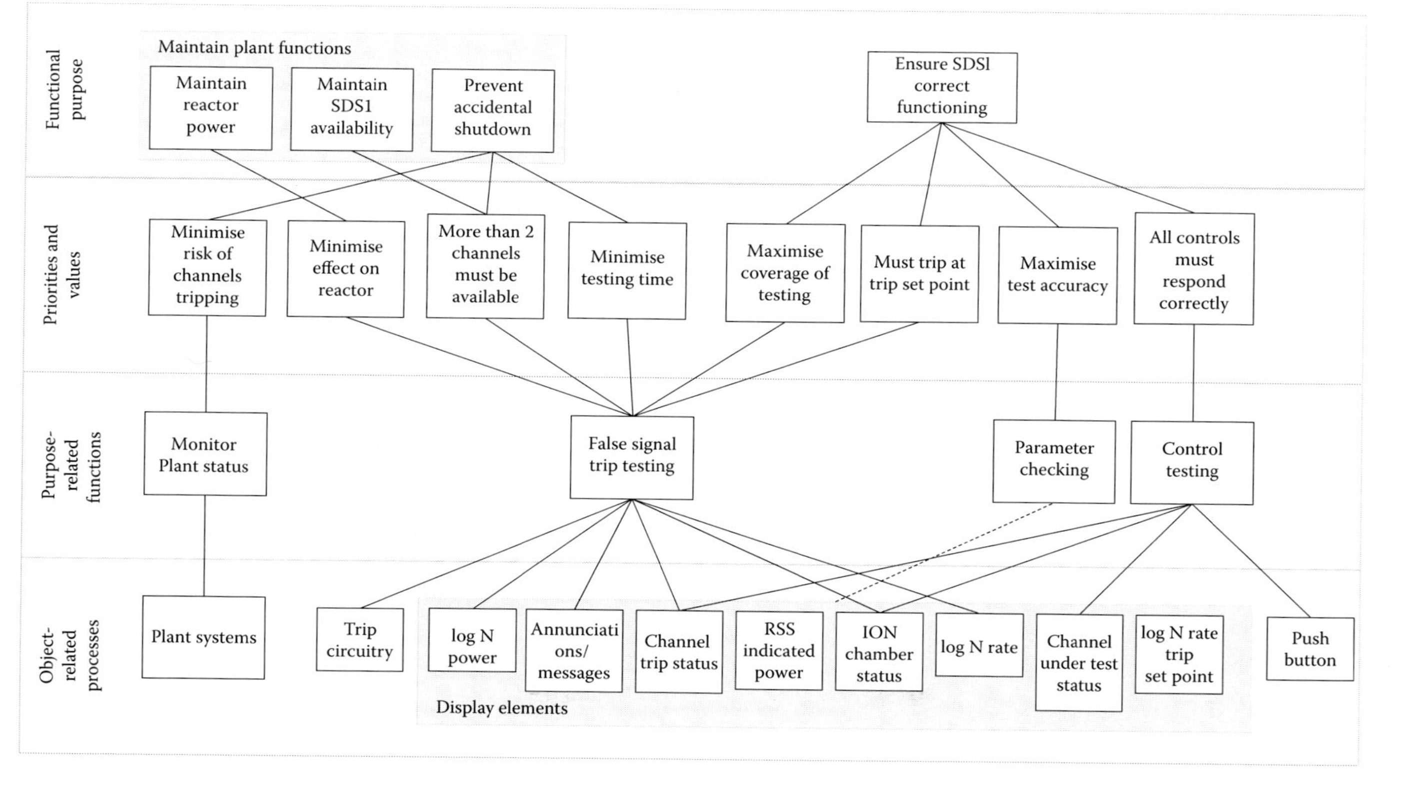

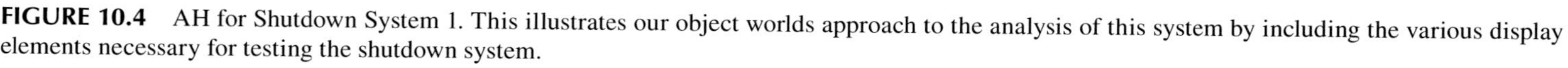
FIGURE 10.4 AH for Shutdown System 1. This illustrates our object worlds approach to the analysis of this system by including the various display elements necessary for testing the shutdown system.

cause of an operational fault or alarm. We adapted the Elix and Naikar (2008) question terminology from their work on decision ladders and applied it to IFMs. This terminology involves displaying the string of questions operators would ask themselves while moving through a process. By using the question terminology, we were able to limit the exit paths from each box to a '*yes*' response, a '*no*' response, or in some cases, an '*unsure*' response. This allowed for various paths to all reach the same end state even under different intermediate circumstances.

The number of strategies depicted by an IFM can quickly become excessive if each possible combination of intermediate states is considered a distinct strategy. Therefore, it is more efficient to examine specific locations where discrete strategies can diverge. For example, in the IFM depicting operator strategies for assessing the cause of an active alarm that is presented in Figure 10.5, there are two principal areas where an operator can employ different strategies. The first is in the organisation of alarms on the display and the second is in the type of search for the cause of the alarm (Kim and Kim, 2014). Each of the orange boxes at these locations can have its own IFM, which would depict the different paths through the individual strategies. This illustrates the idea of different levels of specificity similar to that of the AH. Again, the level of detail of analysis should be determined by the individual study's specific needs.

10.3 RESULTS: EID DISPLAYS

Each of the three phases of our CWA manifests in different ways in the ecological interfaces that we designed. The WDA is most noticeable in the individual graphical forms while the ConTA and StrA will be more evident in the configuration of these individual forms, and the overall structure of the interface.

One of the underlying ideas driving EID is the idea of representing functionally related information together (Vicente and Rasmussen, 1992). Whereas mimic displays represent a system based on the actual physical layout of the system, ecological designs represent a system according to how the system behaves and the underlying physical properties driving that behaviour. To illustrate how our CWA led to the graphical forms, we will present an example of a display we developed for the pressuriser.

The pressuriser is essentially a large tank in the CANDU 6® that controls the pressure in the HTS of the nuclear plant. The HTS is responsible for transporting the heat generated in the reactor to the steam generators. If the pressure in the HTS falls, the pressuriser will add inventory to the system. If the pressure in the HTS rises above the set point, the pressuriser will absorb excess inventory. As is the case with any tank in the nuclear power plant, the level in the pressuriser is critical. There are three fundamental parameters governing the behaviour of the level in the pressuriser: input flow, output flow and reactor power. This relationship is depicted in the *abstract function* level of the AH for the pressuriser and led directly to the development of the display presented in Figure 10.6a. This figure stacks trend information for level, reactor power and flow and by grouping this information, operators are able to see how each parameter changes in response to changes in the other two. Figure 10.6b adds to the accessibility of parameter relationship information by mapping the

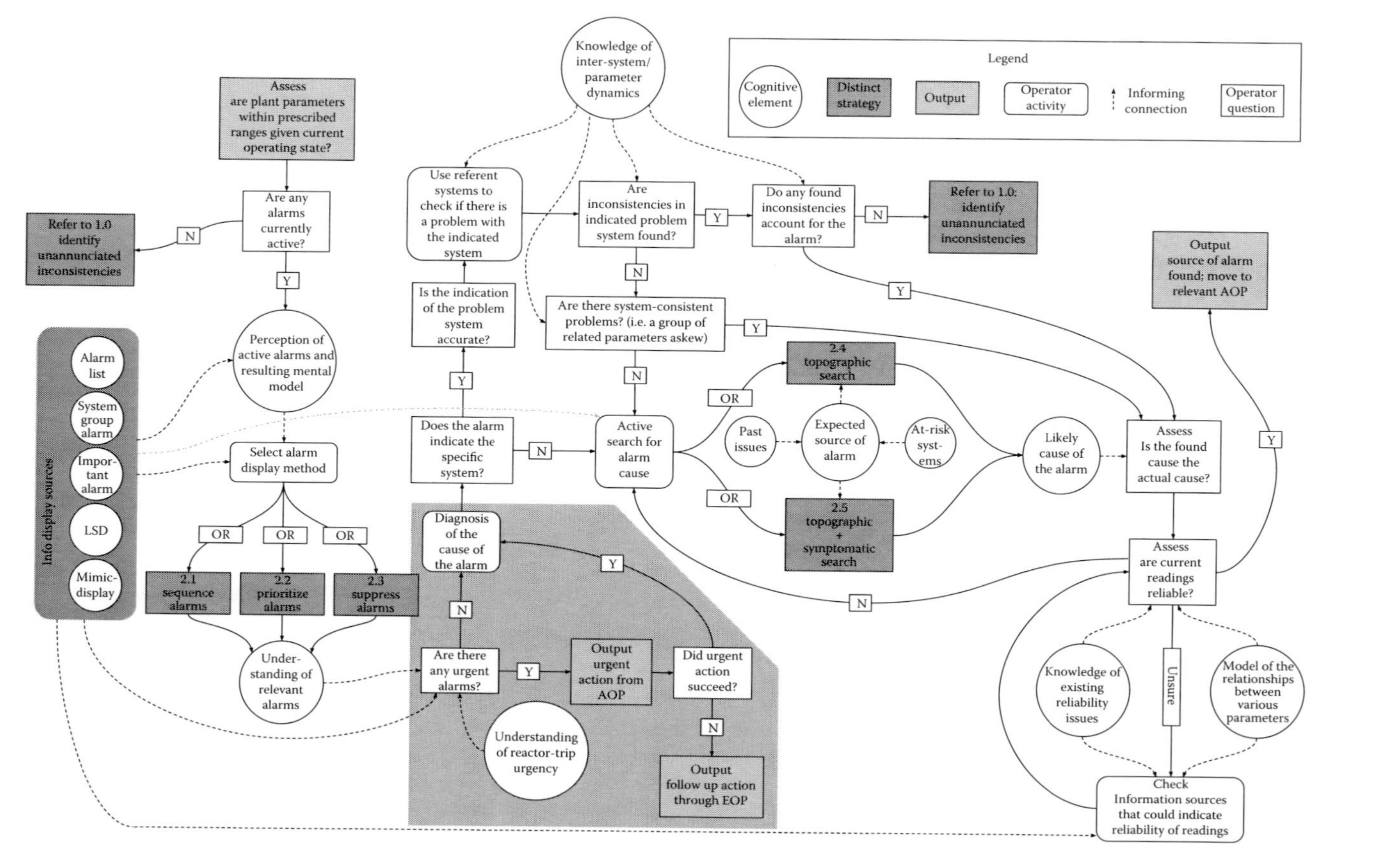

FIGURE 10.5 IFM depicting search strategies for the source of an active alarm. Adapted from Kim, D. Y. and Kim, J. 2014. *Journal of Nuclear Science and Technology*, 51(10), 1288–1310.

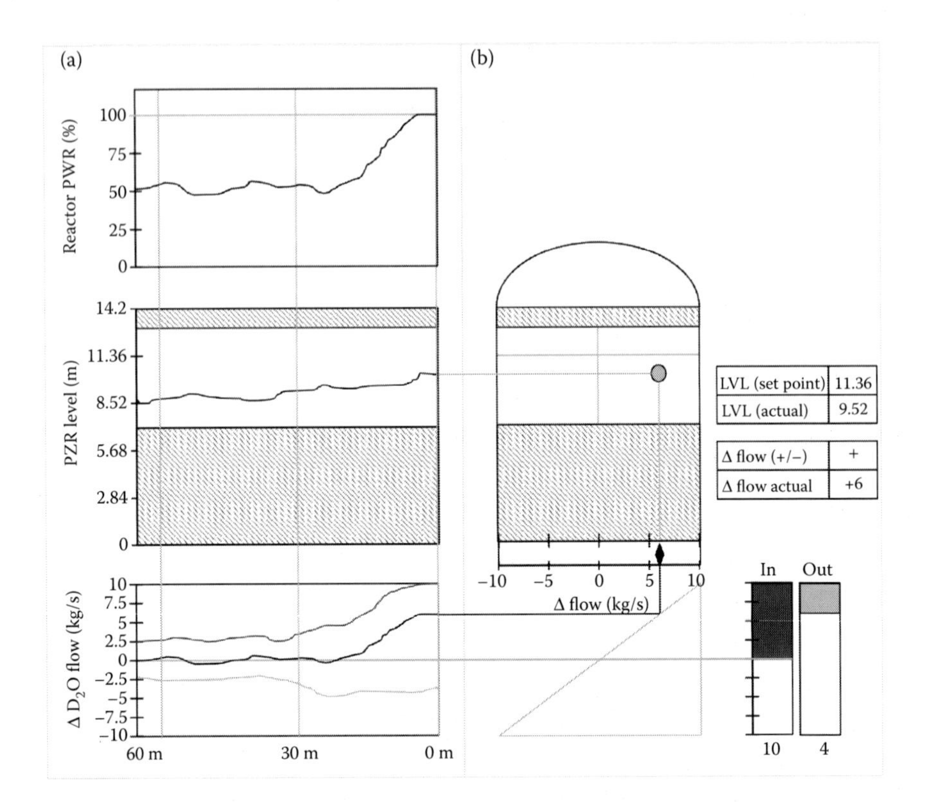

FIGURE 10.6 (a) Ecological display depicting the stacked trend charts for the parameters related to the pressuriser. (b) This illustrates the necessary control actions to return the pressuriser level back to the set point. Note that the current ΔD_2O level taken from the lowest trend chart in (a) is reflected off of the triangle in (b) to provide the x-axis coordinate for the ΔFlow-Level chart above.

pressuriser's delta flow against its level. By doing this, we are able to illustrate the control actions necessary for maintaining the set point in the pressuriser. The green point on this plot should either be in the upper-left quadrant, or the lower-right. If the system is not behaving properly, the point will be in the lower-left quadrant, which means that the tank is emptying out when the level is below the set point, or in the upper-right, which means that the tank is filling when the level is above the set point. By configuring all of these parameters and including the time trends, the operator is able to directly perceive the relationships between important parameters related to the pressuriser.

10.4 DISCUSSION

This article presented a methodological case study for conducting a CWA that was focused by our specific experimental needs. We detailed the major steps that we took along the way to the development of that experiment in an effort to show the practical implications within the realm of experimentation of each stage of CWA. Because we have yet to conduct our experiment, our findings are purely subjective methodological recommendations based on our personal experience with CWA until we are able to support these findings with the upcoming experiment.

The most valuable measure that we undertook in focusing our analysis and design was to maintain a clear understanding of our experimental goals, which were to evaluate the efficacy of LSDs and to assess the impacts of different informational structures when they are placed on these LSDs. Although we maintained these high-level goals throughout our study, our understanding of what was required to achieve them changed over time. Therefore, in order to achieve these overarching goals, we modified some of our sub-goals. For example, although we wanted to represent the plant as accurately as possible, this was not the best option for an evaluation using novice participants. We therefore shifted our focus on representing the system as accurately as possible, to focus more on representing the operations of that system as accurately as possible.

Although the understanding of your experimental goals and constraints will undoubtedly change throughout the course of a multi-year project, there are actions that can be performed early in the project that can facilitate the ease of accommodation. One of these actions for us was to initially over-analyse the system. Although we had a clear understanding of our goal and had clearly defined the boundaries of our analysis ahead of time, we analysed many of the systems beyond what we thought would be required by the final experiment. This allowed for two things. First, it garnered the experimenters a better understanding of the systems that they were working with. This was imperative for our project since the experimenters were new to the nuclear domain. Second, it allowed for later selective simplification as the experimental constraints were better understood. This selective simplification is difficult without an initially complex analysis.

The WDA proved to be invaluable in this effort to analyse our systems, which is consistent with past literature that has shown that it is an effective tool for generating a comprehensive set of information requirements (e.g. Vicente and Rasmussen, 1992). However, the usefulness of the WDA and AH has often overshadowed how useful ConTA and StrA can be. The IFMs that were generated through StrA provided a level of understanding with respect to the actual operations of the plant in various scenarios that could not be attained through WDA. This operator-centric information enhanced the findings from our WDA by relating them to operator behaviour. For example, in the active alarm IFM (see Figure 10.6), an important activity that we discovered was comparing a given parameter to referent systems. The configuration of the displays presented in Figure 10.6a illustrates how this activity can be designed directly into the configuration of graphics in EID. This demonstrates how StrA enabled us to make use of the relationship between the individual graphics that were generated from our WDA. Because the usefulness of StrA is far reaching, we recommend that it be conducted at levels of specificity as determined by the study's needs.

The major limitation of our study is that expert operators are not available for our testing. This has had a significant impact both on our analysis work and on our experimental and display design. As described earlier, this forced us to reduce the complexity of the scenarios and the displays. However, while it limited some of the representativeness of our system, it also allowed us to use participants that with few if any preconceived notions of the interface. This could result in less bias during experimentation and allow the differentiation in performance between the two interfaces to be more accurate.

The next step in our study is to perform pilot testing and to run the experiment. We will apply the flexible yet guided approach that we have used in our analysis and design phases to our experiment, as it is inevitable that our understanding of the experimental constraints will change as we get deeper into experimentation.

REFERENCES

Burns, C. M. and Hajdukiewicz, J. R. 2004. *Ecological Interface Design*. Boca Raton, FL: CRC Press.

Burns, C. M., Skraaning, G., Jamieson, G. A., Lau, N., Kwok, J., Welch, R. and Andresen, G. 2008. Evaluation of ecological interface design for nuclear process control: Situation awareness effects. *Human Factors*, 50, 663–679.

Davey, E. 2000. Process monitoring during normal operation at Canadian nuclear power plants. *Proceedings of the Human Factors and Ergonomics Society Annual Meeting*, 44, 811–815.

Elix, B. and Naikar, N. 2008. Designing safe and effective future systems: A new approach for modeling decisions in future systems with Cognitive Work Analysis. *Proceedings of the 8th International Symposium of the Australian Aviation Psychology Association*.

Fushiki, S. 2013. Radiation hazards in children – Lessons from Chernobyl, Three Mile Island and Fukushima. *Brain & Development*, 35, 220–227.

Hurlen, L., Skraaning, G., Myers, B., Jamieson, G. A. and Carlsson, H. 2015. The plant pane: Feasibility study of an interactive large screen concept for process monitoring and operation. *NPIC & HMIT 2015*. Charlotte, NC.

Juhasz, M. and Soos, J. K. 2007. Impact of non-technical skills on NPP teams' performance: Task load effects on communication. *Joint 8th IEEE HFPP/13th HPRCT*. pp. 225–232.

Kilgore, R. M., St-Cyr, O. and Jamieson, G. A. 2008. From work domains to worker competencies: A five-phase CWA for air traffic control. In A. M. Bisantz and C. M. Burns (Eds.). *Applications of Cognitive Work Analysis*. Boca Raton, FL: CRC Press, pp. 15–46.

Kim, D. Y. and Kim, J. 2014. How does a change in the control room design affect diagnostic strategies in nuclear power plants. *Journal of Nuclear Science and Technology*, 51(10), 1288–1310.

L-3 MAPPS Inc. 2011. Galerie de photos de centrales [online]. http://www.mapps.l-3com.com/photo_gallery_power_fr.html. Accessed 1 September 2015.

Lau, N., Jamieson, G. A. and Skraaning, G. 2012. Situation awareness in process control. NPIC & HMIT 2012. San Diego, CA, pp. 1511–1523.

Lau, N., Jamieson, G. A. and Skraaning, G. 2013. Distinguishing three accounts of situation awareness based on their domains of origin. *Proceedings of the Human Factors and Ergonomics Society*, 2013, 220–225.

McIlroy, R. C. and Stanton, N. A. 2015. Ecological interface design two decades on: Whatever happened to the SRK taxonomy? *IEEE Transactions on Human-Machine Systems*, 45, 145–164.

Meshkati, N. 1998. Lessons of Chernobyl and beyond: Creation of the safety culture in nuclear power plants. *Proceedings of the Human Factors and Ergonomics Society Annual Meeting*, 1, 745–749.

Mumaw, R. J., Roth, E. M. Vicente, K. J. and Burns, C. M. 2000. There is more to monitoring a nuclear power plant than meets the eye. *Human Factors*, 42, 36–55.

Myers, W. P. and Jamieson, G. A. 2013. *Operating Experience Review of Large-Screen Displays in Nuclear Power Plant Control*. Toronto, Canada: University of Toronto Cognitive Engineering Laboratory, CEL-13-01.

Owre, F., Kvalem, J., Karlsson, T. and Niglwing, C. 2002. A new integrated BWR supervision and control system. IEEE Human Factors Meeting. Scottsdale, AZ, pp. 41–47.

Rasmussen, J. 1983. Skills rules, knowledge: Signals, signs, and symbols and other distinctions in human performance models. *IEEE Transactions on Systems, Man, and Cybernetics*, SMC-13, 257–267.

Roth, E. M., Lin, L., Thomas, V. M., Kerch, S., Kenney, S. J. and Sugibayashi, N. 1998. Supporting situation awareness of individuals and teams using group view displays. *Proceedings of the Human Factors and Ergonomics Society (1998)*. pp. 244–248.

Suzuki, A. 2014. Managing the Fukushima challenge. *Risk Analysis*, 34, 1240–1256.

Torenvliet, G. L., Jamieson, G. A. and Chow, R. 2008. Object worlds in work domain analysis: A model of naval damage control. *IEEE Transactions on Systems, Man, and Cybernetics*, 38, 1030–1040.

Vicente, K. and Rasmussen, J. 1992. Ecological interface design: Theoretical foundations. *IEEE Transactions on Systems, Man and Cybernetics,* 22, 1–18.

World Nuclear Association. 2015. *World Energy Needs and Nuclear Power* [online]. http://www.world-nuclear.org/info/Current-and-Future-Generation/World-Energy-Needs-and-Nuclear-Power/. Accessed 1 September 2015.

Section IV

Design and Evaluation

11 The Cognitive Work Analysis Design Toolkit

Gemma J. M. Read, Paul M. Salmon and Michael G. Lenné

CONTENTS

11.1 INTRODUCTION

The importance of design and systems thinking within the human factors and ergonomics (HFE) discipline is well-recognised (Dul et al., 2012; Norros, 2014). Recognising the need for systems-based approaches to design, the cognitive work analysis (CWA) framework of methods was put forward to support the analysis of complex sociotechnical systems with the specific aim of improving system design (Vicente, 1999). Yet while CWA remains a popular approach for analysis and design (Read et al., 2012), there remain questions regarding the extent to which CWA analyses directly inform design (Lintern, 2005; Jenkins et al., 2010; Mendoza et al., 2011). While the utility and uniqueness of the approach for understanding complex system performance is well-established, the extent to which outputs are used directly in the design process is unclear and there is no typical approach evident for using CWA to inform design (Read et al., 2015b).

In response to this practical limitation of the framework, we developed the cognitive work analysis design toolkit (CWA-DT). In developing the CWA-DT we drew upon a related field, the sociotechnical systems theory approach, as well as an analysis of what the CWA framework provides to design, and consideration of common approaches to design. In this chapter, we will describe the approaches underpinning the CWA-DT, detail its structure and demonstrate three case studies which demonstrate its application within different domains.

11.1.1 The Sociotechnical Systems Approach

The sociotechnical systems theory approach (Trist and Bamforth, 1951; Cherns, 1976) provides theoretically consistent principles that can be combined with CWA to augment the process aspects of the CWA framework (see Read et al., 2015d for further discussion). Like CWA, the sociotechnical systems theory approach intends to design systems (typically organisations) that exhibit adaptive capacity, that is, systems that can adapt to external environmental conditions and disturbances by adopting alternative means to achieve goals.

A core feature of the sociotechnical systems theory approach is the participation of stakeholders within the system (users, supervisors, managers and maintainers) in the design process. Participation is not just important to ensure that user issues are understood and to gain buy-in from stakeholders, but because design is not a defined process with a clear end point. Instead, it is an extended social process that continues well beyond initial implementation (Clegg, 2000). After the experts leading the change process have departed and users begin to interact with the new technology, changes are inevitably required, adaptations made and new functionality discovered in the processes or structures within the real work environment. Where those conducting this continuing design process understand the underlying design principles and reasoning around initial design choices such on-going adaptations are more likely to align well with the overall objectives.

The sociotechnical systems theory approach is also strongly values driven and its proponents have emphasised the need to treat humans as the solution to problems rather than the cause (i.e. due to 'human error'). Clegg (2000) notes specifically an overemphasis of technological solutions to solve design problems (including designing out humans), rather than the adoption of solutions that instead support human decision making. Other values are associated with ideas around quality of working life (that workers should be provided with variety in their work, with reasonable challenge, with a sense of contributing to wider goals, etc.) and with responsibility for design decisions involving not only the direct system stakeholders but including the wider social, economic and environment impacts within the community, similar to notions of corporate social responsibility.

Table 11.1 provides a list of the values, process principles and content principles arising from the sociotechnical systems theory literature (Cherns, 1976, 1987; Davis, 1982; Clegg, 2000; Sinclair, 2007; Walker et al., 2009). A fuller description of how the values and principles were identified is available in a previous publication (Read et al., 2015d). The CWA-DT has been developed to facilitate the process principles and with the values in mind. It also aims to ensure that the content principles are met in the design concepts developed.

TABLE 11.1
The Values and Principles of Sociotechnical Systems Theory

Values	Process Principles	Content Principles
• Humans as assets • Technology as a tool to assist humans • Promote quality of life • Respect for individual differences • Responsibility to all stakeholders	• Adoption of agreed values and purposes • Provision of resources and support • Adoption of appropriate design process • Design and planning for the transition period • Documentation of how design choices constrain subsequent choices • User participation • Constraints are questioned • Representation of interconnectedness of system elements • Joint design of social and technical elements • Multidisciplinary participation and learning • Political debate • Design driven by good solutions, not fashion • Iteration and planning for ongoing evaluation and re-design	• Tasks are allocated appropriately between and amongst humans and technology • Useful, meaningful and whole tasks are designed • Boundary locations are appropriate • Boundaries are managed • Problems are controlled at their source • Design incorporates the needs of the business, users and managers • Intimate units and environments are designed • Design is appropriate to the particular context • Adaptability is achieved through multifunctionalism • System elements are congruent • Means for undertaking tasks are flexibly specified • Authority and responsibility are allocated appropriately • Adaptability is achieved through flexible structures and mechanisms • Information is provided where action is needed

11.1.2 Information Provided by CWA

Each phase of CWA and its associated tools provide a vast amount of information about the constraints that limit activity within the system that could be used to inform design. For example, the work domain analysis (WDA) identifies the constraints associated with the functionality of physical objects in the system, the contextual activity template (CAT) identifies situational constraints on activity and the worker competencies analysis phase identifies the constraints of actors which affect their ability to undertake activity. Acknowledging the importance of a constraints-based approach to design, the CWA-DT incorporates activities to support discussions around constraints and how they might be manipulated (i.e. strengthened and removed) to improve design goals such as safety, efficiency or effectiveness.

In addition to information about system constraints, CWA provides a unique perspective on system functioning compared to other HFE methods such as task analysis methods. CWA users who participated in an online survey about their use of the framework for design noted the rich understanding of the system that they gained from conducting the analysis and emphasised that conducting the analysis gave them insights which contributed to design (Read et al., 2015d). This notion of insight is used in the CWA-DT as an important tool for negotiating the gap between analysis and design.

11.1.3 Human-Centred and Participatory Design

Broadly, HFE design approaches subscribe to the human-centred design philosophy. Human-centred design (used here interchangeably with the term user-centred design) focuses design activity on understanding the needs and preferences of users, as well as their limitations, and designing to suit these. HFE knowledge and methods are used in human-centred design to uncover and understand user needs, capabilities and limitations. The international standard on user-centred design for computer-based interaction systems (ISO, 2010) incorporates the following principles which can also be usefully applied beyond computer-based systems:

- The design is based on an explicit understanding of users, tasks and environments.
- Users are involved throughout design and development.
- The design is driven and refined by user-centred evaluation.
- The design process is iterative.
- The design addresses the whole user experience.
- The design team includes multidisciplinary skills and perspectives.

Both CWA and the sociotechnical systems perspective might be considered human-centred approaches as they place humans at the centre of the design process. However, they represent an enhanced approach to the type of human-centred design processes envisaged by the ISO standard. Specifically, they explicitly take a systems perspective, not referring to users (which may narrow the remit of design) but instead considering all actors (both human and technical) and the interactions between them. Further, the sociotechnical systems theory approach encourages more than a consideration of users and stakeholders, or even their involvement. It requires that decision-making authority around design be bestowed upon users and stakeholders to promote their ownership of the outcomes of the design process (Clegg, 2000).

11.1.4 Design Thinking

An approach that is increasingly being employed within the human-centred design paradigm is design thinking. This involves the application of design processes and design skills to promote innovation within organisations. It aims to assist non-designers to think like professional designers and to focus on the needs of the users of the product, service and system being designed (Brown, 2008). Design thinking is proposed not as a method, but rather a way of making design accessible to non-professional designers. Edward De Bono introduced the term lateral thinking to clarify that one did not need to be a 'creative' person to generate novel and useful designs (De Bono, 1992).

A design thinking publication by Liedtka and Ogilvie (2010) provides a process encompassing 10 tools and methods structured into four questions. This follows a standard design process structured through the use of the tools that begins by exploring the existing situation (what is?), then diverges to generate a wide range of new ideas (what if?), subsequently converging to refine a selection of these ideas (what wows?) and finally, evaluating and refining the ideas further ready to test them in the market (what works?). Based on previous proposals to use a selection of these tools (such as a design brief document) to develop design concepts on the basis of CWA (G. Lintern, personal communication, May 24, 2012), these were adopted for the CWA-DT. Another approach that could be considered to follow the design thinking movement is the Design with Intent toolkit, created by Lockton et al. (2010) to provide design patterns that can be used in design for behaviour change. The toolkit provides cards based on eight different lenses or perspectives on design, both environmental and cognitive. The toolkit provides information about each lens and guidance on how the cards can be used to generate design ideas and has been adopted as a design tool for use within the CWA-DT.

These design thinking approaches are consistent with the co-creation approach to design and their format as toolkits or 'how-to' guides further inspired a toolkit format for the CWA-DT.

11.2 STRUCTURE OF THE CWA-DT

The CWA-DT was developed to assist users of CWA to apply the outputs of the framework to the design of sociotechnical systems. The toolkit intends to support the creative process of designing elements of a sociotechnical system (such as equipment, environments, processes, organisational structures, and strategies) in an integrated manner that takes account of HFE considerations. The toolkit is intended to be flexible, and users are therefore encouraged to apply those aspects that are most useful given the purposes, scope and constraints of their design activity.

The toolkit provides written guidance, templates and tools that follow 11 steps associated with CWA-based design. A description of each of the steps is provided in Table 11.2.

While the 11 steps are presented in sequential order, it is intended that there be interconnections between non-adjacent steps and iteration forward and backward through the process. Suggested interconnections are illustrated in Figure 11.1. The process recognises analysis, design and evaluation as part of an iterative process of learning about the domain, considering potential changes and their impact and improving knowledge throughout. Further, the figure shows that the aspect of participation and engagement, which is central to the sociotechnical systems approach and human-centred design, permeates the entire process.

TABLE 11.2
The 11 Steps of the CWA-DT

Step	Description	Tools, Templates and Guidance Provided
1. Participation and engagement	Given that the users and system stakeholders are responsible for the on-going adaptation and re-design of the system, they must have ownership of the design process. Ownership requires more than participation in the design process, it requires genuine engagement. Further, control over the process and decision making authority should remain with the users and stakeholders. The toolkit envisages a workshop approach to engagement with users and stakeholders for design; however, engagement may take other forms such as a series of meetings, webinars or alternative types of engagement.	• Guidance on selecting facilitators, creating appropriate an appropriate physical and social environment and involving the right design participants in the process.
2. Analysis planning	Planning the analysis process ensures clarity for the design team and stakeholders about the purpose of the project and ensures appropriate boundaries are drawn for the analysis. The key activity is the development of an *Analysis brief.*	• *Analysis brief* (adapted from Liedtka and Ogilvie, 2010) which identifies the project need, analysis deliverables, key stakeholders, target users, project constraints, etc. • Questions for conducting a context/problem analysis, including a *Stakeholder needs analysis template* to consider the characteristics of users and stakeholders. • Guidance on the selection of appropriate CWA phases.

(*Continued*)

TABLE 11.2 (*Continued*)
The 11 Steps of the CWA-DT

Step	Description	Tools, Templates and Guidance Provided
3. Analysis process	The analysis process begins with data collection activities (e.g. document review, subject matter expert interviews and observation). Subsequently, the appropriate CWA phases are applied and outputs developed, reviewed and refined. During the analysis process, 'insights' about the functioning of the system are recorded by the analyst/s. The different categories of insights include: *Assumptions* – The underlying hypotheses, expectations and beliefs upon which the system, or part of the system, is based. Assumptions could relate to how the system functions or how people are expected to behave within the system. Assumptions may be correct or incorrect, regardless of which they should be made explicit as they are a key influence on system design. *Leverage points* – Aspects of a system which if changed in a small way can produce big changes across the system. There may be evidence within the analysis that suggests there is a leverage point that is under-utilised or the identification of a potential leverage point. *Metaphors* – Metaphors and analogies promote thinking about how to apply existing ideas in new situations. Metaphor involves the comparison, interaction or substitution of two subjects on a symbolic level. An insight might involve, for example, realising that there are similarities in two domains (i.e. scheduling in manufacturing and health care) which can be used to generate design ideas. *Scenario features* – The data collection and analysis activities include rich contextual information about the domain being analysed which can be captured to communicate to others involved in the design process. Scenario features could include a type of actor, attributes of an actor, a type of task, an environmental disturbance or influence, etc.	• Guidance for navigating the literature on conducting CWA (referencing influential texts and papers on the topic). • Guidance about insights and their use in design. • *Insights template* which enables documentation of the insight, the thought processes leading to its identification, and how it might be used in the design process. • Insight prompt questions. • *Constraints template.* • *Stakeholder object world template.*

(*Continued*)

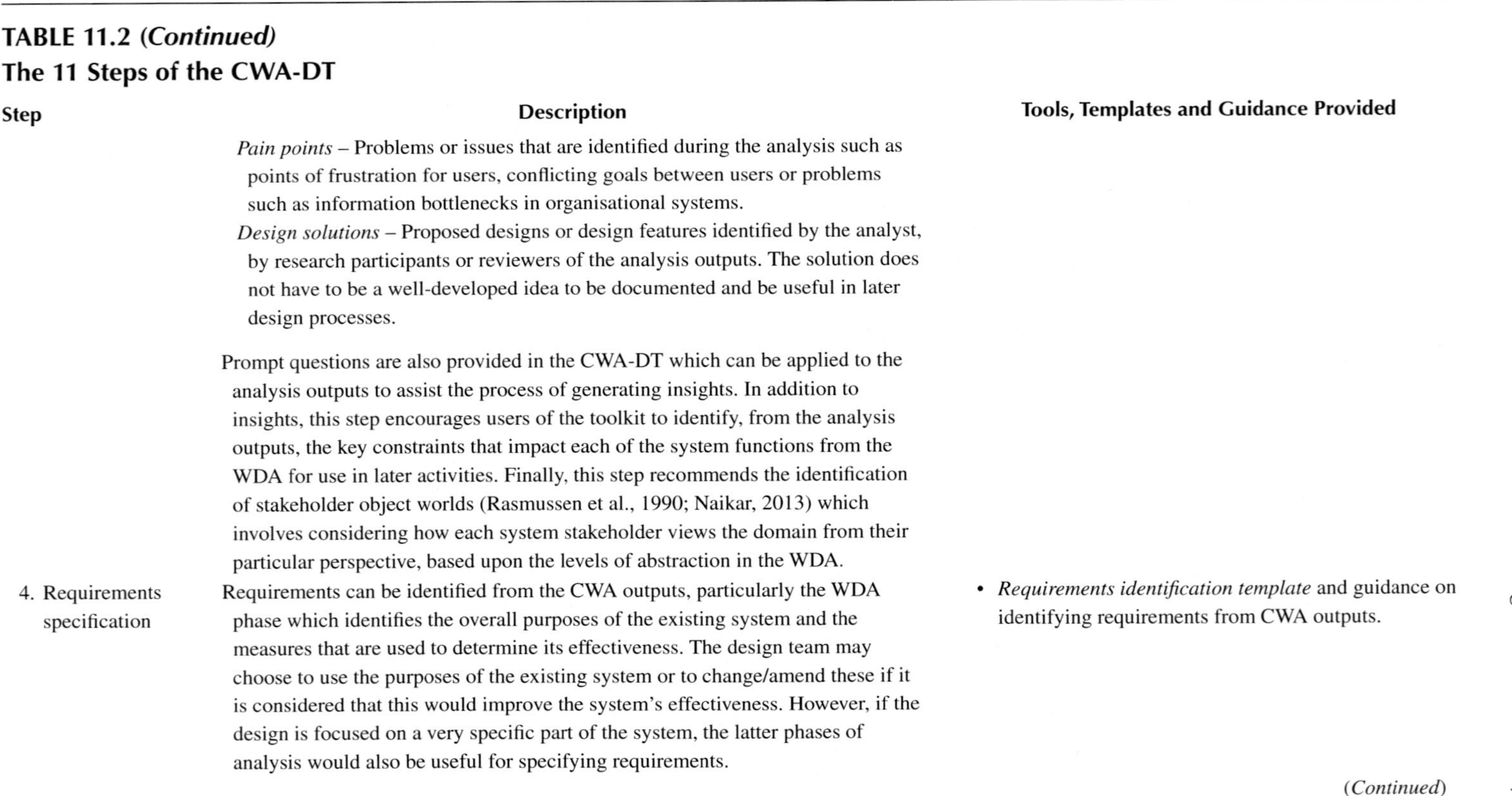

TABLE 11.2 (*Continued*)
The 11 Steps of the CWA-DT

Step	Description	Tools, Templates and Guidance Provided
	Pain points – Problems or issues that are identified during the analysis such as points of frustration for users, conflicting goals between users or problems such as information bottlenecks in organisational systems. *Design solutions* – Proposed designs or design features identified by the analyst, by research participants or reviewers of the analysis outputs. The solution does not have to be a well-developed idea to be documented and be useful in later design processes. Prompt questions are also provided in the CWA-DT which can be applied to the analysis outputs to assist the process of generating insights. In addition to insights, this step encourages users of the toolkit to identify, from the analysis outputs, the key constraints that impact each of the system functions from the WDA for use in later activities. Finally, this step recommends the identification of stakeholder object worlds (Rasmussen et al., 1990; Naikar, 2013) which involves considering how each system stakeholder views the domain from their particular perspective, based upon the levels of abstraction in the WDA.	
4. Requirements specification	Requirements can be identified from the CWA outputs, particularly the WDA phase which identifies the overall purposes of the existing system and the measures that are used to determine its effectiveness. The design team may choose to use the purposes of the existing system or to change/amend these if it is considered that this would improve the system's effectiveness. However, if the design is focused on a very specific part of the system, the latter phases of analysis would also be useful for specifying requirements.	• *Requirements identification template* and guidance on identifying requirements from CWA outputs.

(*Continued*)

TABLE 11.2 (*Continued*)
The 11 Steps of the CWA-DT

Step	Description	Tools, Templates and Guidance Provided
5. Design planning	Design planning involves the development of documentation to drive and scope the design process. Further, this step also includes the selection of the most appropriate design activities to undertake for sharing the analysis findings with users/stakeholders, for idea generation and for design concept definition (e.g. synthesising ideas into more holistic concepts). In addition, this step involves developing the design materials such as scenarios, inspiration cards, etc. based on the insights documented and planning for events such as workshops, meetings, etc.	• *Design brief* template which identifies the design scope, project planning for design, the design specifications and relevant design requirements from sociotechnical systems theory. • *Design criteria* template which identifies criteria for success for the designs produced, taken from the values and priority measures of the WDA. • *Design tool selection matrix* to assist selection of appropriate design tools. • One-page guides for tools to communicate results, generate ideas and synthesis ideas (e.g. scenarios, personas, inspiration card exercises, lateral thinking exercises, affinity diagramming, etc.). • Sample workshop plans. • Guidance on workshop delivery.
6. Concept design	The purpose of this step is to use creative and divergent thinking to identify many design ideas which are then synthesised into one or more design concepts. The CWA-DT encourages involvement of users and stakeholders in this step through workshops using design activities that promote creativity and innovation.	
7. High-level evaluation and concept selection	Design concepts can be evaluated at a high-level using the CWA outputs enabling the design team to determine the effects of the change on the system. Changes could be benefits or unanticipated negative effects. The results of this initial evaluation provide a basis for the selection, rejection or refinement of design concepts.	• Guidance for using CWA outputs for evaluating design concepts. • *Design concept summary* template.

(*Continued*)

TABLE 11.2 *(Continued)*
The 11 Steps of the CWA-DT

Step	Description	Tools, Templates and Guidance Provided
8. Detailed design	In this step, the design concepts are further developed and decisions are made about the details of the design as it would be implemented. There are many ways in which the detailed design phase may proceed. The CWA-DT recommends a collaborative approach with users and stakeholders is continued. This could involve workshops or design sessions where activities such as prototyping are completed collaboratively. Alternatively, the design team might create mock-ups or scenarios of use that are presented to users and stakeholders for input.	• Guidance on rapid prototyping and the use of scenarios in detailed design. • Reference to HFE standards, guidelines and literature for detailed design of different system aspects (e.g. function allocation, interface design, job/task design).
9. Evaluation and design refinement	This step is concerned with evaluating the proposed design/s to determine their performance against the design criteria. The empirical evidence gained can support the selection of a particular design over others or enable a case to be made to funders to commit to constructing and/or implementing the proposed design. As such, this step encompasses more rigorous testing than the high-level evaluation (Step 7).	• Guidance on evaluation processes including inspection-based evaluation/expert review, user testing against performance measures, simulation and modelling, and safe-to-fail experiments.
10. Implementation	This step involves the final construction and implementation of the design into the real world. The CWA-DT does not focus on specific processes for creating the final design as it will vary considerably between projects. From a sociotechnical systems theory perspective, the participation and engagement aspects of the design process should ensure a smooth implementation process if done appropriately.	• Guidance on developing an implementation plan and guidance on communication during implementation.
11. Testing and verification	This step involves testing whether the design, as implemented, has met the requirements specified in Step 4 via the criteria specified in the *Design criteria* document (Step 5). It can encompass immediate testing as well as long-term monitoring and on-going testing and evaluation to ensure the design remains appropriate to the needs of the domain.	• *Design verification template* which can be used to record the results of tests against the design criteria, noting whether the tests meet pre-determined acceptance measures.

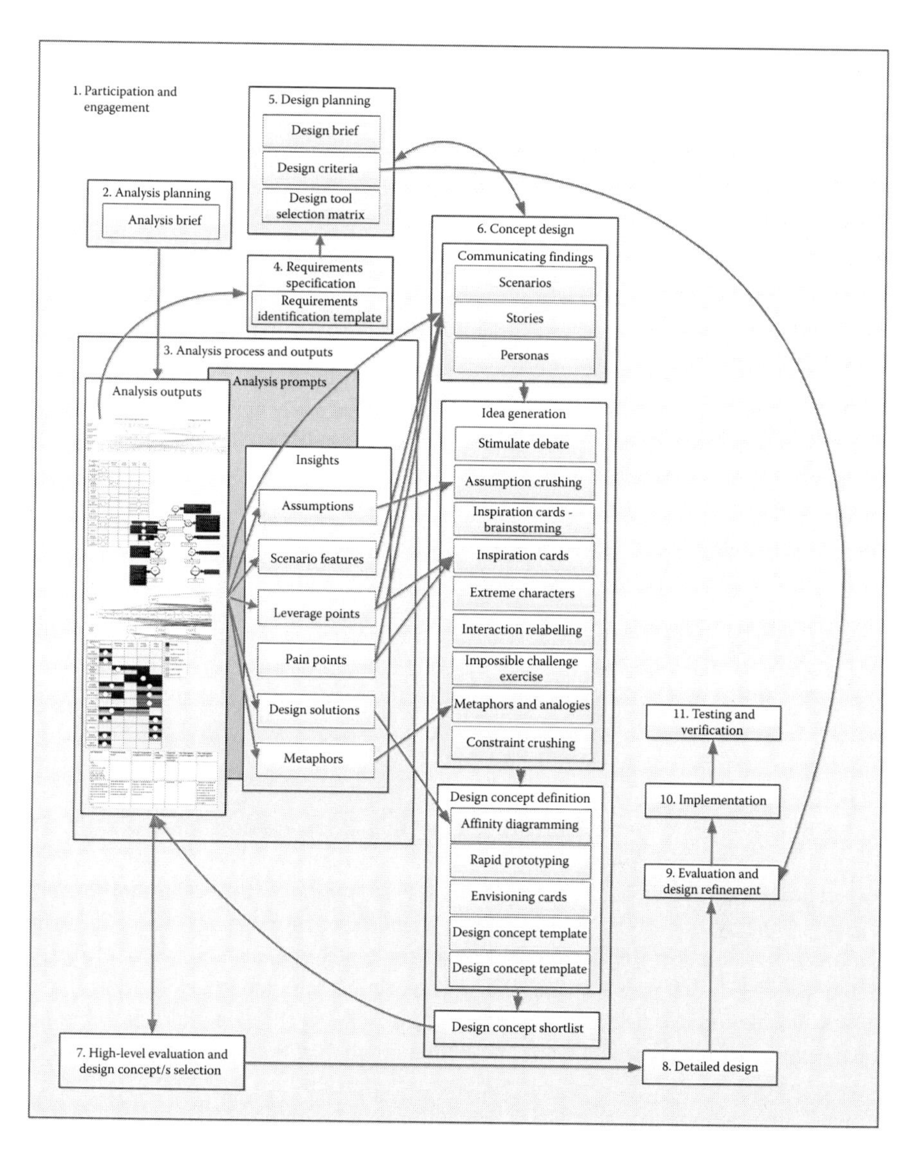

FIGURE 11.1 Overview of the CWA-DT process.

As noted previously, users may choose to use all of the steps within the toolkit, or only the tools and templates that they consider will be beneficial for their design purposes.

11.3 APPLYING AND EVALUATING THE CWA-DT

The CWA-DT has been applied in a number of domains to provide some initial evidence of its effectiveness and validity. These applications have occurred in the domain of public transport ticketing systems (Read et al., 2015c) rail level crossings (RLXs; Read et al., 2016) and road intersections (Read et al., 2015a). A summary of the analysis and design approach undertaken in each application is provided in this section. A comparison of the design approaches taken across the three case studies is shown in Table 11.3 to illustrate the different ways in which CWA and the CWA-DT can be successfully applied.

11.3.1 Case Study 1: Public Transport Ticketing

This case study was concerned with designing a public transport ticketing system for adoption in an Australian city. Prior to applying CWA, an analysis brief was developed. This case study was intended to be illustrative of the method, rather than having key stakeholder support to implement the findings into practice, however, the recommendations may be of use to those designing and implementing new ticketing systems. The analysis brief documented that an Australian state government had recently implemented a smartcard-based ticketing system for public transport. This system has been subject to some criticism from its users regarding aspects such as the speed of processing, usability and convenience of ticket purchasing. It then outlined a fictional scenario that following a change of government, the new minister for transport (not involved in any past decisions about the system) calls for a review of the current ticketing system to identify and learn from the issues experienced and to design a new system that meets the goals of government and users, appropriate for a metropolitan public transport system in 5–10 years time.

Following the application of the five phases of CWA (see Table 11.3 for details), a design brief was developed. This outlined the high-level design requirements, drawn from the functional purposes of the WDA. These included that the system should collect revenue; collect data (journey details) on public transport usage; promote respect of the system by its users; support users to reach their destination and support users to comply with the transport ticketing legislation. The design brief further stated that the design process would be driven by the sociotechnical systems theory values. Developed alongside the design brief, the design criteria document stated that the design concepts produced would be evaluated against the sociotechnical systems content principles (i.e. that tasks are allocated appropriately between and amongst humans and technology; that useful, meaningful and whole tasks are designed). It also stated a range of system-specific evaluation criteria that were drawn from the values and priority measures of the WDA. These included that the design concept would minimise intentional and unintentional fare evasion; minimise operational costs; support a high level of transaction accuracy and security; support efficient use by users and efficient operation of public transport services.

Three design workshops were held over a short time period with a small number of system users who undertook various design activities from the CWA-DT. An example of a design concept resulting from the workshops was the *Smartcard becomes a smartphone* concept. In this design concept, users could wave their smartphone near a reader and a software application on the phone would communicate with the card reader. The application would display informative messages and include the ability for users to recharge, would provide notifications if the balance is low and

TABLE 11.3
Summary of How the CWA-DT Was Applied across Three Case Study Design Processes

	Public Transport Ticketing (Read et al., 2015c)	Rail Level Crossings (Read et al., 2016)	Road Intersections (Read et al., 2015a)
Analysis outputs used	• Abstraction hierarchy • CAT • Decision ladders • Information flow charts • Strategies analysis diagram • SOCA-CAT • Skills, rules and knowledge inventory	For each type (active and passive) for all road users: • Abstraction hierarchy • CAT • Decision ladders • Strategies analysis diagram • SOCA-CAT Additional outputs focused on pedestrians: • Abstraction hierarchy • CAT • Decision ladders • Strategies analysis diagram • Skills, rules and knowledge analysis	• Abstraction hierarchy Additional HFE analyses: • EAST • Perceptual cycle model
Participation/ engagement activities	Three short workshops: 1. Idea generation 2. Design refinement 3. Design concept selection Participants were system users.	Two full-day workshops: 1. Idea generation 2. Design refinement and concept selection Participants were academics, user representatives and industry representatives.	A two-day workshop involving idea generation and design refinement. Participants were academics, users and industry representatives.

(Continued)

TABLE 11.3 (*Continued*)
Summary of How the CWA-DT Was Applied across Three Case Study Design Processes

	Public Transport Ticketing (Read et al., 2015c)	Rail Level Crossings (Read et al., 2016)	Road Intersections (Read et al., 2015a)
Tools for communicating the findings	• Scenarios	• Powerpoint presentation of key analysis findings • Personas • Scenarios	• Powerpoint presentation of key analysis findings • Constraint crushing exercise
Tools for generating design ideas	• Assumption crushing • Inspiration cards	• Lateral thinking exercises • Assumption crushing • Metaphors (e.g. 'separation') • Inspiration cards	• Lateral thinking exercises • Assumption crushing • Metaphors (e.g. 'team') • Inspiration cards • Constraint crushing exercise
Tools for defining design concepts	• Affinity diagramming	• Design concept template	• Design concept template • Refining design concepts through sociotechnical systems principles
Number of design concepts generated	Two	Four	Three

enable secure payment before, during or within 24 hours after travel providing more flexibility regarding when payment is made. The smartphone application would collect optional personal information in return for entry into prize draws or rewards. Transport for low-income passengers would be subsidised by enabling passengers to donate winnings from prize draws that would be turned into travel money for people in need. Random surveys would be available via the application covering topics such as customer experience/feedback on public transport services, particular reasons for travel and travel patterns, etc. A reporting system would also be integrated within the application enabling users to report issues such as disrespectful behaviour, vandalism, faulty equipment and safety hazards. Users would be rewarded for reporting issues and feedback would be provided about the action taken in response to the report. The reporting system would also be available through a website enabling those without the application to report issues or concerns.

11.3.2 Case Study 2: Rail Level Crossings

The design brief for this work noted the longstanding safety problems associated with crashes between trains and road users at RLXs and described the aim of the process as the development of novel design concepts that will improve safety at RLXs. As part of a research grant involving relevant industry partner organisations, the intention was to create designs that could have real world application. Given the two general types of RLXs common in Australia; *active* (with boom barriers and/or flashing lights and audible bells) and *passive* (static warning signage only), the analysis was conducted separately for each type.

The high-level design requirements were extracted from the functional purposes of the WDA. These were that the design should protect road users; protect rail traffic; provide access across the rail line; maintain priority access for rail traffic; minimise delays to the rail network and minimise delays to the road network. The design brief also stated that the design process would be driven by the values of sociotechnical systems theory (humans as assets, technology as a tool to assist humans, etc.).

The design criteria document stated that the design concepts would be measured against the values and priorities of the WDA, being the extent to which they minimise collisions; minimise trauma, injuries and fatalities; minimise near miss events; minimise risk; maximise efficiency; minimise road rule violations; maximise reliability and maximise conformity with standards and regulations. It also stated that the designs would be considered successful if they meet the sociotechnical content principles and an additional project-specific criterion around cost-effectiveness of the designs (i.e. lifecycle cost of the design against savings in relation to injuries, damage and delays).

Two full-day workshops were held involving representatives from road and rail stakeholder organisations. One of the design concepts developed was titled the *Mercedes metropolitan rail level crossing.* This design concept included a number safety interventions in addition to the standard controls provided at active RLXs (e.g. flashing lights, boom barriers and automatic gates over pedestrian paths). The additional warnings included traffic lights, the enhancement of the boom barrier to resemble a more solid structure, in-road lights within the painted stop line on the road and pedestrian gates that remain closed by default and can be opened only when no train is approaching. The design concept also incorporated the linking of traffic lights at the RLX with those upstream and downstream from the site to manage traffic entering and exiting the crossing. Further, it included changes to the location of train station platforms adjacent to the RLX such that they are staggered diagonally (one on each side of the roadway, rather than directly across from one another) to enable the train to stop at the platform after having traversed the RLX. This removes the need for the crossing warnings to be activated while passengers are boarding and alighting. Specifically for pedestrians, the design included shelters over the waiting areas with some areas incorporating a ticket machine and/or community hub display that can be used to access community information and news while pedestrians wait for the train to pass. Cafes near the crossing would have electronic displays that provide train information and information about the crossing (recent near misses, performance, etc.) to encourage conversations about the crossing. Finally, a supervisor would be present at the crossing during peak times to monitor and manage road user behaviour, providing the ability to respond to abnormal or emergency situations.

11.3.3 Case Study 3: Designing Road Intersections

In contrast to the previous case studies, this application of the CWA-DT extended its use to incorporate the application of HFE methods in addition to CWA. Specifically, findings from analyses using the event analysis of systemic teamwork (EAST) approach (Walker et al., 2006) and Neisser's (1976) perceptual cycle model were used to inform the design documents and design process.

The design brief, based on the previous analyses, highlighted the issues associated with road crashes being a result of incompatible situation awareness between different types of road users (e.g. drivers, cyclists and motorcyclists). The aim of the design process was then to develop intersection designs that would promote compatible situation awareness amongst road users. The context for the design was identified as an urban environment in Australia, with mixed residential, retail/business land use around the intersection. The scope of the design process was design of the roadway or road infrastructure rather than considering the design of vehicles, in-vehicle devices, training and licensing. Both blue sky designs and designs appropriate for retrofit to existing intersections were within scope.

Three sets of evaluation criteria for determining the success of the design concepts were identified in the design criteria document. The first was the extent to which they address key target behaviours identified in the EAST and schema analyses as being desirable to encourage the promotion of situation awareness through design. These included that the design influences drivers to look for cyclists, motorcyclists and pedestrians; the design influences drivers to look where cyclists, motorcyclists and pedestrians may be; the design ensures that drivers perceive cyclists, motorcyclists and pedestrians; the design influences cyclists and motorcyclists to engage in predictable behaviour and the design ensures that drivers gain experience of cyclists, motorcyclists and pedestrians. The second set of design criteria were drawn from the WDA developed for intersections. These included that the design should minimise collisions, injuries, fatalities; minimise risk; minimise violations, minimise time taken to traverse the intersection; optimise flexibility and maximise reliability. It was noted that the safety-related measures should be considered more important due to their importance in relation to the overall project. The final set of evaluation criteria incorporated the principles of sociotechnical systems theory.

A two-day workshop was held with road safety stakeholders from academia and industry. An example of the design concepts developed from this process was the *Self-regulating intersection.* This concept was based on the principles of a roundabout, however, rather than a traditional roundabout, it incorporates a large oval-shaped median strip in the centre of the intersection so that motorised traffic would be unable to perform a standard right-hand turn. Instead, when traffic from each intersecting road is given priority to enter the intersection, they would move around the median strip in the same direction, and exit where they wish. Cyclists have the option to either move with the motorised traffic or to 'cut through' via dedicated lanes available through the central median strip. Within the intersection, no lane markings are provided with the intention to promote connectedness between road users and require them to negotiate their way through with other road users (similar to a shared space). Filtering lights would allow vehicles to enter the intersection in a steady stream and once in the intersection the traffic stream would self-regulate the speed of the intersection which would be expected to be slow (i.e. 20 km per hour).

11.4 EVALUATION RESULTS

Following the workshops for each of the three case studies described above, participants were asked to rate their agreement against evaluation statements to gauge their reactions to the design process. Overall, these early evaluation results have been encouraging. For example, across all applications of the toolkit, participants agreed or strongly agreed that the workshops facilitated creativity. To illustrate, all ticketing system participants agreed the process facilitated creativity (Read et al., 2015c); 95% of the RLX design participants agreed that the workshop activities facilitated them to generate a variety of different kinds of ideas (Read et al., 2016) and all participants in the intersection design workshop agreed that the process assisted them to generate novel ideas, to think about the design problem in a different way and that they felt creative during the workshop (Read et al., 2015a).

Participants in the workshops also provided feedback relating to the validity of the process. In the ticketing system group, all participants agreed that the process was valid (Read et al., 2015c), in the RLX case study 90% of participants agreed that the process produced effective designs to improve human behaviour at RLXs and 100% of participants agreed that the process provided answers to relevant design problems (Read et al., 2016). Those in the intersection design workshop all agreed that the process would be useful for other safety-related projects and that the process provided answers to relevant design problems (Read et al., 2015a).

Overall, the evaluation results gained from participants have suggested their satisfaction with the process providing an indication that the process has an acceptable level of face validity.

11.5 CONCLUSION

The CWA-DT has been developed to provide a practical resource for HFE practitioners to assist them to design based on the outputs of CWA. Three case study applications described here have shown how the toolkit can be useful in planning and conducting a design process involving users and stakeholders. The case study applications have shown that the toolkit can be used flexibly and tailored to the needs of a particular design project and the early evaluation results have suggested that the process is generally acceptable to users and system stakeholders.

The methodological implications of this work lie in the extension to CWA ensuring theoretical consistency through the adoption of the values and principles of the sociotechnical systems approach. However, there is a need to further investigate the effectiveness of the toolkit. An important area for further evaluation is the validity of the outcomes of the design process. That is, to what extent do the designs meet the criteria set for their evaluation based on rigorous evaluation methods such as user testing with prototypes, simulation and modelling or safe to fail experiments in the real world (i.e. using step 9 of the toolkit). Furthermore, there is a need to test the toolkit in domains outside of transport to determine its more general applicability. Finally, there is a need to test the usability of the toolkit for HFE professionals not involved in its development. We would be interested to understand whether the CWA-DT is useful, usable and otherwise meets the needs of HFE professionals in the quest to design safer, more efficient and more effective sociotechnical systems.

ACKNOWLEDGEMENTS

The research discussed in this chapter was funded by an Australian Research Council Linkage Grant (ARC, LP100200387) to the University of Sunshine Coast, Monash University and the University of Southampton, in partnership with the following partner organisations: the Victorian Rail Track Corporation, Transport Safety Victoria, Public Transport Victoria, Transport Accident Commission, Roads Corporation (VicRoads) and V/Line Passenger Pty Ltd. and by the Australian Research Council Discovery Scheme (grant number DP120100199). Paul Salmon's contribution to this chapter was funded through his Australian Research Council Future Fellowship (FT140100681).

REFERENCES

Brown, T. 2008. Design thinking. *Harvard Business Review*, 6, 84–92.

Cherns, A. 1976. The principles of sociotechnical design. *Human Relations*, 29, 783–792.

Cherns, A. 1987. Principles of sociotechnical design revisited. *Human Relations*, 40, 153–161.

Clegg, C. W. 2000. Sociotechnical principles for system design. *Applied Ergonomics*, 31, 463–477.

Davis, L. E. 1982. Organization design. In G. Salvendy (Ed.). *Handbook of Industrial Engineering*. New York, NY: Wiley.

De Bono, E. 1992. *Serious Creativity: Using the Power of Lateral Thinking to Create New Ideas*. New York, NY: HarperBusiness.

Dul, J., Bruder, R., Buckle, P., Carayon, P., Falzon, P., Marras, W. S., Wilson, J. R. and Van Der Doelen, B. 2012. A strategy for human factors/ergonomics: Developing the discipline and profession. *Ergonomics*, 55, 377–395.

ISO. 2010. *Human Centred Design for Interactive Systems*. Geneva, Switzerland: International Organization for Standardization (ISO).

Jenkins, D. P., Salmon, P. M., Stanton, N. A. and Walker, G. H. 2010. A new approach for designing cognitive artefacts to support disaster management. *Ergonomics*, 53, 617–635.

Liedtka, J. and Ogilvie, T. 2010. *Designing for Growth: A Design Thinking Tool Kit for Managers*. New York: Columbia Business School Publishing.

Lintern, G. 2005. Integration of cognitive requirements into system design. *Proceedings of the Human Factors and Ergonomics Society 49th Annual Meeting*, 49, Orlando, FL, pp. 239–243.

Lockton, D., Harrison, D. and Stanton, N. A. 2010. The design with intent method: A design tool for influencing user behaviour. *Applied Ergonomics*, 41, 382–392.

Mendoza, P. A., Angelelli, A. and Lindgren, A. 2011. Ecological interface design inspired human machine interface for advanced driver assistance systems. *Intelligent Transport Systems, IET*, 5, 53–59.

Naikar, N. 2013. *Work Domain Analysis: Concepts, Guidelines and Cases*. Boca Raton, FL: Taylor and Francis Group.

Neisser, U. 1976. *Cognition and Reality: Principles and Implications of Cognitive Psychology*. San Francisco: Freeman.

Norros, L. 2014. Developing human factors/ergonomics as a design discipline. *Applied Ergonomics*, 45, 61–71.

Rasmussen, J., Pejtersen, A. M. and Schmidt, K. 1990. *Taxonomy for Cognitive Work Analysis*. Roskilde, Denmark: Risø National Laboratory.

Read, G. J. M., Salmon, P. M. and Lenne, M. G. 2012. From work analysis to work design: A review of cognitive work analysis design applications. *Proceedings of the Human Factors and Ergonomics Society Annual Meeting*, 56, Boston, MA, pp. 368–372.

Read, G. J. M., Salmon, P. M. and Lenné, M. G. 2015a. The application of a systems thinking design toolkit to improve situation awareness and safety at road intersections. *Procedia Manufacturing*, 3, 2613–2620.

Read, G. J. M., Salmon, P. M. and Lenné, M. G. 2015b. Cognitive work analysis and design: Current practice and future practitioner requirements. *Theoretical Issues in Ergonomics Science*, 16, 154–173.

Read, G. J. M., Salmon, P. M. and Lenné, M. G. 2016. When paradigms collide at the road-rail interface: Evaluation of a sociotechnical systems theory design toolkit for cognitive work analysis. *Ergonomics*, 11, 1–23.

Read, G. J. M., Salmon, P. M., Lenné, M. G. and Jenkins, D. P. 2015c. Designing a ticket to ride with the cognitive work analysis design toolkit. *Ergonomics*, 58, 1266–1286.

Read, G. J. M., Salmon, P. M., Lenné, M. G. and Stanton, N. A. 2015d. Designing sociotechnical systems with cognitive work analysis: Putting theory back into practice. *Ergonomics*, 58, 822–851.

Sinclair, M. A. 2007. Ergonomics issues in future systems. *Ergonomics*, 50, 1957–1986.

Trist, E. L. and Bamforth, K. W. 1951. Some social and psychological consequences of the longwall method of coal-getting: An examination of the psychological situation and defences of a work group in relation to the social structure and technological content of the work system. *Human Relations*, 4, 3–38.

Vicente, K. J. 1999. *Cognitive Work Analysis: Toward Safe, Productive, and Healthy Computer-Based Work*. Mahwah, NJ: Lawrence Erlbaum Associates.

Walker, G. H., Gibson, H., Stanton, N. A., Baber, C., Salmon, P. and Green, D. 2006. Event analysis of systemic teamwork (EAST): A novel integration of ergonomics methods to analyse C4i activity. *Ergonomics*, 49, 12–13.

Walker, G. H., Stanton, N. A., Salmon, P. M. and Jenkins, D. P. 2009. *Command and Control: The Sociotechnical Perspective*. Aldershot: Ashgate.

12 Cognitive Work Analysis for Systems Analysis and Redesign

Rail Level Crossings Case Study

Paul M. Salmon, Gemma J. M. Read, Michael G. Lenné, Christine M. Mulvihill, Nicholas Stevens, Guy H. Walker, Kristie L. Young and Neville A. Stanton

CONTENTS

12.1 INTRODUCTION

Rail level crossings are 'at grade' intersections comprising rail vehicles and their infrastructure, and other travel modes, usually roads. Typically the train has priority and other types of traffic have to be deconflicted with the train's movement. This is achieved by inducing other non-rail traffic to stop and wait until the train has passed. In engineering terms achieving this deconfliction is a simple problem and involves the use of various risk controls such as signs, road markings, warning lights and barriers.

Unfortunately, these controls are not always effective, leading to collisions between trains and vehicles or trains and pedestrians. Despite a high focus on rail level crossing safety in recent years, collisions continue to occur at unacceptable rates and with unacceptable consequences. In addition to analyses of high profile, high fatality incidents (e.g. Salmon et al., 2013), much has been written on the numbers and outcomes in terms of fatalities and injuries in different jurisdictions (see Evans, 2011; Hao and Daniel, 2014; Salmon et al., 2016). It is clear that collisions at rail level crossings represent a significant problem for which existing solutions are not working.

This chapter argues that a new approach is needed to identify and remove issues that contribute to collisions at rail level crossings. Specifically, it is argued that the cognitive work analysis framework (CWA; Vicente, 1999) provides a suitable approach for analysing and redesigning rail level crossing systems. Accordingly, we present an overview of a rail level crossing design lifecycle process that involved applying CWA first to analyse existing rail level crossing systems, and then to generate, evaluate and refine new rail level crossing design concepts. The process was adopted as part of a major research programme currently being undertaken in Victoria, Australia. The overall aim of the research programme is to develop and evaluate new rail level crossing system designs underpinned by systems thinking. To demonstrate, we provide an overview of the process and discuss selected outputs from each of the following phases: systems analysis; generation of design concepts, evaluation of design concepts and refinement of design concepts.

12.2 COGNITIVE WORK ANALYSIS

An overview of CWA and its different phases is presented in Section I. To support the aim of understanding rail level crossing system behaviour and then developing new rail level crossing design concepts, CWA was used as part of an overall rail level crossing systems analysis and design process, presented in Figure 12.1. This involved four key phases beginning with data collection, followed by the application of CWA, the development of design concepts and the evaluation and refinement of those concepts.

In addition to the CWA framework, the design process utilised the recently developed CWA-Design Toolkit (CWA-DT, see Chapter 11; Read et al., 2016a,b). The CWA-DT was developed to assist CWA users to identify design insights from CWA outputs and to use these insights within a participatory design process. It promotes

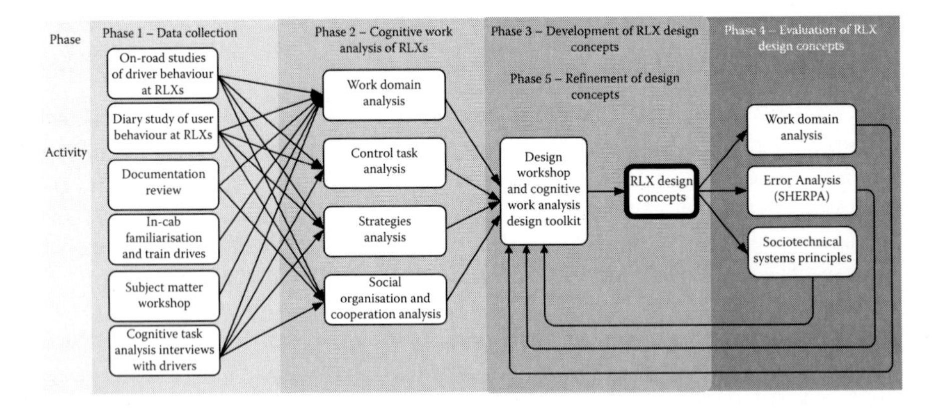

FIGURE 12.1 An overview of the CWA-based design lifecycle (*RLXs*, rail level crossings).

the collaborative involvement of experts (i.e. ergonomics professionals, designers and engineers), stakeholders (i.e. company representatives, supervisors and unions) and end users (i.e. workers or consumers) to solve design problems, based on insights gained through CWA.

Phases 1 and 2: Data Collection and Cognitive Work Analysis of Rail Level Crossings – Phases 1 and 2 culminated in an in-depth CWA of rail level crossings in Victoria, Australia. The analysis focused on both 'active' and 'passive' rail level crossings. Active crossings are so-called because they have 'active' warning devices that provide a warning of an approaching train, including flashing lights, boom gates and warning bells. Passive crossings, on the other hand, do not have active warnings and rely on static warnings of the presence of rail level crossing only (e.g. road signs, road markings and rail level crossing markers).

In phase 1, a series of data collection efforts were undertaken in order to gather detailed information to support the CWA of rail level crossings (see Salmon et al., 2016). These included

- On-road studies of driver behaviour at rail level crossings incorporating driver verbal protocols
- Critical decision method (CDM; Klein and Armstrong, 2005) interviews with drivers regarding decision making at rail level crossings
- A diary study of user behaviour (drivers, pedestrians, cyclists and motorcyclists) at rail level crossings with and without trains approaching
- Subject matter expert workshops
- Documentation review including a review of current rail level crossing design standards, accident reports, etc.
- Analysis of selected rail level crossing collisions
- In-cab train rides

12.2.1 Work Domain Analysis

The data derived from these activities was first used to support the development of work domain analysis (WDA) models. A summary of the active rail level crossing WDA is presented in Figure 12.2.

A discussion of the WDA is presented in Salmon et al. (2016). For the purposes of this chapter it is worth drawing out some of the key findings. First, at the functional purpose level it is notable that rail level crossings appear to have competing purposes. For example, maintaining priority access for rail traffic whilst minimising delays to the road network is difficult if not impossible to achieve. Second, at the values and priorities level the extent to which stakeholders possess the data required to measure different priorities is questionable. For example, it is debatable whether the road and rail sectors possess accurate data on the level of risk associated with different rail level crossings. Likewise, there are challenges in gaining an accurate picture of the number of road rule violations and near misses at rail level crossings. An assumption then is that road and rail organisations may not fully understand the extent to which rail level crossings are meeting key values and priorities. Third, the generalised function level gives an indication of the complexity of rail level crossings

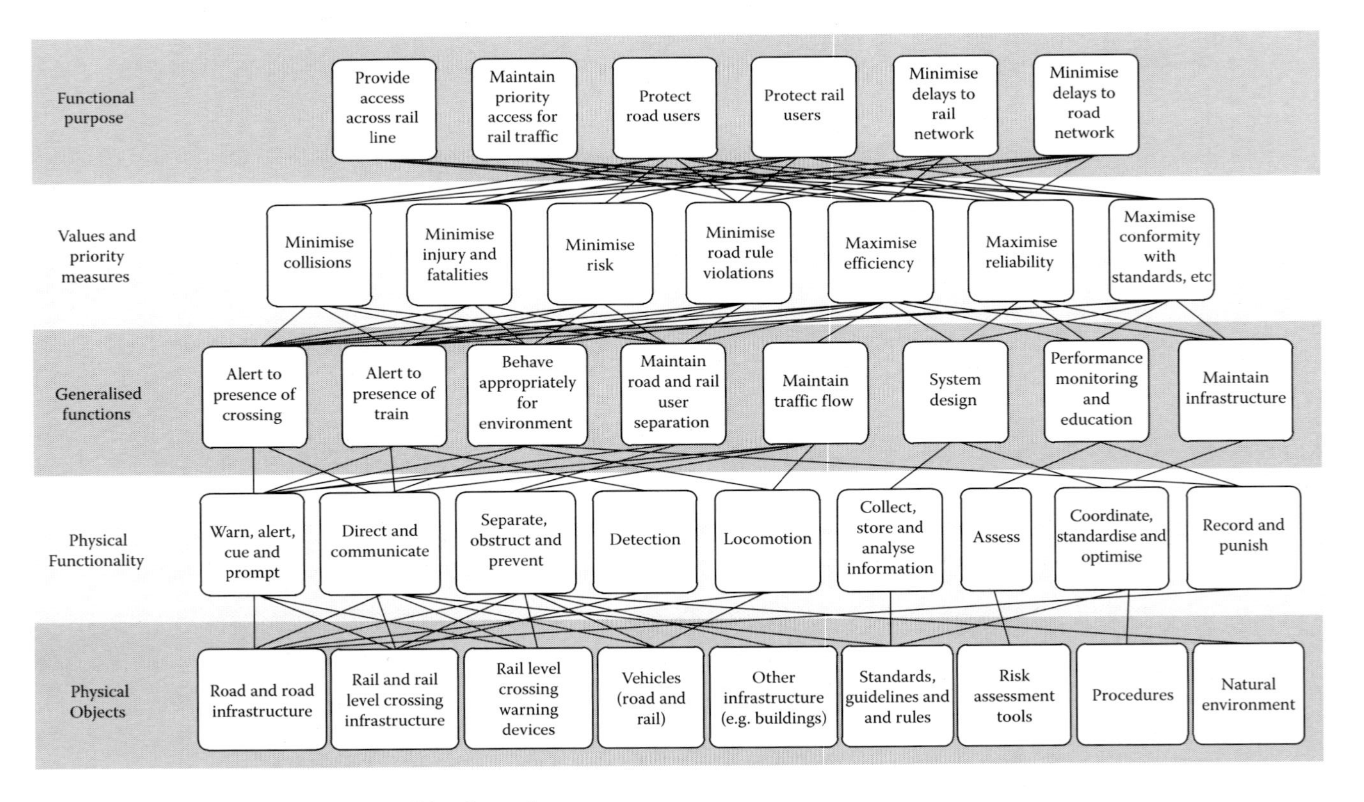

FIGURE 12.2 Summary of WDA of active rail level crossings.

and the collisions that occur. This level shows clearly that there are various ways in which the system can fail and create rail level crossing collisions. For example, the crossing failing to alert the road user to the presence of a train represents one failed function that can cause a collision. A road user not behaving appropriately (i.e. using a mobile phone whilst on approach) is another direct pathway to a potential collision. On the contrary, collisions may be an emergent property of the failure of various different functions. For example, poor maintenance could lead to the active warnings not working and road users not being warned of the approaching train. This diverse set of collision pathways ensures that various defences are required to prevent collisions. Another important aspect of this level is that it shows how functions physically and temporally distant from rail level crossings themselves have a bearing on behaviour and safety at the crossing. For example, functions such as 'system design' and 'performance monitoring and education' can conceivably play a role in creating or indeed preventing collisions even though the particular activity might occur or fail to occur days, weeks, months, or even years before an incident (Salmon et al., 2013). An example would be where performance monitoring and education does not communicate to users a series of near miss incidents at a particular crossing. Fourth and finally, the lower two levels of the WDA show the physical objects along with their affordances. At the bottom level, physical objects were grouped into the following categories: road and road infrastructure (e.g. the road, kerb and lane markings), rail and rail level crossing infrastructure (e.g. tracks, whistleboard and train detection systems), rail level crossing warning devices (e.g. flashing lights, early warning signage and rail level crossing markers), vehicles (e.g. cars, trucks and trains), other infrastructure (e.g. buildings), standards, guidelines and rules (e.g. road rules, road and rail level crossing design standards), risk assessment tools (e.g. rail level crossing risk assessment tools), procedures (e.g. safety and maintenance procedures) and the natural environment (e.g. vegetation and weather conditions).

12.2.2 Control Task Analysis

The control task analysis (ConTA) phase analyses the control tasks that are performed to support the system in meeting its functional purposes. Rasmussen's decision ladder method provides one approach for analysing in-depth the decisions that users make during these control tasks. The present analysis focussed specifically on the 'stop or go' decision whereby users decide whether they should proceed through the crossing or stop at the crossing and wait for an approaching train to pass. This decision is the key decision involved in safely negotiating rail level crossings when a train is approaching and is one for which an inappropriate decision can lead to a collision or near miss with a train.

Using the data from the on-road studies, CDM interviews and diary study we applied the decision ladder to understand the 'stop or go' decision from the point of view of different users, including drivers, pedestrians, cyclists and motorcyclists.

Initially, a generic decision ladder for the 'stop or go' decision was populated based on the data (see Figure 12.3). This involved taking data from the diary study and mapping them onto the appropriate sections of the decision ladder. For example, responses to the question 'what information did you use to make your decision?'

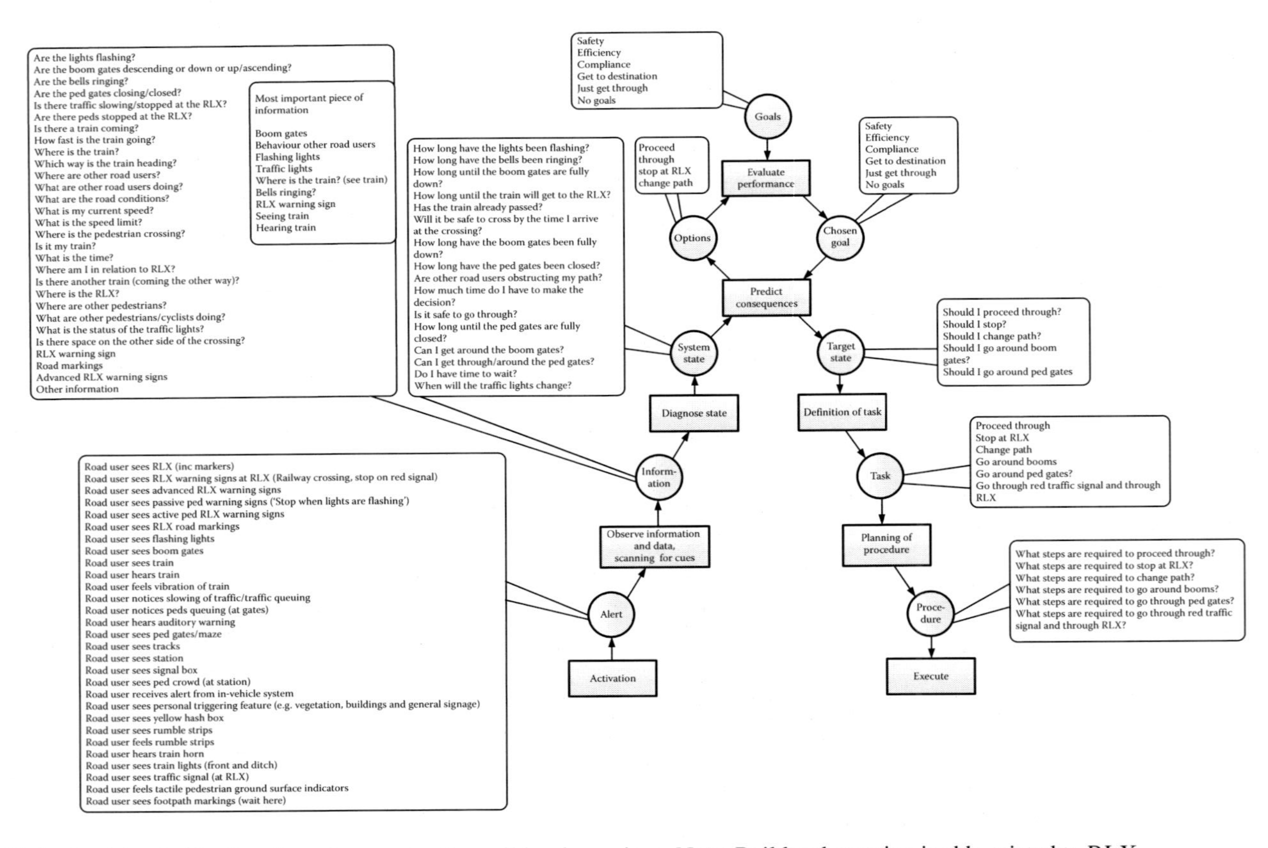

FIGURE 12.3 Decision ladder for all road users at active rail level crossings. Note: Rail level crossing is abbreviated to RLX.

were added to the 'Information' component of the decision ladder. The decision ladder presented in Figure 12.3 therefore represents an overview of the possible decision-making processes adopted by users during encounters with a train.

The decision ladder shows that there are multiple sources of information that road users and pedestrians reported using to inform their 'stop or go' decision (the information and system state components in Figure 12.3). In addition to expected sources of information, such as signage, flashing lights and boom gates, other pieces of information reportedly used included the behaviour of other road users, judgements about own behaviour (such as 'what is my current speed?') and personal triggering features such as a particular environmental features (e.g. houses and vegetation).

When asked what the most important piece of information was in determining whether to stop or go, participants described a range of sources including boom gates, behaviour of other road users, flashing lights, traffic lights, where the train is, ringing bells, rail level crossing warning signs, and seeing or hearing the train. The options identified by participants as available to them on approach to the crossings included to proceed through, stop at the crossing or change path and the goals influencing behaviour were safety, efficiency, compliance, getting to desired destination or 'just to get through'. The remainder of the decision ladder depicts the procedures available to cross and the users' choice of procedure. The procedures available included: to stop, to proceed through, to change path (in order to avoid the crossing), to go around the boom gates, to go around the pedestrian gates or to pass through the traffic lights and then the crossing.

Overall, the decision ladder analysis demonstrated the complexity of decision making for rail level crossing users which involved key differences between types of road user (e.g. driver, cyclist and pedestrian), between individuals, and between individuals in different situations. The overriding conclusion from the decision ladders was that the user decision-making process differs significantly, both within and across different user groups (e.g. drivers vs. pedestrians).

12.2.3 Strategies Analysis

The strategies analysis followed Cornelissen et al.'s (2013) strategies analysis diagram (SAD) methodology, whereby verbs (e.g. drive, ride, walk, queue and anticipate) and criteria (e.g. weather, visibility, distraction and impairment) are added to the WDA in order to identify the range of strategies possible for different road users in RLX environments. In addition, the strategy identification process was supported by the data derived from the on-road study verbal protocols, the CDMs and the diary study responses.

The SAD generated a number of important insights. First, multiple strategies were identified for each type of road user; essentially there are many ways in which different users can pass through rail level crossing environments. Particularly important for design efforts was the finding that different users use a diverse set of physical objects to support passage through the crossing. For example, drivers can receive a warning of an approaching train from multiple sources, including the train itself, the level crossing warning devices and the behaviour of other road users. Whilst this points to a high level of redundancy within the system and also the multiple roles of objects within the system, it also highlights that one design solution will likely not be suitable for all users. Second, an important gap in current rail level crossing warnings is the failure to provide specific information regarding approaching trains (e.g. time to arrival at crossing, speed). Currently the only information typically provided is that a train is approaching; however, a number of the strategies identified would be better supported through the provision of more useful information, such as time to arrival, number of trains approaching and time that user will be delayed at the crossing. Indeed, other strategies are currently adopted by users to gather this information, such as pedestrians looking down the track and attempting to gauge time to arrival. A third and final important insight from the SAD was that there are a range of potential conflicts when different users adopt different strategies to pass through crossings. For example, strategies adopted by one form of user (e.g. cyclists crossing on the footpath to avoid traffic) can impede or prevent a strategy for another form of user (e.g. pedestrians attempting to pass through via the footpath).

12.2.4 Social Organisation and Cooperation Analysis

The social organisation and cooperation analysis (SOCA) involved mapping different human (e.g. drivers and pedestrians) and non-human actors (e.g. rail level crossing warning devices and detection systems) onto the WDA and decision ladder to identify how functions, affordances, decisions and strategies are currently allocated across actors, and also to identify how they could be allocated in order to identify potential redesign recommendations.

The outputs of the SOCA provide an exhaustive description of who and what currently do what, and could do what, in rail level crossing systems. For example, for the generalised function, 'alert to presence of rail level crossing' the following actors currently contribute: detection and alert systems (e.g. rail level crossing signage), regulators/authorities (through providing road and rail infrastructure) and the physical infrastructure (the rail level crossing itself). The formative analysis shows which of the actors identified could potentially contribute to the functional purposes and functions following system redesign. For example, for the functions, 'alert to presence of rail level crossing' and 'alert to presence of a train' the 'road user' group was included since an in-vehicle display or intelligent transport system could potentially provide a warning of an approaching rail level crossing or train.

Phase 3: Development of New Rail Level Crossing Design Concepts – Following interpretation of the CWA a series of initial rail level crossing design concepts were developed through a workshop incorporating the CWA-DT (see Chapter 11). Eighteen participants participated in the design workshop. Participants were invited as representatives of rail level crossing stakeholder organisations (i.e. government departments, regulators, road authorities, road user peak bodies and transport investigators) or as interested persons with a professional interest in the research (i.e. HFE professionals, researchers and designers).

The design workshop was delivered over 2 days and involved participants generating design concepts and solutions for improving behaviour and safety at rail level crossings. Workshop activities were based on the insights identified from the CWA outputs and included an idea generation phase (using targeted approaches such as assumption crushing and metaphor-based design), concept design and concept prioritisation. The workshop culminated in 11 rail level crossing design concepts prioritised by the workshop participants in order of those thought likely to be most effective in improving safety.

Phases 4 and 5: Evaluation and Refinement of Design Concepts – The highest ranked seven design concepts of the 11 created were selected by the research team for further evaluation. The four remaining design concepts were not selected based on the research teams' judgements regarding low alliance with systems thinking, practicality and likelihood of implementation.

The initial evaluation process involved three core activities: insertion of design concepts into the WDA abstraction hierarchy, identification of potential user errors and evaluation against sociotechnical systems values and principles. This enabled the evaluation to consider the extent to which the designs aligned with systems thinking along with the likely impact and emergent behaviours associated with each design concept.

Insertion of the concepts into the rail level crossing abstraction hierarchy involved adding the features of each design concept (e.g. in-vehicle display, new road markings, new warning signs and rumble strips) into the physical object level of the abstraction hierarchy presented in Figure 12.2, removing any nodes as appropriate, and then remodelling the means–ends relationships. For example, a new in-vehicle display would link upward to the physical functionality affordances of visual and auditory warning of approaching rail level crossing and approaching trains, which in turn would link up to the generalised functions of alerting users to the presence of crossing and presence of train, and so on. This process essentially enabled the initial WDA to be modified to describe in-depth each of the seven rail level crossing design concepts.

The modified WDAs were then used to assess the potential impacts of each new rail level crossing design concept. This was achieved by reviewing the modified abstraction hierarchies and summing the following:

- New nodes in the abstraction hierarchy, for example, the new physical object 'optimal speed to avoid train in-vehicle display' would 'communicate optimal speed' and 'provide distance to rail level crossing' notification.
- Support for existing nodes within the abstraction hierarchy, for example, the new physical object 'in-vehicle warning display' would provide support for the existing functions of 'alert user to the presence of rail level crossing' and 'alert user to the presence of train'.
- Appropriate restriction, for example, the new physical object 'default closed pedestrian gates' would appropriately restrict (pedestrian) traffic flow which in turn would support the function of 'maintain road and rail user separation'.
- Negative influence on nodes within the abstraction hierarchy, for example, the new physical object 'speed limit reduction sign' would have the effect of slowing traffic through rail level crossings which in turn could potentially negatively influence the 'maximise efficiency' value and priority measure.

Identification of user errors for each design concept was achieved by applying the systematic human error reduction and prediction approach (SHERPA, Embrey, 1986) to predict the likely errors that would arise when users interacted with the new rail level crossing. Initially, a hierarchical task analysis (HTA) was created for rail level crossings generally. Then one analyst modified the HTA for each design concept and applied SHERPA to identify the credible errors that could occur. The output was a series of likely errors for each concept, including a description of each error and the associated consequences, ratings of likelihood (low, medium and high) and probability (low, medium and high) and potential remedial measures. Metrics such as the number of existing potential errors reduced by the new design concept as well as new errors introduced were calculated.

The evaluation against sociotechnical systems theory values and principles involved the research team considering each design concept and providing a rating of 1 to 3 (low, medium and high) in relation to the following values and principles:

- Tasks are allocated appropriately between and amongst humans and technology.
- Useful, meaningful and whole tasks are designed.
- Boundary locations are appropriate.
- Boundaries are managed.
- Problems are controlled at their source.
- Design incorporates the needs of the business, users and managers.
- Intimate units and environments are designed.
- Design is appropriate to the particular context.
- Adaptability is achieved through multifunctionalism.
- Adaptability is achieved through flexible structures and mechanisms.
- Information is provided where action is needed.
- Means for undertaking tasks are flexibly specified.
- Authority and responsibility are allocated appropriately.
- System elements are congruent.

The ratings were then aggregated giving each design concept a total score. Based on the three evaluation activities five railway level crossing design concepts were selected for further refinement.

12.2.5 Refinement of Railway Level Crossing Design Concepts

The five selected railway level crossing design concepts were further refined through the conduct of a workshop involving ten participants. Prior to the workshop, participants were provided with a written summary of the top five ranked design concepts and a summary of the overall evaluation findings. At the start of the workshop, a detailed overview of each concept was provided. Following this participants engaged in a series of CWA-DT activities designed to refine each concept. This included reviewing suggested design improvements identified during the evaluation process, identifying additional design improvements and conducting an evaluation and final ranking of concepts following the inclusion of design improvements deemed to be practical and appropriate. The output of the concept refinement workshop was six refined design concepts (three for urban environments and three for rural environments). Example concepts are presented below in Figure 12.4.

12.3 DISCUSSION

Rail level crossings pose a significant threat to road and rail safety; exactly the sort of problem that systems thinking frameworks such as CWA are suited to solving. The aim of this chapter was to present an overview of how the authors have applied CWA to understand the problem of collisions at rail level crossings and then design new solutions to remove them. Accordingly, we have provided an overview of a rail level crossing analysis and design process underpinned entirely by CWA. Specifically, the CWA framework and its accompanying design tool, the CWA-DT, were used to analyse existing rail level crossing systems and develop and evaluate a series of new rail level crossing design concepts underpinned by systems thinking principles.

An increasing number of researchers are arguing for a so called systems approach to be taken when attempting to improve safety and behaviour in transport settings (e.g. Read et al., 2013; Salmon et al., 2013; Salmon and Lenné, 2015; Wilson and Norris, 2005). Previously an engineering based approach has been adopted, primarily focussing on the introduction of new warning devices and providing important information regarding their likely effectiveness upon implementation (e.g. Lenné et al., 2011; Tey et al., 2014). Applying CWA to understand the problem of rail level crossing collisions and then to design and evaluate solutions represents a novel systems approach to a longstanding issue. There are many and varied important findings that emerged from the overall process, including insights on why collisions occur and on what is required to prevent them in future. For a discussion the reader is referred to Salmon et al. (2016). To facilitate further systems thinking applications in transport analysis and design efforts the discussion here focusses on the strengths and weaknesses that emerged when applying CWA in the present study.

A key strength of CWA when applied to analyse existing rail level crossing systems as part of a design process was its depth and explanatory power. CWA was useful in that it examined the behaviour of multiple end users (e.g. drivers, pedestrians, cyclists, motorcyclists and train drivers) and the factors influencing their behaviour. In addition, the analysis covered the overall system and its constraints (WDA), user decision making (ConTA), the different ways in which users can behave in rail level crossing environments (strategies analysis) and the contribution of different actors, human and non-human, to rail level crossing system behaviour (SOCA). Together this provided a highly exhaustive analysis and ensured that the new design concepts consider multiple end users, rather than one group in isolation, and also multiple aspects of behaviour. Indeed, a previous criticism of analyses and safety interventions in road and rail safety has been that they tend to focus on one user group in isolation. Another important strength of CWA is that it can focus on overall systems, rather than, in this case, merely the physical manifestation of the rail level crossing. The present analysis was notable in that it identified factors outside of the rail level

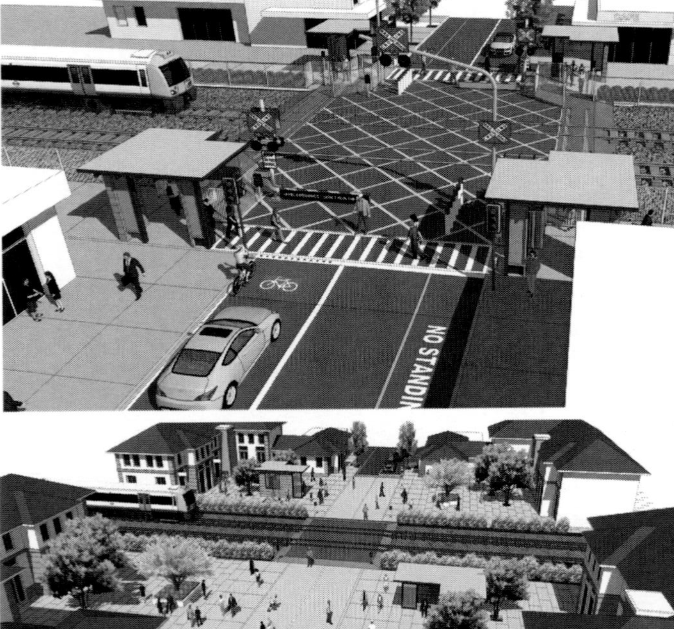

High technology crossing

Key features

- Camera enforcement, linked to traffic lights – with automatic fines via number plate recognition.
- Runway red lights in the stop line (in-road studs) that light up when a train is approaching.
- 'skirt' underneath boom gate to increase its physical presence and use this to provide safety messages and possibly for advertising to reduce costs.
- Traffic lights at the crossing with coordination to control up and downstream traffic
- A loop in the road to detect traffic queues on the far side of the crossing linked to an active 'Hold' (or 'Keep tracks clear') sign for approaching traffic, with 'clear tracks' sign displayed to traffic that might be on the crossing
- 'Break the chain' promotion/awareness campaign to discourage queuing on the crossing
- A 'no standing' area on road on exit side of crossing to provide space for vehicles caught on the crossing
- An emergency lane on exit side of crossing, with active bollards linked to the warnings to avoid everyday use of the lane
- A default closed pedestrian gate that is locked when train is approaching
- An all-across traffic phase for pedestrians – with official diagonal paths across the RLX
- Fences used to channel pedestrians towards the gates, so they don't use the road to cross
- A crossing supervisor is present at locations and times of higher risk (e.g. crossings with many near misses, peak times, etc.). Supervise the crossing, can assist in emergencies and reports maintenance and safety issues to management.

Community crossing

Key features

- A city square/courtyard feel with cafes, meeting areas and community information booths (community focused transit-orientated design)
- The stop line is moved back for road traffic, with traffic lights to control entry to the shared space
- Road traffic speed is reduced to 20km/h through the shared space area, and the road is raised to further delineate the area
- The train speed is reduced to 20km/h through the RLX
- Crossing supervisors stand in front of the RLX when the train approaches – would advise road users of appropriate behaviour
- Red man standing lights are activated for pedestrian traffic when train is approaching
- A different colour (i.e. orange) is used to delineate the track area/danger zone from the shared space area

FIGURE 12.4 Example rail level crossing design concepts produced from CWA-based design process.

crossings themselves that influence safety, such as design standards, risk assessment processes, and incident reporting systems. Whilst design concepts for the rail level crossing itself were produced, a series of recommendations on modifications to other parts of the system were also developed (e.g. modifications to incident reporting systems and improved risk assessment processes).

The next strength worth discussing is the use of CWA outputs throughout the analysis and design process. CWA was used to analyse existing rail level crossings, drive the design of new rail level crossing design concepts, evaluate the new design concepts, and then finally to drive further refinement of the design concepts. A key issue associated with human factors methods is that they often provide information to inform design processes, but they may not directly inform design. In the present study use of CWA throughout the design process ensured that CWA directly informed the design of new rail level crossing environments and also enabled a level of consistency across the analysis and design lifecycle. Importantly, it ensured that the new design concepts were explicitly linked to the problems identified through the original CWA analyses. Finally, the formative capabilities of CWA ensured that the impact of proposed design changes on system performance was considered and any negative impacts were removed through design concept refinement. This is a highly useful feature of CWA and one that can potentially contribute to huge cost savings in design efforts as very early design concepts (e.g. paper drawings) can be tested for their likely impacts on performance.

It is important to note some difficulties that emerged through the process. A significant level of resources was required to collect appropriate data and undertake each of the phases; certainly more so than a typical human factors study. However, the depth and explanatory power of the analyses produced is a logical benefit of this and justifies the level of resources invested. In addition, there was a high requirement for subject matter expert input throughout the process. Whilst this can be difficult to arrange, in the present study it ensured that the designs were matched to stakeholder needs and also that stakeholders maintained involvement in the entire process. A final difficulty worth noting is that collaborators unfamiliar with CWA may not readily grasp some of the terminology used within CWA phases and methods (e.g. 'constraints' and 'affordances'). In future applications it is important that such terms are clarified and explained early on in the process.

The challenge of embedding sociotechnical systems analysis and design methodologies within design processes has been noted (Eason, 2014). To conclude, it is our opinion that CWA provides a suitable framework for doing so. Moreover, the study presented has highlighted that the utility of CWA can be significantly heightened when it is used throughout the system design lifecycle for analysis, design, and design testing and refinement. Whilst it is useful for analysis alone, often the power of CWA can be lost when it is not used to drive the design process. Accordingly, it is recommended that further similar applications are undertaken across the safety critical domains. Of course, the acid test now lies in whether the designs produced are safer and more efficient than existing rail level crossing environments. Currently the authors are using driving simulation to test the RLX design concepts produced, with initial studies providing promising results (see Read et al., 2016b).

REFERENCES

Cornelissen, M., Salmon, P. M., McClure, R. and Stanton, N. A. 2013. Using cognitive work analysis and the strategies analysis diagram to understand variability in road user behaviour at intersections. *Ergonomics*, 56(5), 764–780.

Eason, K. 2014. Afterword: The past, present and future of sociotechnical systems theory. *Applied Ergonomics*, 45(2), 213–220.

Embrey, D. E. 1986. SHERPA: A systematic human error reduction and prediction approach, *Paper Presented at the International Meeting of Advances in Nuclear Power Systems*, Knoxville, Tennessee.

Evans, A. W. 2011. Fatal accidents at railway level crossings in Great Britain 1946–2009. *Accident Analysis & Prevention*, 43(5), 1837–1845.

Hao, W. and Daniel, J. 2014. Motor vehicle driver injury severity study under various traffic control at highway-rail grade crossings in the United States. *Journal of Safety Research*, 51, 41–48.

Klein, G. A. and Armstrong, A. 2005. Critical decision method. In N. A. Stanton, A. Hedge, K. Brookhuis, E. Salas and H. Hendrick (Eds.). *Handbook of Human Factors and Ergonomics Methods*. Boca Raton, FL: CRC Press, pp. 35.1–35.8.

Lenné, M. G., Rudin-Brown, C. M., Navarro, J., Edquist, J., Trotter, M. and Tomasevic, N. 2011. Driver behaviour at rail level crossings: Responses to flashing lights, traffic signals and stop signs in simulated rural driving. *Applied Ergonomics*, 42(4), 548–554.

Read, G., Salmon, P. M. and Lenné, M. G. 2013. Sounding the warning bells: The need for a systems approach to rail level crossing safety. *Applied Ergonomics*, 44, 764–774.

Read, G. J. M., Clacy, A., Thomas, M., Van Mulken, M. R. H., Stevens, N., Lenné, M. G., Mulvihill, C. M. et al. (2016b). Evaluation of novel urban rail level crossing designs using driving simulation. *Proceedings of the Human Factors and Ergonomics Society*, 60(1), 1921–1925.

Read, G. J. M., Salmon, P. M. and Lenné, M. G. (2016a). When paradigms collide at the road rail interface: Evaluation of a sociotechnical systems theory design toolkit for cognitive work analysis. *Ergonomics*, 59(9), 1135–1157.

Salmon, P. M. and Lenné, M. G. 2015. Miles away or just around the corner: Systems thinking in road safety research and practice. *Accident Analysis and Prevention*, 74, 243–249.

Salmon, P. M., Lenné, M. G., Mulvihill, C., Young, K., Cornelissen, M., Walker, G. H. and Stanton, N. A. 2016. More than meets the eye: Using cognitive work analysis to identify design requirements for safer rail level crossing systems. *Applied Ergonomics*, 53(Part B), 312–322.

Salmon, P. M., Read, G., Stanton, N. A. and Lenné, M. G. 2013. The Crash at Kerang: Investigating systemic and psychological factors leading to unintentional non-compliance at rail level crossings. *Accident Analysis and Prevention*, 50, 1278–1288.

Tey, L. S., Zhu, S., Ferreira, L. and Wallis, G. 2014. Microsimulation modelling of driver behaviour towards alternative warning devices at railway level crossings. *Accident Analysis & Prevention*, 71, 177–182.

Vicente, K. J. 1999. *Cognitive Work Analysis: Toward Safe, Productive, and Healthy Computer-Based Work*. Mahwah, NJ: Lawrence Erlbaum Associates.

Wilson, J. R. and Norris, B. J. 2005. Rail human factors: Past, present and future. *Applied Ergonomics*, 36(6), 649–660.

13 Work Domain Analysis Applications in Urban Planning
Active Transport Infrastructure and Urban Corridors

Nicholas Stevens, Paul M. Salmon and Natalie Taylor

CONTENTS

13.1 INTRODUCTION

Urban planning and design require new ways of interpreting urban form that allows for the understanding of multidisciplinary approaches and cooperative outcomes. This innovative study brings together the disciplines of Human Factors with Urban Design to investigate the form and design of active transport infrastructure (ATI) in urban transport corridors (Stevens and Buksh, 2013). Urban transport corridors are connectors of people and place that go beyond the roadway and consider the dynamics of the adjacent urban form. They are defined by the disparate demands of engineering, urban planning, urban design, property development, and community expectation. Using cognitive work analysis (CWA) the research detailed here investigates and identifies the interdependencies and ability of these urban corridors to operationalise ATI which permits best practice road user hierarchies that give highest priority to walking and cycling.

CWA (Vicente, 1999) is a systems analysis and design framework that identifies the constraints imposed on activities and designs new systems that better support the activities of interest. CWA has been used in a variety of design activities in several domains, including defence, disaster management, process control and road safety (Salmon et al., 2010). This research is an important contribution to the emerging use of complex systems approaches for the examination of urban planning and design challenges (Stevens and Salmon, 2014, 2015; Stevens, 2016). This work applies the first phase of CWA, work domain analysis (WDA), as a tool to interpret the interdependencies of urban form and its component elements. The application and extension of sociotechnical systems theory in an urban setting has allowed for unique insights.

A key aspect of this approach is that it is formative in nature describing what could happen if design modifications are undertaken, rather than providing normative analyses of what should happen. The results of this study allow for a clearer interpretation of the relationships between all of the physical elements of ATI and its intended functional purposes. Through the analysis of ATI this research provides a means to understand the multidisciplinary requirements for establishing urban form which supports safe and accessible use by pedestrians and cyclists.

13.2 BACKGROUND

This chapter proposes that the urban form of our cities needs to be underpinned by a better understanding and prioritisation of community access. It seeks to offer insights into the design of streets which are prioritised as key community conduits and destinations in their own right. It outlines a systems approach which enables multidisciplinary responses to the establishment of ATI which supports and encourages pedestrians and cyclists within the urban corridor.

Public space within cities is a dwindling resource, yet though appropriate design, or indeed redesign, our footpaths and roadways have the capacity to host community life. They offer the greatest opportunity and potential for our neighbourhoods and urban environments to act as social places. Indeed, the role that streets and footpaths should play in contributing to quality of life is firmly entrenched in urban design and urban planning theory (Lynch, 1960; Jacobs, 1961; Appleyard, 1980; Gehl, 2011).

There is a range of literature, plans and strategies that outline the known benefits of active transport, including, reducing vehicular congestion, greenhouse gas emissions and supporting individual and community health (Woodcock et al., 2007; Frank et al., 2010; SCRC, 2011b; Millward et al., 2013). This research instead focuses on the design and form of ATI. Specifically it is concerned with the provision of pedestrian footpaths and on-road cycling facilities within existing urban roadway contexts. Many of these roadways were never designed or intended to provide for active modes of transport and subsequently often do not support safe interactions between different road user groups.

This 'systems thinking' approach to road design is a key requirement for future road systems (Salmon et al., 2012). It is able to consider the relationships between the ideal functional purposes of ATI and highlight the system relationships. While the urban form and roadway contexts are key inputs into the system, the approach specifically identifies the range of interdependencies between functions, purposes and objects within a typical ATI configuration.

The literature on active transport emerges from a range of discipline areas including health promotion and preventative medicine (Frank et al., 2010; Hamer and Chida, 2008); transport geography (Kelly et al., 2011; Millward et al., 2013); and urban planning and design (Forsyth et al., 2007, 2008; Ewing and Cervero, 2010). This research provides new knowledge through the convergence of the Human Factors and Urban Design disciplines.

13.3 URBAN CORRIDORS AS COMMUNITY AND NEIGHBOURHOOD PLACES

The potential of roadways and more specifically existing urban corridors to act as community and neighbourhood places supported by active mobility is recognised in the literature, planning and policy (IPWEA, 2010; SCRC, 2011b; Stevens and Buksh, 2013). Additionally, there are a variety of evaluative techniques that seek to establish and reprioritise active transport, and the amenity to support it, through principles of urban design and health promotion (Cerin et al., 2007; Clifton et al., 2007; Ewing and Handy, 2009; Millward et al., 2013). While much has been done to better understand and evaluate the contributions of built form and community life which encourages active transport there is also a continued acknowledgement that implementation of appropriate strategies needs to be improved (Stevens and Buksh, 2013). This research recognises three main challenges that have inhibited the ability of urban corridors to be significant locations of active mobility.

First, is that past and present design standards for roadway corridors prioritise motor vehicles over all other uses (P&NJ DoT, 2008; TMR, 2013a). The engineering based standards are largely not concerned with the delivery of quality urban design or form, but the efficient and safe movement of motorised traffic. As transport agencies increasingly seek to enable active mobility, the supplied active transport infrastructures are required to be established within an existing roadway context. Footpaths and on-road cycleways are required to 'fit' within the roadway corridor without specific consideration of the needs of active mobility. This post hoc delivery of painted on-road cycle markings and standard concrete footpaths may fulfil departmental and agency policy for implementation, but they are not fit for their purpose.

Second, is that the engineering or urban design guidance which does seek to enable walkable and cyclable neighbourhoods is very rarely considered in an integrated manner. The engineering standards are not interested in ensuring visual amenity and sense of place, and the normative principles of urban design are often difficult to apply in practice. That is, much of the urban design guidance for ATI is often descriptive or illustrative (ITE, 2010; IPWEA, 2010). The provision of key principles and listed objectives of primary considerations; supported by illustrations and elevations of what should be done, may be argued to limit the in-practice uptake of such advice (Stevens and Buksh, 2012).

Third, current active transport 'solutions' rarely seek to exemplify safe, efficient or best practice sustainable transport, nor do they support a high-quality active transport user experience. As such, despite implementation of ATI, increases in activity are often limited (Alfonzo, 2005; Paige Willis et al., 2013). Further, there is very little evidence that such infrastructure considers the implications or interactions between active transport modes which are often also incompatible, for example pedestrians and cyclists.

The safety of active transport users continues to represent a significant issue, both in Australia and worldwide. Of the 1303 crash-related fatalities occurring in Australia during 2012, 174 were pedestrians and 33 were cyclists (BITRE, 2013). The challenge is how to retrofit urban corridors which are safely engineered for active transport and establish urban form which encourages use through contributions such as amenity and sense of place. The identification of design frameworks that can support the development of appropriate active transport infrastructures which prioritise walking and cycling is critical.

The application of systems analysis and design approaches for roadway configuration and evaluation is gaining momentum (Salmon et al., 2014; Cornelissen et al., 2013). This research argues that the CWA framework (Vicente, 1999) provides a suitable design approach for ATI within the urban corridor. The aim is to show how CWA can contribute to the design of safe and efficient active transport environments which provide a high-quality user experience. This chapter outlines an approach whereby it is possible to better understand the role and interactions of ATI within the urban corridor. The aim is to provide a single analysis with the potential to consider both the engineering and technical standards of safe ATI with the urban design opportunities and experiential contributions as high-quality places for walking and cycling. Much of the current literature and policy seeks to deal with these issues independently, or in turn, here they are considered as interdependent.

The study described uses the first phase of the CWA framework, WDA to model the 'ideal' ATI configuration, considering the current practices of retrofitting existing urban corridors.

13.4 URBAN CORRIDORS

The idea of the urban corridor is not new and has historically been used to describe the growth and development between key activity centres (Sargeant et al., 2009). Indeed the concept of the urban corridor has been used a spatial planning unit of analysis in several national and international investigations (Aust. Gov, 2007; OA, 2010; Government of South Australia, 2011).

These transport focussed corridors, within an urban environment, are naturally complex as they transverse across different scales and are influenced by a collection of economic, environmental and social relationships connecting people and place. Hale (2011) and Arrington and Cevero (2008) identify the need to carry out corridor analysis at different scales to better understand how a particular transport initiative, corridor upgrade, or transport orientated development (TOD) proposal relates to the broader urban setting.

Hale (2011) argues that this is a practical step forward and must be carried through the development of analytical tools, ultimately leading toward better outcomes for projects at different spatial scales over time. Urban corridors of any scale or function and their surrounding urban catchment have reciprocal latent and explicit impacts which require further investigation, understanding and management (Stevens and Buksh, 2013).

For the purposes of this paper the *urban corridor* represents a 400 m buffer along road-based transport corridors, often containing public transport (Adams, 2009). These urban corridors are key development areas and community conduits which respond to and consider adjacent site, local and corridor urban context.

13.5 ASSEMBLING THE WORK DOMAIN ANALYSIS FOR ACTIVE TRANSPORT INFRASTRUCTURE

The development of the WDA was based on key design guidelines and standards, best practice documentation and the literature on urban design, active transport, and on-road cycleway and footpath engineering (See Table 13.1). This data represents national standards for footpath, pathway and cycleway design; national and international engineering design guidance; and governmental active transport and urban design policy and protocol. It is indicative of the range of active transport design guidance from across the often disparate disciplines of engineering, transport planning and urban design. Based on these inputs the WDA was developed by one Human Factors analyst and one Urban Planning practitioner.

13.5.1 Active Transport and Urban Design Literature

The WDA has also drawn on a range of urban design and active transport literature to allow for the establishment of high-quality active transport user experiences. It is important to note that the concepts described here also emerge across a range of practice and academic literatures and are by no means exhaustive on the topic of active transport infrastructures or the operationalisation of urban design principles or subjective qualities. Concepts within this study represent those which respond largely to the context of policy and practice in South East Queensland (SEQ) and Australia more generally (Aust Gov, 2011; SCRC, 2011a; PIA, 2010). Additionally there has been the inclusion of key elements and considerations that have emerged within the literature, planning and policy relating to: walkability and active transport (Purciel and Marrone, 2006; Ewing and Handy, 2009; Ewing and Cervero, 2010; Millward et al., 2013); urban thoroughfares and corridors (Premius and Zonneveld,

TABLE 13.1
Inputs of WDA

Australian and Queensland Design Standards
Austroads (2009) '*Guide to Road Design – Part 6A: Pedestrian and Cyclist Paths'*, Austroads Incorporated, Sydney, Australia.
Queensland Department of Transport and Main Roads (2014) *Transport Operations (Road Use Management – Road Rules) Regulation 2009).* Queensland Government, Brisbane.
Queensland Department of Transport and Main Roads (2013a) *Road Planning and Design Manual (2nd Ed).* Integrated Transport Planning Branch, Brisbane, Queensland.
Queensland Department of Transport and Main Roads (2013b) *Queensland Manual of Uniform Traffic Control Devices Part 9 Bicycle Facilities.* Queensland Government, Brisbane.
National and International Engineering Guidelines for Active Transport Infrastructure
Institute of Public Works Engineering Australia (Qld) (2011) '*Complete Streets: Guidelines for Urban Street Design'*, Queensland Division Inc, Fortitude Valley, Queensland.
Institute of Transportation Engineers (ITE) (2010) '*Designing Walkable Urban Thoroughfares: A Context Sensitive Approach'*. Institute of Transportation Engineers, Washington, DC.
US DOT, 2010, 'International Technology Scanning Program: Pedestrian and Bicyclist Safety and Mobility in Europe', US Department of Transportation, National Cooperative Highway Research Program, February 2010, viewed 28th March 2014.
WSDoT (2011) Roadway Bicycle Facilities Design Manual M 22.01.08, Washington State Department of Transport,1520–1, July 2011
National and International Urban Design and Active Transport Policy and Literature
Australian Government (2011) '*Creating Places for People: An Urban Design Protocol for Australian Cities.*' Canberra.
Australian Government (2009) 'Healthy spaces and places. A national guide to designing places for healthy living' Australian Government Department of Health and Aging, Kingston, Canberra.
Davies, L. (2000). '*Urban design compendium.*' London: English Partnerships.
Dixon, K., Liebler, M., Zhu, H. (2008) '*NCHRP Report 612 Safe and Aesthetic Design of Urban Roadside Treatments.* Transportation Research Board, Washington: DC.
Planning Institute of Australia (2010) '*Urban Design Unit Manual. Planning Institute of Australia.*' AKH Consulting.
Stevens, N. and Buksh, B. (2012). *A Leading Practice Framework for Sustainable Transport Corridors: Sustainable Transport Corridor Research.* Final Report to Queensland Department of Transport and Main Roads. University of the Sunshine Coast.
Sunshine Coast Regional Council (2011b), '*Sunshine Coast Active Transport Plan 2011–2031'*. Regional Strategy and Planning Department, Queensland.

2003; ITE, 2010; Stevens and Buksh, 2012); pathway and cycleway design guidelines (Austroads, 2009; WSDoT, 2011); user-orientated spaces (Yücel, 2013) and quality human environments (Gehl, 2011).

13.6 URBAN CONTEXT OF ATI WDA

When considering the urban design and engineering requirements of ATI within urban corridors it is necessary to acknowledge and understand the context within which they occur (Premius and Zonneveld, 2003; Stevens and Buksh, 2012). First, it is important to identify the transport task and roadway context (e.g. highway, arterial road, local road, etc.). Second, is understanding the type of neighbourhood within which they occur (rural, suburban, urban centre, CBD, etc.) and third, is the context of the adjacent built environment within the corridor (residential, commercial, retail, open space, etc.).

ATI may be found in many urban roadway, neighbourhood and built environment circumstances. However, the interdependency and influence of these three contexts has significant implications for both their physical design and also the quality of the user experience. For example pedestrians and cyclists have no role to play in the context of major highways, but may be expected to be key contributors to the success and vibrancy of inner city streets.

It is important to note that the WDA is actor independent and does not explore the behaviour of different users of the system. As such the WDA includes all objects that may be important to some users and of no consequence to others, for example tactile surface indicators or bike racks. The system view as represented by the WDA considers that the constraints within it are the same for all users.

13.7 ROADWAY CONTEXT FOR ACTIVE TRANSPORT INFRASTRUCTURE

This research is principally interested in urban corridors that contain higher order roads that are administered and managed by state governments. As such it has utilised the urban corridor and roadway context established by Stevens and Salmon (2014) which considered the influence of 'multi-modal urban arterial roads' and 'community boulevards'. These roadway types are two of the categories established under the functional hierarchy of strategic roads as defined by the Queensland Department of Transport and Main Roads (TMR, 2009). This functional hierarchy of strategic roads identifies four (4) categories (p66):

1. *High capacity, high-speed motorways and highways* to move large volumes of traffic, including freight traffic, over longer distances;
2. *Multi-modal urban arterial roads* to provide connections within communities and cater for a range of road users, including pedestrians, cyclists, public transport, private vehicles, as well as commercial delivery vehicles ('first and last mile' freight);
3. *Bypass and ring roads* to remove traffic from activity centres; and
4. *Community boulevards* to provide amenity through activity and town centres, designed to cater for low volumes of traffic, with priority given to pedestrians, cyclists and public transport.

13.8 NEIGHBOURHOOD CONTEXT FOR ATI

This study utilises the *SEQ Place Model* for the classification of the physical form and character of the neighbourhood context (CM-SEQ, 2011; Stevens and Salmon, 2014). This model suggests that settlements in SEQ can be understood as a series of place types, each with common characteristics, similar land use mixes and intensities of development. The model represents what may traditionally be recognised as transect planning (Duany and Talen, 2002). The SEQ Place Model (CM-SEQ, 2011 p 62) identifies seven (7) categories of overlapping place type (Figure 13.1):

P1. Natural Places – Areas dominated by the natural environment
P2. Rural Places – Rural production and landscapes, rural living
P3. Rural Townships – The range of smaller rural townships in SEQ
P4. Next Generation Suburban Neighbourhoods – Walkable local areas, which are people (rather than car) focussed and contain a choice of housing types and some other local uses in a mixed use setting
P5. Urban Neighbourhoods – Walkable, high density local areas, which are people focussed and contain a wider choice of other housing types more mixed use than suburban neighbourhoods
P6. Centres of Activity – Vibrant and intense mixed use centres including housing, retail, employment, education and entertainment facilities. Some taller buildings are part of the built form character
P7. CBDs – Central business districts – Centres of production as well as consumption

For this research the place types within which the ATI WDA has been established are P5 Urban Neighbourhoods and P6 Centres of Activity.

13.9 WDA ATI CONTEXT

The formative model developed through the WDA represents ATI located on an urban arterial or community boulevard within an urban neighbourhood or centre of activity which has an adjacent built form context comprised of mixed use commercial, retail and residential uses. The unit of analysis is an urban block, corner to corner of approximately 150 metres in length, and in width, begins at the building line and traverses the footpath and on-road cycleway to the roadway (Figure 13.2). The configuration for on-road cycling is described as a 'wide parking lane with bicycle

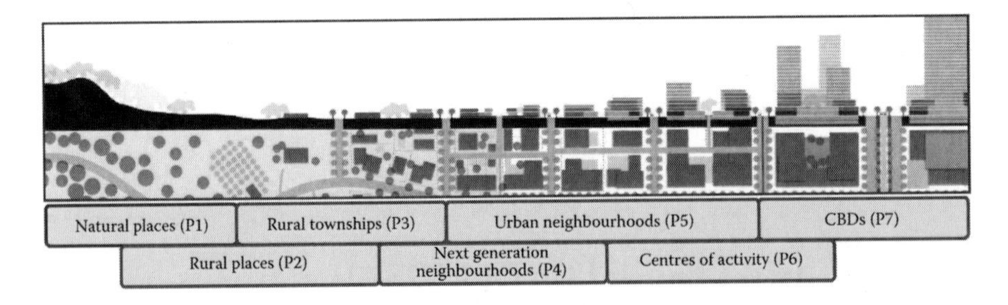

FIGURE 13.1 SEQ Place Model. (Adapted from Council of Mayors South East Queensland [CM-SEQ]. 2011. *Next Generation Planning*. Brisbane: Council of Mayors [SEQ] and State of Queensland. p62).

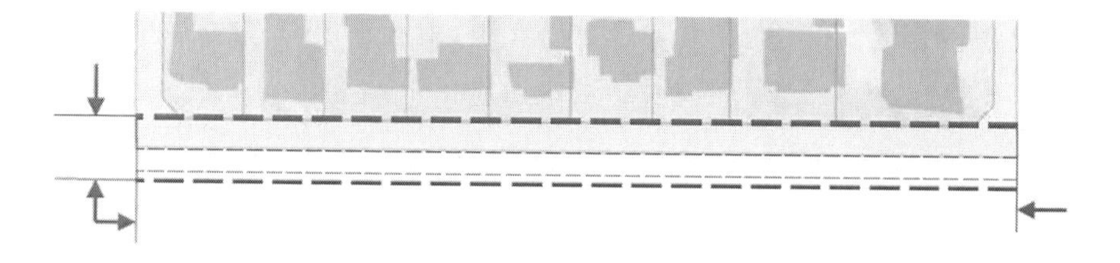

FIGURE 13.2 ATI WDA unit of analysis.

provision' as it is the most commonly delivered by state agencies in SEQ (TMR, 2013b). The WDA considers the functions, purposes and objects within this system of ATI.

13.10 ATI ZONES

As an organising device this research extends the recognised zones of a footpath (Austroads, 2009; ITE, 2010; Stevens and Salmon, 2014) to include the on-road cycleway and the roadway edge. The amount of space allocated to these five (5) zones and the objects located within them has a significant influence on the functions of ATI. The identified zones may be defined as

1. *On-road Cycleway Zone*: This is the distance from the edge of the motorised vehicular travel way to the kerb. It often contains provision for car parking, in addition to the markings required to indicate this space as a cycleway.
2. *Edge/Kerb Zone*: This area is from the face of the kerb and defines the limit of the pedestrian area. It provides the minimum separation between the objects in the footpath and other vehicles in the roadway.
3. *Furnishings Zone*: This zone provides the major buffer between vehicles and pedestrians. It often contains signal poles, seats, landscaping, street trees, parking meters and so forth.
4. *Throughway Zone*: This is the walking zone and the area through which pedestrians usually travel. It is an area that must remain clear and be free of obstructions at all times.
5. *Frontage Zone*: This is the distance from the throughway to the frontage of adjacent property. It is an area which should buffer pedestrians from window shoppers and private business or residential doorway traffic. This zone often contains private street and dining furniture, merchandise displays, private signage, fences and so forth.

The WDA seeks to better understand each of these distinct zones, the relationships between them and the objects they contain. To assist in the analysis of the active transport infrastructure, where appropriate, the physical objects layer within the WDA has been arranged so as to highlight their placement within the zones.

13.11 RESULTS

The system represented by the WDA has been summarised in Figure 13.3. While the *Functional Purposes* and the *Value and Priority Measures* levels of the WDA are presented in their entirety, for efficiency the nodes contained within the *Purpose-Related Functions*, *Object Related Processes*, and the *Physical Objects* levels have been grouped into representative categories.

The following discussion does not endeavour to outline all of the relationships; rather we provide an overview of the interactions between the levels of the abstraction hierarchy to demonstrate the applicability and usefulness of this method.

13.11.1 Functional Purpose

This highest level of the abstraction hierarchy contains the reasons for the existence of the system. When considering ATI, the WDA was required to include both objective and subjective functional purposes. For example the objective purposes of the ATI to 'provide a right of way for active transport', 'provide safe active transport' and 'support efficient active transport', in balance with the best practice and user-orientated functions of 'permit best practice transport hierarchy' and 'provide high quality active transport experience'.

If the analysis had been undertaken simply by considering the function of ATI as 'provide a right of way for active transport' and 'support efficient active transport', the WDA would have yielded a systems view of how ATI is currently implemented – to accommodate these functions within the roadway corridor. Through the inclusion and analysis of the potential functions of ATI, in line with best practice, safety and user experience, the research brings together a means to understand the interfaces and possible outcomes for achieving sustainable active transport.

The important and complex relationships and interdependencies between these functional purposes are established when considering the below levels of values and priority measures; purpose-related functions; physical affordances; and physical objects. It is interesting to note that the five functional purposes described at this level may not necessarily be compatible; that is, it is questionable whether ATI designs can fulfil these purposes without some trade-offs.

13.11.2 Values and Priority Measures

Values and priority measures are positioned on the second level of the WDA and outline the criteria against which to evaluate the achievement of the system's functional purposes. That is, do these inclusions provide an indication of how the functional purposes are being achieved? For example, in this ATI system 'maximise comfort' is an important value and priority measure for the delivery of 'high quality active transport experience'. In turn, 'maximise wayfinding' is one of ten purpose-related functions (next level down) which are required in the system to maximise comfort. Further 'maximise wayfinding' is achieved through six object related processes that are enabled by a variety of physical objects within the footpath. In the WDA system described here the object-related processes and their

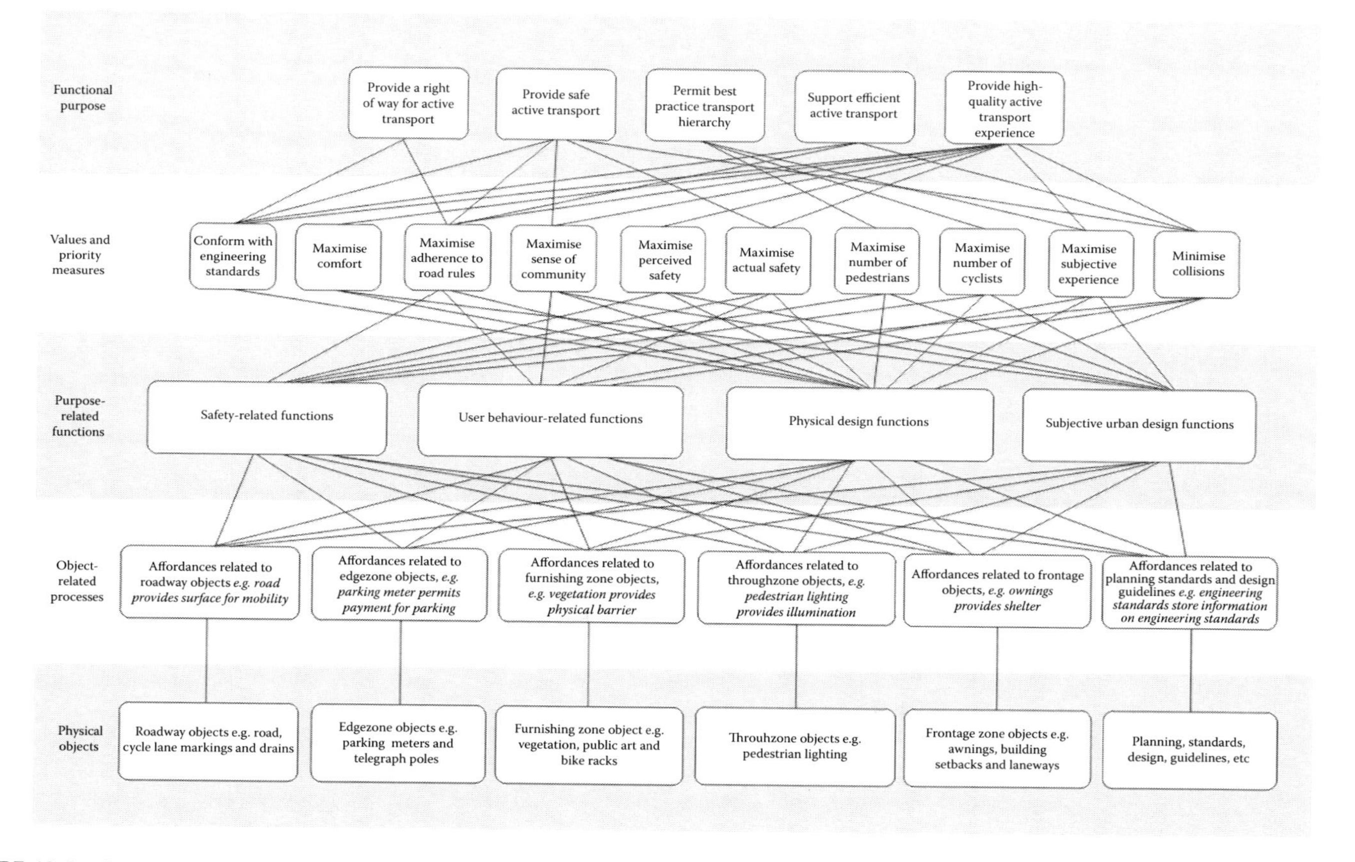

FIGURE 13.3 Summarised WDA.

associated physical objects (in brackets) to 'maximise wayfinding' include: 'provide stop/go information' (traffic signals, regulatory signage); 'landmarks' (public art, trees, building setbacks, community archive and knowledge); 'footpath illumination' (pedestrian lighting); and 'provides wayfinding information' (road markings, cycle lane markings, regulatory signage, street name signage, information signage, tactile surface markers, building number signage) – all necessary processes and objects for maximising wayfinding.

For another example, when considering the value and priority measure of 'maximise actual safety' it is important in this systems approach to be able to offer means by which it may be achieved. It makes sense then that 'maximising actual safety' has connections with purpose-related functions such as: provide protection for the travel way; maximise lighting; minimise obstacles and debris; minimise changes in grade; maintain separation of pedestrians, cyclists and motor vehicles; maximise through zone clearances; minimise crime; provide warnings; design in consideration of the traffic context; maximise quality of the design; and corridor maintenance. These are all sound and adequate indicators of the level of actual safety, in turn supported by object related processes and the objects themselves.

13.11.3 Purpose-Related Functions

The purpose-related functions sit at the middle level of the abstraction hierarchy. This level may be considered to represent the functions that need to be achieved for the system to fulfil its purpose. It is the crucial link between the independent objects and their related processes at the bottom of the hierarchy with the more purpose orientated concepts above. As such the use of the 'means ends' relationships are of significant value here for identifying the interfaces within the system. In the design context, this level represents design and end-user requirements in that they describe what functions need to be supported through design.

For example if it is considered that the purpose-related function (minimise obstacles and debris) is the question '*what*', and the values and priority measures (maximise perceived safety, maximise actual safety and maximise subjective experience) are the question '*why*', then the object-related processes provide '*how*' – prevents roadway water entering footpath, provides bike storage, protects and contains vegetation, stores litter and debris, provides surface for objects, provides access, stores information on maintenance procedures.

13.11.4 Object-Related Processes (Physical Affordances)

The physical affordances of the objects within the system relate to their ability to enable or aid a function. The description of these object-related processes are such that they are generic and not a reflection of the ATI system purposes. When linking the objects to their affordances it is evident that the object-related purposes are often afforded by multiple objects. For example 'provide shelter' is an affordance of 'trees', 'awnings' and 'public transit stop'. While objects which offer the affordance of a 'landmark' may also include 'trees', the affordance is also offered by 'public art' and 'building setbacks' in this WDA.

13.11.5 Physical Objects

At the base of the AH the physical objects that comprise the ATI system are detailed. At this level the objects, where possible, have been represented within the footpath zone in which they occur, and in Figure 13. 3 are summarised as such. This is useful when considering the urban form and socio-spatial aspect of the system and the relationships between physical objects and the interface to the built form and (transport task) roadway contexts. Within the summary WDA there is an additional grouping of objects, which are important components of the ATI system that are not physically located within it. For example road rules; planning regulations; budget and engineering standards are crucial objects within the system, yet physically exist elsewhere. Key to this systems understanding is the recognition that many physical objects fulfil multiple affordances and so contribute to various purpose-related functions. For example a 'planter boxes' not only relates to the affordance 'protect and contain vegetation', but may also act as 'physical buffers or barriers' and 'psychological buffer between the travel way and pedestrians'.

13.12 DISCUSSION

The aim of this article was to demonstrate how CWA, and WDA in particular, provide a design template that will contribute to the design of more appropriate; safer and user-orientated ATI. It details a unique approach whereby it is possible to better understand the contributions that ATI may make in operationalising the hierarchy of best practice sustainable transport in existing urban corridors.

The analysis presented demonstrates the possibilities for the design of ATI that is context sensitive, well-engineered and that also contributes to the subjective experience of pedestrians and cyclists, possibly allowing for greater participation. Through constructing a generic WDA for ATI, this paper has provided a novel design template that may be used to inform the establishment of ATI. Additionally it is also useful to evaluate the extent to which existing ATI within urban corridors achieves anticipated and essential functional requirements.

A key contribution of this method, is that because the overall ATI 'system' is considered, it is possible to introduce different design features and then evaluate them in the context of their interactions with other parts of the corridor and thus evaluate their impact on the overall system itself. For example, in future design efforts one could introduce new objects in the AH and examine the means-ends links with functions, values and priorities, and the systems functional purpose. The analysis presented also reveals the significant influence of the many objects that are not physically contained within the actual footpath environment, such as planning regulations, maintenance and design standards, and significantly for urban form and design, the community archives and knowledge.

This study has allowed a clearer understanding of the contributions that physical objects, both within and beyond the urban corridor, make to the establishment of a leading practice ATI. Further it highlights the interdependency of many issues that have historically been managed in isolation and within discipline areas. For example, the engineering aspects of ATI are largely not expected to contribute to subjective user experience, and the urban design guidance on active transport and public urban environments has often relied on outlining principles without practical direction. This whole of system approach to ATI will enable consideration of this interdependency in future design efforts.

This paper identifies the usefulness of a systems approach in accessing disparate concepts. This system allows for an understanding of the interdependencies of intertwined functional purposes. It allows for multiple disciplines and indeed stakeholders to recognise their place within the system, and the impacts and influences of design decision-making. It makes sense that a systems approach allows for greater insights into the development of urban form. While this method has been applied to the sociotechnical system of ATI, it is envisaged as a useful tool for a range of urban environments. Urban planning needs new ways to interpret built environments and urban form as not only technical systems, but often poorly understood (and designed) socio-spatial systems. Further research in which CWA is applied in urban design applications is therefore recommended.

13.13 FUTURE RESEARCH

The application of the additional phases of CWA will assist in understanding how to best accommodate a variety of uses and users in ATI. Phases 2, 3, 4 and 5 of the framework consider important aspects of performance such as decision-making, strategies and allocation of functions and are important in determining how specific ATI designs might shape pedestrian, cyclist and motor vehicle behaviour. These stages will detail the types of activity permitted in different ATI designs and local considerations and contexts that impact on the placement of physical objects and overall ATI capacity. Further, allocation of functions across different physical objects will also be examined (e.g. whether warnings should be provided to users via road and footpath markings or by signage).

REFERENCES

Adams, R. 2009. Transforming Australian cities for a more financially viable and sustainable future: Transportation and urban design. *The Australian Economic Review*, 42(2), 209–216.

Alfonzo, M. A. 2005. To walk or not to walk? The hierarchy of walking needs. *Environment and Behavior*, 37(6), 808–836.

Appleyard, D. 1980. Liveable streets: Protected neighbourhoods? *The ANNALS of the American Academy of Political and Social Science September*, 451(1), 106–117.

Arrington, G. and Cevero, R. 2008. *Effects of TOD on Housing, Parking, and Travel, TCRP Report 128*. Transportation Research Board, Washington.

Australian Government. 2007. *Brisbane Urban Corridor Strategy – Building our National Transport Future, Auslink, Queensland Government*. Canberra, Australia: Department of Transport and Regional Services.

Australian Government. 2009. *Healthy Spaces and Places, A National Guide to Designing Places for Healthy Living*. Kingston, Canberra: Australian Government Department of Health and Aging.

Australian Government. 2011. *Creating Places for People: An Urban Design Protocol for Australian Cities*. Canberra: Infrastructure Australia.

Austroads. 2009. *Guide to Road Design – Part 6A: Pedestrian and Cyclist Paths*. Sydney, Australia: Austroads Incorporated.

Bisantz, A. M., Roth, E., Brickman, B., Gosbee, L. L., Hettinger, L. and McKinney, J. 2003. Integrating cognitive analyses in a large-scale system design process. *International Journal of Human-Computer Studies*, 58, 177–206.

Bureau of Infrastructure, Transport and Regional Economics (BITRE). 2013. Australian Road deaths Database, http://www.bitre.gov.au/statistics/safety/fatal_road_crash_database.aspx. Accessed 19 December 2013.

Cerin, E., Leslie, E., Toit, L. D., Owen, N. and Frank, L. D. 2007. Destinations that matter: Associations with walking for transport. *Health & Place*, 13(3), 713–724.

Clifton, K., Livi Smith, A. and Rodriguez, D. 2007. The development and testing of an audit for the pedestrian environment. *Landscape and Urban Planning*, 80, 95–110.

Cornelissen, M., Salmon, P. M., McClure, R. and Stanton, N. A. 2013. Using cognitive work analysis and the strategies analysis diagram to understand variability in road user behaviour at intersections. *Ergonomics*, 56(5), 764–780.

Cornelissen, M., Salmon, P. M. and Young, K. L. 2013. Same but different? Understanding road user behaviour at intersections using cognitive work analysis. *Theoretical Issues in Ergonomics Science*, 14(6), 592–615.

Council of Mayors South East Queensland (CM-SEQ). 2011. *Next Generation Planning*. Brisbane: Council of Mayors (SEQ) and State of Queensland.

Davies, L. 2000. *Urban Design Compendium*. London: English Partnerships.

Dixon, K., Liebler, M. and Zhu, H. 2008. *NCHRP Report 612 Safe and Aesthetic Design of Urban Roadside Treatments*. Washington: DC: Transportation Research Board.

Duany, A. and Talen, E. 2002. Transect planning. *Journal of the American Planning Association*, 68(3), 245–266.

Ewing, R. and Cervero, R. 2010. Travel and the built environment: A meta-analysis. *Journal of the American Planning Association*, 76(3), 265–294.

Ewing, R. and Handy 2009. Measuring the unmeasurable: Urban design qualities related to walkability. *Journal of Urban Design*, 14(1), 65–84.

Forsyth, A., Hearst, M., Oakes, J. M. and Schmitz, K. H. 2008. Design and destinations: Factors influencing walking and total physical activity. *Urban Studies*, 45(9), 1973–1996.

Forsyth, A., Oakes, J. M., Schmitz, K. H. and Hearst, M. 2007. Does residential density increase walking and other physical activity? *Urban Studies*, 44(4), 679–697.

Frackelton, A., Grossman, A., Palinginis, E., Castrillon, F., Elango, V. and Guensler, R. 2013. Measuring walkability: Development of an automated sidewalk quality assessment tool. *Suburban Sustainability*, 1, 1–15.

Frank, L.D., Greenwald, M.J., Winkelman, S., Chapman, J. and Kavage, S. 2010. Carbonless footprints: promoting health and climate stabilization through active transportation. *Preventive Medicine* 50(supp. 1), S99–S105.

Gehl, J. 2011. *Life Between Buildings – Using Public Space*. Washington, DC: Island Press.

Hale, C. 2011. New approaches to strategic urban transport assessment. *Australian Planner*, 48(3), 173–182.

Hamer, M. and Chida, Y. 2008. Active commuting and cardiovascular risk: A meta-analytic review. *Preventive Medicine*, 46(1), 9–13.

Heesh, K. C. and Sahlqvist, S., 2013. Key influences on motivations for utility cycling (cycling for transport to and from places). *Health Promotion Journal of Australia*, 227–233.

Institute of Public Works Engineering Australia (IPWEA) (Qld). 2010. *Complete Streets: Guidelines for Urban Street Design*. Queensland: Queensland Division Inc, Fortitude Valley.

Institute of Transport Engineers (ITE). 2010. *Designing Walkable Urban Thoroughfares: A Context Sensitive Approach*. Washington DC: Institute of Transportation Engineers.

Jacobs. 1961. The uses of sidewalks: Safety, in Carmona, M. & Tiesdell, S. (Eds) 2007. *Urban Design Reader.* Oxford, UK: Architectural Press.

Jenkins, D. P., Salmon, P. M., Stanton, N. A. and Walker, G. H. 2010. A new approach for designing cognitive artefacts to support disaster management. *Ergonomics*, 53(5), 617–635.

Jenkins, D. P., Stanton, N. A., Salmon, P. M. and Walker, G. H. 2009. *Cognitive Work Analysis: Coping with Complexity.* Aldershot, UK: Ashgate.

Kelly, C. E., Tight, M. R., Hodgson, F. C. and Page, M. W. 2011. A comparison of three methods for assessing the walkability of the pedestrian environment. *Journal of Transport Geography*, 19(6), 1500–1508.

Lynch 1960. *The Image of the City.* Cambridge Massachusetts: MIT Press.

Millward, H., Spinney, J. and Scott, D. 2013. Active transport walking behaviour: Destinations, durations, distances. *Journal of Transport Geography*, 28, 101–110.

Naikar, N., Hopcroft, R. and Moylan, A. 2005. Work Domain Analysis: Theoretical Concepts and Methodology, *DSTO Technical Report (DSTO-TR-1665).* Edinburgh, Australia: System Sciences Laboratory.

Naikar, N., Pearce, B., Drumm, D. and Sanderson, P. M. 2003. Technique for designing teams for first-of-a-kind complex systems with cognitive work analysis: Case study. *Human Factors*, 45(2), 202–217.

Naikar, N. and Sanderson, P. M. 2001. Evaluating design proposals for complex systems with work domain analysis. *Human Factors*, 43, 529–542.

Olsson Associates (OA). 2010. Urban Corridors, Regional Transit Implementation Plan. Kansas City, MO: Prepared for Mid-America Regional Council, March 2010.

P&NJ DoT. 2008. New Jersey & Pennsylvania Department of Transportation, *Smart Transportation Guidebook Planning and Designing Highways and Streets that Support Sustainable and Livable Communities.* March 2008.

Paige Willis, D., Manaugh, K. and El-Geneidy, A. 2013. Uniquely satisfied: Exploring cyclist satisfaction. *Transportation Research Part F: Traffic Psychology and Behaviour*, 18, 136–147.

Planning Institute of Australia. 2010. *Urban Design Unit Manual.* Brisbane, Australia: AKH Consulting.

Premius, H. and W. Zonneveld. 2003. What are corridors and what are the issues? Introduction to special issue: The governance of corridors. *Journal of Transport Geography*, 11(3), 167–177.

Purciel, M. and Marrone, E. 2006. *Observational Validation of Urban Design Measures for New York City. Field Manual.* New York: Active Living Research Program.

Salmon, P. M., Lenné, M. G., Walker, G. H., Stanton, N. A. and Filtness, A. 2014. Exploring schema-driven differences in situation awareness between road users: An on-road study of driver, cyclist and motorcyclist situation awareness. *Ergonomics*, 57(2), 191–209.

Salmon, P. M., McClure, R. and Stanton, N. A. 2012. Road transport in drift? Applying contemporary systems thinking to road safety. *Safety Science*, 50(9), 1829–1838.

Salmon, P. M., Stanton, N. A., Jenkins, D. P. and Walker, G. H. 2010. Hierarchical task analysis versus cognitive work analysis: comparison of theory, methodology, and contribution to system design. *Theoretical Issues in Ergonomics Science*, 11(6), 504–531.

Sargeant, B., Mitchell, N. and Webb, N. 2009. *Place Making in the Urban Corridor', International Cities Town Centres & Communities Society, ICTC 2009.* Geelong, Australia: Deakin University Campus, 27–30 October, 2009.

South Australian Government. 2011. *Urban Corridor Zone, Technical Information Sheet 10.* Adelaide, Australia: South Australian Planning Policy Library, September 2011.

Stevens, N. and Buksh, B. 2012. *A Leading Practice Framework for Sustainable Transport Corridors: Sustainable Transport Corridor Research*, Final Report to Queensland Department of Transport and Main Roads. Queensland, Australia: University of the Sunshine Coast.

Stevens, N. and Buksh, B. 2013. A leading practice framework for sustainable urban transport corridors. *Proceedings of the 2013 Planning Institute of Australian National Congress*, 24–27 March 2013, Canberra, Australia.

Stevens, N. and Salmon, P. 2014. Safe places for pedestrians: Using cognitive work analysis to consider the relationships between the engineering and urban design of footpaths. *Accident Analysis and Prevention*, 72, 257–266.

Stevens, N. J. 2016. Sociotechnical urbanism: New systems ergonomics perspectives on land use planning and urban design. *Theoretical Issues in Ergonomics Science*, 17(4), 443–451.

Stevens, N. J. and Salmon, P. M. 2015. New knowledge for built environments: Exploring urban design from socio-technical system perspectives. In *Engineering Psychology and Cognitive Ergonomics*. pp. 200–211. New York: Springer International Publishing.

Sunshine Coast Regional Council (SCRC). 2011a. *Sunshine Coast Sustainable Transport Strategy 2011–2031*. Maroochydore, Australia: Integrated Transport Planning Branch, February 2011.

Sunshine Coast Regional Council (SCRC). 2011b. *Sunshine Coast Active Transport Plan 2011–2031*. Maroochydore, Australia: Integrated Transport Planning Branch, February 2011.

TMR. 2009. *Queensland Department of Transport and Main Roads, Connecting SEQ 2031: An Integrated Regional Transport Plan for South East Queensland*. Brisbane: Queensland Government.

TMR. 2013a. *Queensland Department of Transport and Main Roads, Road Planning and Design Manual (2nd Ed)*. Brisbane: Integrated Transport Planning Branch, Queensland Government.

TMR. 2013b. *Queensland Department of Transport and Main Roads, Queensland Manual of Uniform Traffic Control Devices Part 9 Bicycle Facilities*. Brisbane: Queensland Government.

TMR. 2014. *Queensland Department of Transport and Main Roads, Transport Operations (Road Use Management – Road Rules Regulation 2009)*. Brisbane: Queensland Government.

US DOT. 2010. *International Technology Scanning Program: Pedestrian and Bicyclist Safety and Mobility in Europe, US Department of Transportation*. Washington DC: National Cooperative Highway Research Program, February 2010.

Vicente, K. J. 1999. *Cognitive Work Analysis: Toward Safe, Productive, and Healthy Computer-Based Work*. Mahwah, NJ: Lawrence Erlbaum Associates.

Washington State Department of Transport (WSDoT). 2011. Roadway Bicycle Facilitates, In Division 15 – Pedestrian and Bicycle Facilities – Design Manual, WSDoT, pp. 1520–1532. Washington, DC.

Watson, M. O. and Sanderson, P. M. 2007. Designing for attention with sound: Challenges and extensions to ecological interface design. *Human Factors*, 49(2), 31–346.

Woodcock, J., Banister, D., Edwards, P. and Prentice, A. M. 2007. Energy and transport. *The Lancet*, 370,9592, 1078–1088.

Yücel, G. F. 2013. Street furniture and amenities: Designing the user-oriented urban landscape. In Murat Ozyavuz (Ed.) *Advances in Landscape Architecture*, http://www.intechopen.com/books/advances-in-landscape-architecture/street-furniture-and-amenities-designing-the-user-oriented-urban-landscape. Accessed 10 September 2013.

14 Use and Refinement of CWA in an Industrial, Automotive Design, Context

Ida Löscher and Stas Krupenia

CONTENTS

14.1 INTRODUCTION

In this chapter we describe how the cognitive work analysis (CWA) framework can be used in an industrial organisation. The work reported is part of an ongoing collaboration between Scania Commercial Vehicles, a manufacturer of premium heavy trucks and buses, and Uppsala University. As is presented throughout this book, CWA is a powerful analysis tool that can be used to understand the variety of constraints shaping activity within the work system. While few (if any) figures are quoted, CWA is assumed to be a resource heavy analysis method. Furthermore, the link between the analysis and design phases are largely unclear and unstructured (Read et al., 2012). Consequently, within profit sensitive industrial organisations, the use of CWA is risky – a known (or likely) high effort cost, and a vague pathway to design changes. Additionally, because CWA outcomes have the potential to (positively) influence multiple aspects of design (e.g. hardware, software, interfaces, etc.), implementation of the analysis results may require the involvement of multiple organisational units and necessitate some amount of organisation change (another risky activity). Unless some of these risks can be reduced, CWA is unlikely to be adopted on a large-scale throughout industry.

We undertook a first step toward understanding the value of CWA for industry, and reducing some of these foreseen (or predicted) risks. The CWA was conducted within the scope of an externally funded project, thus transferring some of the analysis costs outside of the industrial organisation. The CWA was conducted within the scope of the 'Methods for Designing Future Autonomous Systems' (MODAS) project (Krupenia et al., 2014). The MODAS project had three goals; to develop a method for designing future (highly autonomous) systems, to apply this method and build a prototype truck cab for a highly autonomous truck and to assess the final cab design. The analysis from the CWA framework was one of several analysis and design tools included in the final MODAS methodology.

In the current chapter, we first present the long haul trucking case study, with focus on the methods used, and results obtained. In the second part of the chapter, we present a discussion on how CWA can be adapted to be of greater value to profit margin sensitive organisations. In this second part, we also reflect on some of the shortcomings experienced with CWA in the context of the use case. We conclude with some recommendations for future work that could support the further industrial uptake of the CWA framework.

14.2 METHOD

Compared to other forms of truck driving (e.g. distribution, construction and mining), long haul driving is characterised by longer distances between collection and drop-off locations (driving radius of greater than 160 km; Ministry of Jobs, Tourism and Skills Training, 2010), fewer stops per maximum legislated driving time period, and increased probability of living in the vehicle for some amount of time (see Figure 14.1). These features have important consequences for the vehicle characteristics purchased by customers. For example, a long haul truck is more likely to have a larger cabin with additional comfort and 'home

FIGURE 14.1 Exterior and interior of Scania R series for long haul driving.

living' features (such as beds and microwaves) as compared to a short haul city distribution truck. The CWA was conducted for, and is valid for, long haul truck driving only.

14.2.1 Case Study Methodology Description

The long haul truck was analysed using the first three phases of the CWA framework (Vicente, 1999; see Chapter 1); work domain analysis, control task analysis and strategies analysis. The data collection and modelling for the three CWA phases are described in the following three sections, and the fourth section includes a description of the design solution influenced by the analyses. Finally, although very approximate, it may interest the reader to note that completion of the analysis defined below required about 1,200 hours.

14.2.2 Work Domain Analysis

The work domain analysis completed for the long haul truck (Bodin, 2013) was modelled in an abstraction hierarchy (AH; see Chapter 1). The AH was constructed using data collected during three types of data collection sessions; interviews, observations and verbalisation interviews, see Figure 14.2.

The interview study included experienced long haul drivers and a senior technical adviser (engineer). Answers from the semi open-ended interview questions were used iteratively to construct the first version of the AH. The early AH was used together with the questions suggested by Naikar et al. (2005; also in Jenkins et al., 2009) as a base for the interviews. The AH was then further developed between the interviews based on the informants' comments.

The observation study was conducted to gain a more thorough understanding of the long haul truck. For the observation study, four long haul drivers were observed during one work day each, resulting in a total distance driven of 2500 km. Three video cameras captured video data during 16 hours of driving. Further, the driver was asked to 'think out loud' during four ten-minute sessions of driving. The route included a mix of highway and country road driving, and variations in traffic density and weather conditions. Notes taken during the observation and the video data were

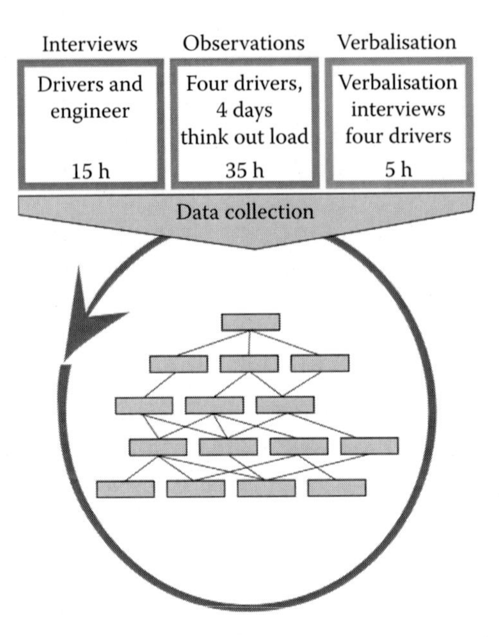

FIGURE 14.2 Illustration of the data collection and modelling of the AH.

used to further develop the AH. The video data was also used to prepare the third type of data collection sessions – the verbalisation interviews.

The verbalisation interview was similar to the collegial verbalisation method (Jansson et al., 2006). For the current study, the same drivers that participated in the observation study were used for the verbalisation interviews. Due to practicalities (driving schedules), interviews occurred one to two and a half weeks after the observation study. For the interviews, interesting episodes from the video data were extracted and grouped according to the captured situation and shown to each driver. The drivers were asked to 'think out load' or explain what was happening in the situation in the video. After each interview the new insights about the functions of the technical parts or values in the work system were modelled in the AH.

14.2.3 Control Task Analysis

The MODAS project control task analysis (Bodin, 2013; Bodin and Krupenia, 2016) was used to model the long haul truck driver's work in a contextual activity template (CAT; Naikar et al., 2006). The CAT included different situations related to long haul driving, and information about the work functions that either 'can', 'usually' or 'never' occur in the situations. The information modelled in the CAT was obtained from the earlier (WDA) data collection parts, and two additional interview studies. By studying the video data collected during the observation study, the to-be-included CAT situations were identified.

Because both the object-related processes and the purpose-related functions from the AH appeared important for development purposes, the former for prioritising overall work, and the latter for their close connection to the technical system, two CATs were created. One CAT included functions from the object-related processes level of the AH and was called the object-related CAT. The other CAT included the purpose-related functions and was called the purpose-related CAT. Each activity (the junction between a situation and a purpose-related function or an object-related function) was given a score, such that 0 denoted that the activity never occurs, 1 denoted that the activity can occur and 2 denoted that the activity usually occurs.

The scores in the object-related CAT were set by structured interviews with the four drivers who also participated in the observation study. Each informant was asked to complete the scores for half of the activities, making each activity scored twice. When the scores from the two drivers where conflicting, or when the scores were conflicting with the analysts' understanding from the observation study, the activity was further investigated by an interview with an additional driver and further telephone interviews with the original drivers from the observation study. These scores were used to create the object-related CAT.

By using the mean-end links between the object-related and purpose-related functions in the AH, the object-related CAT was truncated to construct the purpose-related CAT (Bodin and Krupenia, 2015). It is important to consider that the CAT scores are ordinal, and therefore the score for the purpose-related function cannot be immediately calculated. Instead the truncated scores need to be set qualitatively using the domain knowledge by looking at the scores for the connected object-related processes and consider their relation to the purpose-related function, see Figure 14.3.

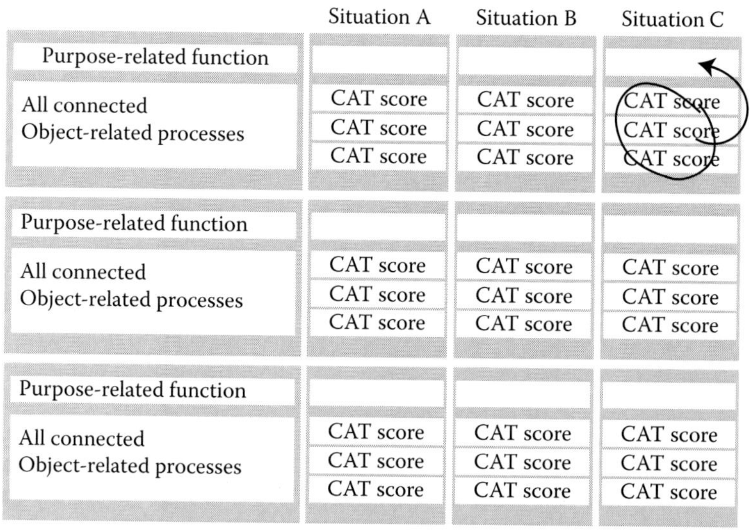

	Situation A	Situation B	Situation C
Purpose-related function			
All connected Object-related processes	CAT score CAT score CAT score	CAT score CAT score CAT score	CAT score CAT score CAT score
Purpose-related function			
All connected Object-related processes	CAT score CAT score CAT score	CAT score CAT score CAT score	CAT score CAT score CAT score
Purpose-related function			
All connected Object-related processes	CAT score CAT score CAT score	CAT score CAT score CAT score	CAT score CAT score CAT score

FIGURE 14.3 Illustration of the CAT score truncation procedure.

In the current study the CATs were extended to include a priority score for each of the activities in the CAT (Bodin and Krupenia, 2015). The CAT scores included how frequently the functions are exploited in the situations. Information about the frequency of the situations was also needed for the prioritisation. Therefore structured interviews with ten drivers from different European driving companies were conducted. Each interviews lasted 30 minutes. Informants were asked what proportion of their working day is usually spent in each of the situations. Answers were given in parts of work time, or time in hours or minutes, and together with the interviewer transferred to the percentage of the work time, which was later transferred to the score from 1 to 10. At the end of the interview informants were also asked two open-ended questions about what situations they think should be the focus of future development, and if they can think of any problem connected to any particular situation.

In addition to the CAT score and the situation frequency, it was also necessary to identify how essential each function was for system performance. This was rated by the researchers, using the approach by Birrell et al. (2011) which involves giving scores based on the number of connections to higher level nodes in the AH. The three scores (situation frequency, function priority and contextual activity frequency) are multiplied to create an activity prioritisation score.

14.2.4 Strategies Analysis

The goal of the strategies analysis was to understand the strategies that can be used to perform tasks in a future system that manages the demands imposed by a future traffic environment. It is difficult to analyse systems that are not yet developed and for which no users have experience. To overcome this issue the formative strategies analysis approach by Hassall and Sanderson (2014) was modified (Löscher et al., 2017).

The approach by Hassall and Sanderson (2014) consists of a preparatory phase and an application phase. The preparatory phase includes eight categories of strategies that can be applied in any domain. Those strategies are; (1) 'Avoidance' which means to delay or not perform the task, (2) 'Intuitive' which is the category for routine tasks that are automatically executed, (3) 'Option-based' where a possible action is selected by reasoning, evaluation, or 'rules-of-thumb' from alternatives and criteria, (4) 'Cue-based' were possible actions are selected by considering evidence from the environment, (5) 'Compliance' were rules and procedures are followed, (6) 'Analytical' were the selection of strategy is based on fundamental principles of the work system, mental simulation of outcomes or other reasoning and (7) 'Arbitrary-choice' which regards tasks that are uncommon and no consideration of options take place (Hassall and Sanderson, 2014).

The application phase includes four steps were the first step is to choose activities to analyse. The second step is to consider the time pressure, difficulty level and risk level associated with the activity. The last two steps are to identify likely categories of strategies and factors influencing strategy selection and when a change of strategy occurs.

In Löscher et al. (2017), the predefined categories of strategies were used to facilitate a focus on a variety of strategies rather than the one that comes first to mind.

As a part of the preparation phase these categories were also elaborated on (by the research team) with an initial set of truck driver strategies. These pre-worked examples were used as explanation tools when required by respondents in the (driver) application phase.

The application phase started with choosing the activities to be included in the study. To identify the focus areas, knowledge from the previous CWA phases and the work within the MODAS project were used. The functions 'Vehicle Control', 'Logistics', 'Cargo Transport' and 'Fuel Efficient Driving' were associated with many activities with high activity prioritisation scores, meaning they are essential for system performance and are frequently occurring.

The functions were reformulated as tasks, both for how they are conducted today and how they could be conducted in the future. The future tasks were described based on the expectation that the future driver will be supported by more automation regarding operational control. To let the participants discuss different variations of the task and to focus on different contexts, a range of possible events that were identified within the MODAS project as being particularly demanding were used. Using these contexts was also a way to keep the discussion on a concrete level, even when many parameters regarding the work in the future system are flexible.

Because of uncertainty regarding the future, and that all (acceptable) strategies were to be regarded in the system design, a smaller emphasis was placed on the second step of the application phase – identifying categories of strategy that are more or less likely to occur. Instead, all categories of strategies were used as a tool to identify more than just the obvious strategies.

The two last steps of the application phase of the strategies analysis conducted by Löscher et al. (2017) were completed via four workshops each involving three to four long haul truck drivers. In the first workshop, the categories of strategies from Hassall and Sanderson (2014), and the set of initial strategies were presented. Further, the initial assumptions about the future system and driver environment were explained. Possible strategies that could be used in the future system were elicited from the discussion with the drivers. The three other workshops brought the different strategies alive through discussing interface ideas. For this, the third group received earlier groups' ideas to evaluate and elaborate upon.

14.2.5 The MODAS Design

A design concept of the driver environment in a self-driving long haul truck was developed within the MODAS project. The concept included visual and auditory displays. Many researchers, developers and designers were involved in the project, and the designers received input from multiple studies during the project (Fagerlönn et al., 2015). The MODAS project design was built using the analysis results from the three first phases of the CWA framework as well as user iterations based on the GMOC model (goal, mental model, observability, controllability; Tschirner, 2015).

Figure 14.4 illustrates the contribution from the different project activities on the final design concept. In Figure 14.4 it is shown that the work domain analysis gave

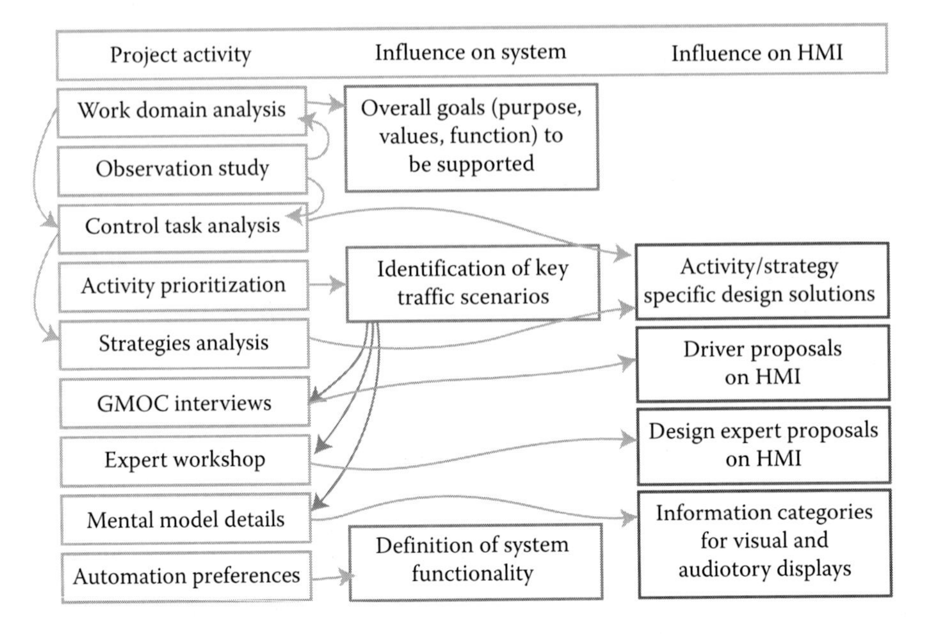

FIGURE 14.4 Overview of project activities effect on other activities, HMI and system design.

input about the overall goals to be supported, while the control task analysis and the strategies analysis gave information regarding activities or strategies that influence design solutions.

14.3 FINDINGS

14.3.1 Presentation and Description of CWA Outputs

14.3.1.1 The Abstraction Hierarchy

An AH for long haul trucks was modelled from the collected data. A description of how the AH was modelled is described in Chapter 1. In Figure 14.5 we present the four highest levels of the AH, thus the lowest level of the AH (physical objects) was excluded. The functional purpose is described as 'Goods Distribution Via road Transportation' and the three value and priority measures are 'Effectivity and Efficiency', 'Safety' and 'Comfort'. The value and priority measure 'Effectivity and Efficiency' includes the system decompositions 'Efficient Relocation' and 'Work & Cargo Transportation', which represents both the efficiency when moving to a new location and the ability to transport cargo. The value and priority 'Safety' covers both the safety for the driver, other road users, the truck and the cargo, and is divided into five different system decompositions regarding; when the truck stands still, avoiding accidents, the truck's sustainability, safety during an accident and avoiding work injuries.

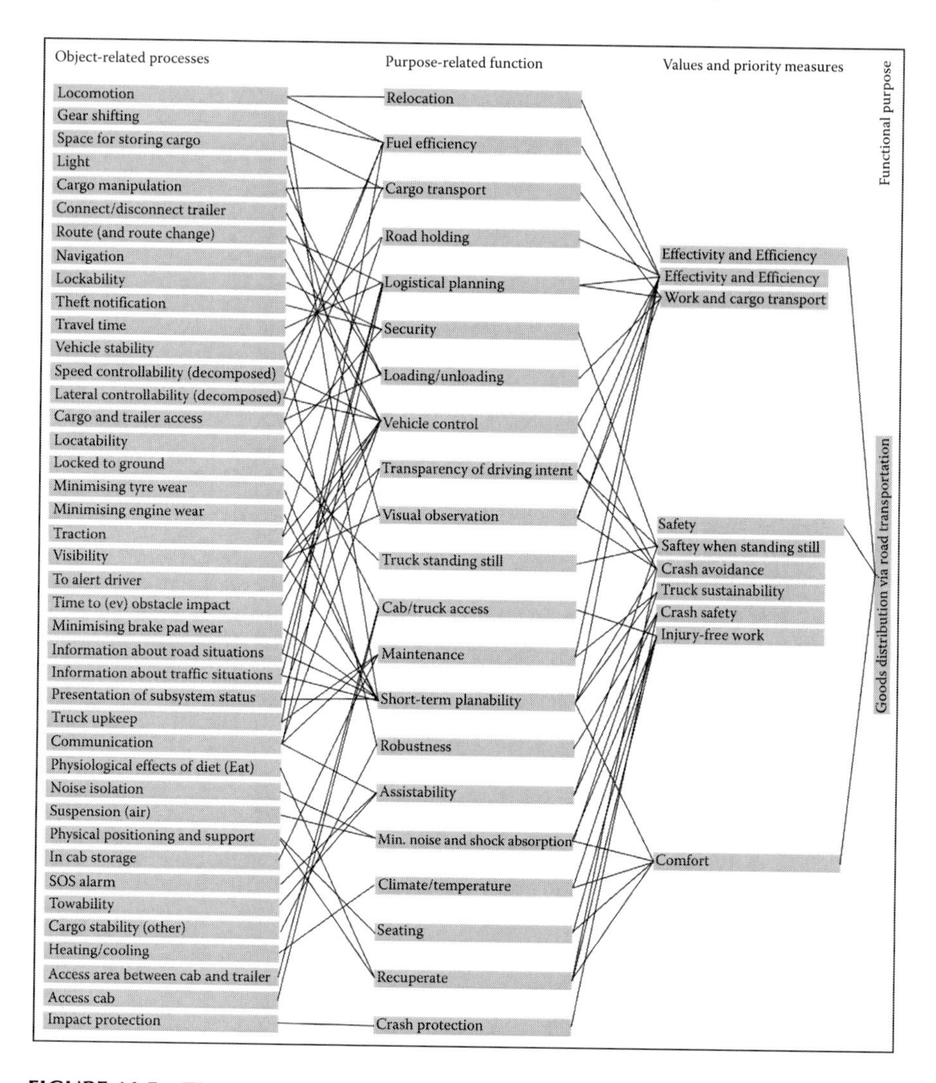

FIGURE 14.5 The top four levels of the AH for long haul trucks.

14.3.1.2 The Contextual Activity Templates

The object-related CAT included 2184 possible activities, where 1034 usually occur, 463 can occur and 687 do not occur. The purpose-related CAT is smaller because of the lower quantity of purpose-related functions, and including 882 possible activities of which 459 usually occur, 170 can occur and 253 do not occur.

Figure 14.6 shows the object-related CAT with the selected activities CAT score written in black instead of being represented by figures (as per Naikar et al., 2006). The activity prioritisation scores are the values in white. Figure 14.7 includes an excerpt of the purpose-related CAT showing the activity CAT scores and their activity prioritisation scores.

	Highway	Country road	Urban driving	Break during the way	Sleep in truck	Road work	Traffic jam	Ferry	Refuel	Truck break down	Function priority
Cargo manipulation	0 / 0	0 / 0	0 / 0	0 / 0	0 / 0	0 / 0	0 / 0	0 / 0	0 / 0	0 / 0	x3
Travel time (estimated)	18 / 2	27 / 2	12 / 2	9 / 2	0 / 0	0 / 0	0 / 0	2 / 1	2 / 1	0 / 0	x3
Speed controllability	18 / 2	27 / 2	12 / 2	0 / 0	0 / 0	3 / 2	6 / 2	0 / 0	0 / 0	0 / 0	x3
Lateral controllability	18 / 2	27 / 2	12 / 2	0 / 0	0 / 0	3 / 2	6 / 2	3 / 2	0 / 0	0 / 0	x3
Traction	18 / 2	27 / 2	12 / 2	1 / 1	6 / 1	3 / 2	6 / 2	3 / 2	0 / 0	0 / 0	x3
Visibility	54 / 2	81 / 2	36 / 2	14 / 1	0 / 0	9 / 2	18 / 2	9 / 2	5 / 1	5 / 1	x9
Info. about traffic and road situations	27 / 1	41 / 1	36 / 2	0 / 0	0 / 0	9 / 2	18 / 2	0 / 0	0 / 0	5 / 1	x9
Communication	54 / 2	81 / 2	36 / 2	14 / 1	18 / 1	9 / 2	18 / 2	5 / 1	5 / 1	9 / 2	x9
Situations frequency	x3	x4.5	x2	x1.5	x2	x0.5	x1	x0.5	x0.5	x0.5	

FIGURE 14.6 Sample from the object-related CAT including the situations (top), object-related processes (left), CAT scores (in black), function priority scores (right), situations frequency scores (bottom) and the activity prioritisation scores (in white).

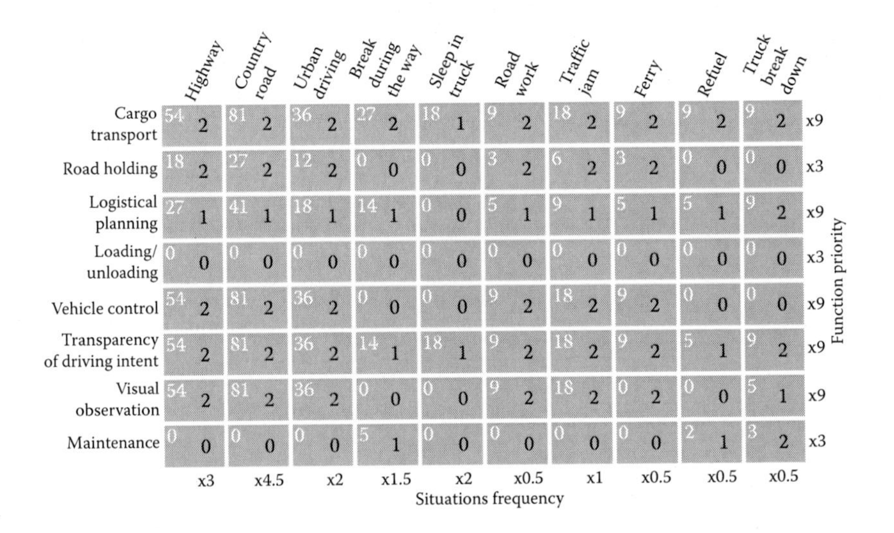

	Highway	Country road	Urban driving	Break during the way	Sleep in truck	Road work	Traffic jam	Ferry	Refuel	Truck break down	Function priority
Cargo transport	54 / 2	81 / 2	36 / 2	27 / 2	18 / 1	9 / 2	18 / 2	9 / 2	9 / 2	9 / 2	x9
Road holding	18 / 2	27 / 2	12 / 2	0 / 0	0 / 0	3 / 2	6 / 2	3 / 2	0 / 0	0 / 0	x3
Logistical planning	27 / 1	41 / 1	18 / 1	14 / 1	0 / 0	5 / 1	9 / 1	5 / 1	5 / 1	9 / 2	x9
Loading/ unloading	0 / 0	0 / 0	0 / 0	0 / 0	0 / 0	0 / 0	0 / 0	0 / 0	0 / 0	0 / 0	x3
Vehicle control	54 / 2	81 / 2	36 / 2	0 / 0	0 / 0	9 / 2	18 / 2	9 / 2	0 / 0	0 / 0	x9
Transparency of driving intent	54 / 2	81 / 2	36 / 2	14 / 1	18 / 1	9 / 2	18 / 2	9 / 2	5 / 1	9 / 2	x9
Visual observation	54 / 2	81 / 2	36 / 2	0 / 0	0 / 0	9 / 2	18 / 2	0 / 2	0 / 0	5 / 1	x9
Maintenance	0 / 0	0 / 0	0 / 0	5 / 1	0 / 0	0 / 0	0 / 0	0 / 0	2 / 1	3 / 2	x3
Situations frequency	x3	x4.5	x2	x1.5	x2	x0.5	x1	x0.5	x0.5	x0.5	

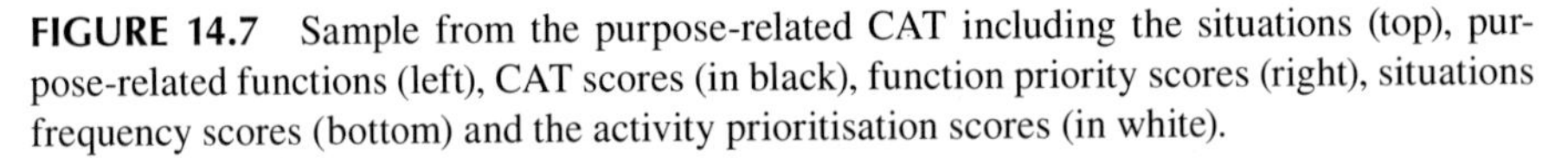

FIGURE 14.7 Sample from the purpose-related CAT including the situations (top), purpose-related functions (left), CAT scores (in black), function priority scores (right), situations frequency scores (bottom) and the activity prioritisation scores (in white).

14.3.1.3 Strategies for Long Haul Driving

The strategies for vehicle control are described in Table 14.1, and categorised according to the strategy categories defined by Hassall and Sanderson (2014). The elicited strategies are briefly described, and the table also includes the information whether the strategy is for the future or the current system and the strategy prompts. Most of

TABLE 14.1
The Identified Strategies for 'Vehicle Control'

Strategy Category	Elicited Strategy	In Current or Future System	Strategy Prompt
Avoidance	Use cruise control and Eco Roll	Current	Highway driving
	No supervision	Future	Trust
	Let passenger supervise	Future	Tired driver
Intuitive	Emergency breaking/steering	Current	Obstacles close in front
	Vehicle placement in lane	Current	Driving
	Intuitively take over control	Future	Emergency, automation failure
Option-based	Either release the acceleration, hit the brakes, or swerve	Current	Obstacles close in front
	Drive safe, efficient or fast	Current	Time pressure, visible police cars, etc
	Choose between alternatives of drive safe, efficient, fast or in a cargo specific mode	Future	Policy, status, time pressure, cargo (hanging, liquid, livestock or fixed)
Cue-based	Turning left/right, changing lane	Current	Curves, needed lane changes
	The driver uses cues in the environment to drive and turn	Current	Narrow place
	Speed limit signs, GPS	Both	Information input
	Reads cues from traffic far in front	Both	Visible traffic
	See where other vehicles will drive depending on how they behave (to know if the truck behaviour is sufficient)	Both	For example when changing lanes in high traffic
	Use environmental cues to know what happens further ahead	Both	Obstacles
	Compare environment with what the system detects	Future	Driving, wants to know that the automated systems are functioning
	Use information from truck to know what the truck will do next	Future	Lane change, overtaking other vehicles
Compliance	Keep within speed limits	Current	When driving
	The driver is following some rules of thumbs to drive and turn	Current	Narrow place
	Using knowledge about how the trailer acts when turning to drive and turn	Current	Narrow place
	Make sure the truck drives within speed limit	Future	When driving Temporary speed limits
	Take over control	Future	Traffic rules (local or terrain specific) against automation Automation or sensor failure
Analytical	Deciding to take over control or not by reasoning whether the truck can handle an upcoming situation	Future	Novel situations

FIGURE 14.8 Heads-up display for controlling a self-driving truck.

the elicited strategies are categorised as belonging to the 'Cue-based' category, and half of them are expected to be used both in the current and future systems.

The strategies analysis also resulted in information visualisations generated by the drivers. These were described both via text and by the sketches created during the workshops.

14.3.2 Heads-Up Display for Control of a Self-Driving Truck

On the basis of the work described above, a concept heads-up display (HUD) was created. The HUD was composed of five 'panels', (1) strategic display (top), (2) entertainment (bottom), (3) convoy display (right), (4) bird's eye view/tactical display (left) and (5) augmented reality overlay (centre). The strategic display supports planning while driving, and extends the strategy of using environmental cues to interpret events further ahead. The convoy display supports communication with other drivers during convoy driving. The augmented reality (e.g. the yellow line markings in Figure 14.8) and the bird's eye view are ways to visualise what the system has detected. These latter displays support the cue-based strategy for vehicle control where the driver compares what is seen in the environment to what the systems detects. The blue line (see Figure 14.8) shows where the truck plans to drive, and is an example of information about what the truck will do next augmented on the environment.

14.4 DISCUSSION

When applying the CWA framework in a certain domain, project and industry, the method needs to be adjusted due to the immediate constraints and needs. When adapting the methods to fit the constraints there are some practical and methodological implications. Some of the ones we faced are described in the following two sections.

14.4.1 Practical Implications

14.4.1.1 CWA in Industry

The CWA framework is comprehensive and using it provides important insights. However, the framework is effort intensive, and requires much work to collect and analyse data. To rationalise the investment, the CWA results should live beyond one project. Ideally, the knowledge gained via the CWA should be made available for multiple product (development) teams and be renewed with new insights or technology advances. To make this possible the analysis needs to be, and appear to be, useful for other possible stakeholders.

For the use case described previously, the CWA was conducted to support the design of a future truck concept. For this use case, it was not straightforward to know how all the information modelled in the AH and the CATs should be presented to the designers, and how it actually could give input to a specific design solution. One innovative solution was to facilitate handover sessions between analysis and designers. Following key stops of the analysis phase, a dedicated handover workshop conducted during which the analyst handed over their results to the design team in a way that was practically useful for the designers. Results from the analysis were discussed item by item using mutually understandable terms. As mentioned, for the strategies analysis handover, this provided the MODAS project designers with more concrete directions on what to design.

To make the CWA useful for other stakeholders, and to share the analysis results with the correct persons in the right way, we needed to investigate who within the industrial organisation would benefit from using the CWA models, and how the models could be used. To investigate how different parts of the organisation can use the AH and the CATs, an exploratory study was conducted with persons representing three work roles in the organisation; group managers, object owners and property owners (Bodin and Krupenia, 2015). Group managers supervise groups of around 10–30 persons, and are the formal lowest organisational unit. Object owners are responsible for an object in the vehicle, and is the lowest organisational decomposition of technical systems. Property owners are responsible for product quality and could involve the quality arising from the integration of multiple products developed by multiple groups.

Four group managers, four object owners and four property owners were interviewed individually to explore how they could benefit from using the WDA and CATs for strategic planning and for product development planning over different time horizons. After a brief explanation of the CWA framework in general, the AH and the CATs were explained. Each informant rated to what extent the models could support their strategic and product planning for time horizons ranging from 6 months to 10 years. The AH was rated first and then the CATs (object-related and purpose-related) in counterbalanced order. For the analysis, a large emphasis was placed on the participants' comments about how they expected they could, or could not, use the models. Participants were asked to explain their answers and suggest alternative ways of using the models.

From this study, we found that the AH was seen as useful for visualising relations between sub-systems and abstract functions of the system. The difference in rated usefulness for the object-related CAT and the purpose-related CAT was small, but the qualitative feedback given shows that the models would likely be used in different ways. Some participants perceived the object-related CAT as easier to interpret, which seems to be connected to the closeness to the physical form of the technical system (something typically kept in focus during development). All four property owners, who have responsibility for a higher level function of the truck system, argued for the purpose-related CAT's usefulness over the longer term because of the time needed to make larger product changes or because the purpose-related functions are more likely to be valid for a longer time compared to the object-related processes (due to technical developments).

The main concern with the models was that they were hard to interpret (especially the more abstract parts) and were somewhat overwhelming due to the large amount of data represented – people therefore found it difficult to obtain a good overview. Further technical development was seen as a threat to the long-term usefulness of the models, because rapid technical advances could outdate the models. This scepticism depended on whether the model was considered as a description, or a process where the documents are supposed to be updated, or even used as a tool to analyse what project updates would mean or to analyse future scenarios.

14.4.1.2 Focusing Analysis Effort

Five of the twelve informants mentioned that the CATs could be valuable tools for prioritising development or that the accompanying CAT prioritisation score would be valuable (Bodin and Krupenia, 2015). Note that prior work by the authors – the activity prioritisation score (Bodin and Krupenia, 2016) was not included or mentioned during these interviews. Spontaneous suggestions about how to complete this prioritisation was via event frequency (focus development on situations where the functions can be used), or via information in the models that was unexpected.

As described above, the need for prioritisation was foreseen when the CATs were modelled. A high number of activities made it unreasonable to further investigate all activities via task and strategies analysis, at least in the first stage. To randomly select a few activities was judged as insufficient; hence a deliberate way of prioritising and selecting was needed. Therefore the activity prioritisation score was developed as guidance when focusing further analysis efforts, that is, to fit the scope of the analysis to the size of the project and the available resources.

Different criteria could be used for prioritisation, depending on the aim of the project. In this project where a first concept for a future system was developed, the score from the CAT (can, usually, or do not occur), situation frequency and connections of the functions to higher level purposes (with a score developed by Birrell et al., 2011) was seen to be central. Other projects or other stages of product development may require other criteria. An alternative to prioritisation could be to define the activities in less detail, thus reducing the overall number of activities. The reason for not doing that in the study described in this chapter was to avoid excluding important detail. Instead the list with highest prioritised activities was used as a guide when selecting what to include in the strategies analysis. Instead of selecting the activities with top activity prioritising score the scores was used to see which functions were parts of activities (a function in a situation) with the highest activity prioritising scores. The functions were selected and described as tasks conducted today and by the future driver. Here it was seen that the activities with higher abstraction level, from the purpose-related CAT, was more suited for guiding the choice of what to include in the strategies analysis because they were less narrowly defined. The purpose-related functions were also merged and complemented with challenging events to bring in the situation aspects and to support a rich discussion of the selected tasks in preparation for the strategies analysis. Further research could define a procedure of mapping activities together for further analysis, which could be used as an alternative to prioritisation.

14.4.2 Methodological Implications

14.4.2.1 The Activity Prioritisation – A Violation of the Fundament of CWA?

The implications of the activity prioritising goes beyond practical needs to also include methodological concerns regarding compromising the purpose and the strength of the CWA framework. As described in Bodin and Krupenia (2016) and the associated critique (Burns and Naikar, 2016), a primary concern of prioritisation is that the CWA framework is often used for analysing safety critical socio-technical systems, and prioritising activities for focus of further analysis excludes other activities from analysis. The decision to include frequency as a prioritisation factor may not be suitable in some projects, because uncommon situations can be more demanding for the worker and need to be considered during the development process.

The motivation for prioritising is that resources are often limited (not only in an industrial context) and the CWA analysis can grow too large to complete in full, at least as a first step. Alternatives to this more resource heavy approach would be to avoid using the CWA framework, to exclude detail in the earlier stage of the analysis, or to select the activities to be analysed in an arbitrary manner. We argue, however, in favour of using the CWA framework but to be aware, and selective, of what aspects of the analysis are carried over to later analysis phases.

Even if time and resources was unlimited and no prioritising method is used, it is still necessary to assume that not everything will be covered by any analysis of tasks and strategies. It is impossible to include all possible tasks and strategies in any analysis, due to the nature of unanticipated events which always exist in complex systems.

14.4.2.2 Relevance to the Development Work Gives Descriptive Elements in the Analysis

A concern brought up by the informants in the study about the usefulness of the CWA models (Bodin and Krupenia, 2015) was that technical development will outdate the models. If the models are describing the ecology of the task and visualising the work domain, can technical development really outdate the models? We would say that this is the case in this study, because the CWA was based on the description of the existing system. While CWA is suited for handling novelty by giving the possibility of designing a new system from the ground up, based on the constraints imposed by the work itself, CWA can also (as it is here) be used as a way to describe the current systems' relations to the higher purposes imposed by the work domain (Vicente, 1999). For example lateral controllability (object-related process) and vehicle control (purpose-related function) are needed to transport goods to the right location, while the physical form 'steering wheel' is the name of the control today which might not be the same even in the near future. With this reasoning we can say that the top levels of the AH are more stable over time, while the two bottom levels are largely influenced by the current system. The desire for the model to be useful for development, and at the same time maintain a concrete level description on the lower abstraction levels, was the reason for the descriptive nature.

Over a longer period of time, and when environments change, a change in the purpose and values in the work domain could occur. The constraints in the project described here were that we assumed a future truck to handle future traffic demands. New technology gives new opportunities and the work itself will change. We used CWA to analyse the current system and in that way to also understand the work domain and control tasks. Similar values will be of importance in the future, and similar functions or tasks will need to be performed, even when the drivers work tasks might change.

While the AH and the CATs will be outdated over time because of technical development, the higher levels of the AH and the purpose-related CAT can be assumed to be more stable. As some of the informants in the interview study about the CWA usefulness (Bodin and Krupenia, 2015) commented, the CWA models can be useful during technological changes and visualise important information needed when taking decisions about product updates.

14.5 CONCLUSION

While the CWA framework is useful for analysing complex sociotechnical systems, and can be used in an industrial context, further work is needed to better fit the framework within industrial requirements and constraints and to reduce the risks associated with the effort invested and outcomes attained. Having methods or processes by which different stakeholders can use, and maintain, the models after their initial construction would significantly justify the initial resource investment in CWA. Further approaches to focus the analysis effort and scale the CWA to a smaller project could increase the possibility to use the CWA framework in industry.

ACKNOWLEDGEMENTS

The authors would like to thank the following key contributors to the MODAS project; Anders Jansson, Anton Axelsson, Johanna Vännström, Victor Ahlm, Johan Fagerlönn, Jon Friström, Stefan Larsson, Rickard Leandertz, Stefan Lindberg, Matteo Manelli, Daniele Nicola and Anna Sirkka. We would also like to thank the Scania Transport Laboratory for their collaboration on this work. Parts of the work were funded by the Strategic Vehicle Research and Innovation (FFI) funding scheme to the MODAS project (2012-03678).

REFERENCES

Birrell, S. A., Young, M. S., Jenkins, D. P. and Stanton, N. A. 2011. Cognitive work analysis for safe and efficient driving. *Theoretical Issues in Ergonomics Science*, 13(4), 430–449. doi: 10.1080/1463922X.2010.539285.

Bodin, I. 2013. Using cognitive work analysis to identify opportunities for enhancing human-heavy vehicle system performance (*Masters dissertation*). Stockholm, Sweden: KTH Royal Institute of Technology.

Bodin, I. and Krupenia, S. 2015. Supporting industrial uptake of cognitive work analysis. *Proceedings of the HFES 2015 Annual Meeting*, 59(1), 170–174.

Bodin, I. and Krupenia, S. 2016. Activity prioritization to focus the control task analysis. *Journal of Cognitive Engineering and Decision Making*, 10(1), 91–104.

Burns, C. M. and Naikar, N. 2016. Prioritization: A double-edged sword?. *Journal of Cognitive Engineering and Decision Making*, 10(1), 105–108.

Fagerlönn, J., Lindberg, S., Sirkka, A., Krupenia, S., Larsson, S., Nicola, D. and Manelli, M. 2015. Deliverable 5.3: Third Version of the Auditory and Visual Displays. MODAS Project Deliverable [2012-03678].

Hassall, M. E. and Sanderson, P. M. 2014. A formative approach to the strategies analysis phase of cognitive work analysis. *Theoretical Issues in Ergonomics Science*, 15(3), 215–261. doi: 10.1080/1463922X.2012.725781.

Jansson, A., Olsson, E. and Erlandsson, M. 2006. Bridging the gap between analysis and design: Improving existing driver interfaces with tools from the framework of cognitive work analysis. *Cognition, Technology & Work*, 8(1), 41–49.

Jenkins, D., Stanton, N., Salmon, P. and Walker, G. 2009. *Cognitive Work Analysis: Coping with Complexity: Human Factors in Defence [Elektronisk resurs]*. Farnham, UK: Ashgate Publishing Limited.

Krupenia, S., Selmarker, A., Fagerlönn, J., Delsing, K., Jansson, A., Sandblad, B. and Grane, C. 2014. The 'Methods for Designing Future Autonomous Systems' (MODAS) project: Developing the cab for a highly autonomous truck. In N. Stanton, S. Landry, G. Di Bucchianico and A. Vallicelli (Eds), *Advances in Human Aspects of Transportation*, Part II, pp. 70–81. Louisville, KY: AHFE Conference.

Löscher, I., Axelsson, A., Vännström, J. and Jansson, A. 2017. Eliciting strategies in a revolutionary design: Exploring the hypothesis of predefined strategy categories. *Theoretical Issues in Ergonomics Science*, 1–17. doi: 10.1080/1463922X.2017.1278805.

Ministry of Jobs, Tourism and Skills Training, Canada 2010. Employment Standards Branch Factsheet: Truck Drivers. http://www2.gov.bc.ca/gov/content/employment-business/employment-standards-advice/employment-standards/factsheets/truck-drivers

Naikar, N., Hopcroft, R. and Moylan, A. 2005. Work domain analysis: Theoretical concepts and methodology (No. DSTO-TR-1665). Victoria, Australia: Commonwealth of Australia.

Naikar, N., Moylan, A. and Pearce, B. 2006. Analysing activity in complex systems with cognitive work analysis: Concepts, guidelines and case study for control task analysis. *Theoretical Issues in Ergonomics Science*, 7(4), 371–394.

Read, G. J. M., Salmon, P. M. and Lenné, M. G. 2012. From work analysis to work design: A review of cognitive work analysis design applications. *Proceedings of the Human Factors and Ergonomics Society Annual Meeting*, September 2012; vol. 56, 1: pp. 368–372.

Tschirner, S. 2015. The GMOC Model: Supporting Development of Systems for Human Control (*Doctoral dissertation*). Uppsala: Acta Universitatis Upsaliensis.

Vicente, K. J. 1999. *Cognitive Work Analysis: Toward Safe, Productive, and Healthy Computer-Based Work*. Mahwah, NJ.: Lawrence Erlbaum Associates.

15 Using Work Domain Analysis to Design a Rotary Wing Head-Up Display

Neville A. Stanton, Katherine L. Plant, Aaron P. Roberts, Catherine Harvey and T. Glyn Thomas

CONTENTS

15.1 INTRODUCTION

The current operational environment for helicopters varies greatly with role, but helicopters generally operate outside of direct air traffic control, at low altitudes and under visual flying conditions (British Helicopter Association, 2014). These operational advantages mean that helicopters are used in operational contexts that are not suitable for fixed wing aircraft, including medical rescue over land, search and rescue over water or mountains, rapid corporate passenger transfer, oil platform transfer, police search, television broadcasting and firefighting. Helicopters tend to perform take-off and landing manoeuvres that are unlike fixed wing aircraft, generally being steep and with greatly reduced landing distances at both managed and unmanaged landing sites (Brook and Dodge, 2014). This benefit of helicopter flight is also the helicopter's greatest hazard as these manoeuvres can be dangerous and in remote locations. A primary cause of rotary wing accidents is degraded visibility caused by poor weather. A DVE is one in which ocular visibility is reduced due to light levels (e.g. night), weather phenomena (e.g. fog) or atmospheric conditions (e.g. dust) (Hart, 1988). Baker et al. (2011) found that bad weather was the second most common precipitating factor (after mechanical failure) for fatal and nonfatal crashes in their analysis of 178 Gulf of Mexico helicopter accidents. Due to the nature of helicopter operations, in remote locations and in emergency operational contexts, the likelihood of encountering a DVE can be significant. Aside from the increased safety risk, a DVE presents one of the most disruptive factors in civil aviation and is a leading contributing factor to flight delays at major commercial airports (Allan et al., 2001; Pejovic et al., 2009).

HUDs have been commonly used in military aviation for a number of years but are increasingly being used in commercial flight operations and for other applications such as driving (Harris, 2011; Jakus et al., 2014). A HUD interposes images on a transparent layer between the pilot and windshield, allowing the pilot to simultaneously look at the HUD symbology and the outside world. This allows the pilot to fly 'eyes out' rather than switching attention to head-down displays (HDDs) inside the cockpit. Conformal symbology is often included on HUDs to increase the realism of the presented information. Conformal, or scene-linked, symbology allows information to be displayed at a static position relative to the real world; an example of this would be the outline of a helipad displayed on the HUD which remains overlaid on top of the view of the helipad in the outside world, regardless of what part of the HUD the pilot is looking at. The objective of a HUD is to replicate the operational benefits of clear-day flight operations regardless of the actual outside visibility condition, thus increasing operational capacity and reducing accidents caused by low visibility conditions. The presentation of information via a HUD in a manner that does not require the pilot to divert visual attention and cognitive resources into the cockpit has the ability to optimise workload and enhance situation awareness (Fadden et al., 1998; Snow and Reising, 1999; Snow and French, 2002). Conformal symbology leads to faster detection of changes in symbology and improved flight path tracking accuracy (Fadden et al., 1998; Snow and French 2002). Current HUDs offer a very limited field of view for conformal symbology and the pilot's head must be located in the eye-box to ensure the visibility of all HUD symbology. Due to these constraints it has been argued that HUDs are unsuitable for helicopter operations, however the HUD concept tested here was developed using an extended field of view (180°) to consider how future cockpits might present information, for example, windshield displays. Furthermore, the HUD was developed in a full colour HUD to represent the potential of future cockpits. Dudfield (1991) demonstrated that a coloured, compared to monochrome, HUD reduced subjective workload although there was no beneficial effect on performance.

Helicopter flight operations represent a complex sociotechnical system made up of numerous interacting parts, both human and non-human, operating in dynamic, ambiguous and safety critical domains. The complexities in these systems present an opportunity for cognitive work analysis (CWA) to assist in the modelling and analysis. The attraction of using this method to assist in the development of HUD concepts is that the analysis offers a formative approach, that is, describing how the system *could* perform, as opposed to analysing what a system current does (descriptive modelling) or should do (normative modelling). Cornelissen et al. (2014) argued that CWA is one of the few formative methods in the ergonomics toolkit and is particularly important as a design method. A description of each phase of CWA is provided in Chapter 1. Whilst each phase of the analysis process builds upon the last, McIlroy and Stanton (2011) argued that not all phases are required to be used equally and it is down to the analyst to decide which phases are necessary to answer the research question under investigation. To address the research aims of this chapter, the first phase of the CWA process, work domain analysis (WDA, see Chapter 1) was utilised to analyse the cognitive work of landing a helicopter in a DVE in order to identify the critical information requirements associated with this task.

15.2 METHOD

15.2.1 Constraints Analysis of Approach and Landing

WDA is represented via an abstraction hierarchy (AH); this models the system at a number of levels of abstraction (see Chapter 1 for a description of each level). The analysis was conducted with the CWA software tool developed by the Human Factors Integration Defence Technology Centre. The AH was populated via consultation sessions with four subject matter experts (SMEs) who were experienced helicopter pilots and had worked in various operational contexts including the military, private passenger transport, and search and rescue. These sessions involved the construction of AHs assisted by experienced Human Factors researchers. This information was supplemented with documentation to support domain familiarisation including flight manuals, simulator observations and videos of DVE landing. The individual AHs were amalgamated into one representative AH (see Figure 15.1) that was verified by a test pilot.

The *functional purpose* captures the reason why the system exists. Here, the functional purpose of the situation is to 'establish hover and land' this encompasses the purpose of the situation as it is the prerequisite to ensure you are ready to land. The next level down, *values and priority measures*, captures the criteria that can be used to determine how well the system is achieving its functional purpose. To ensure a successful approach and maintenance of hover is achieved the values and priority measures were defined as; 'aircraft altitude decreased', 'maximise smooth transition of aircraft', 'minimise time to conduct approach and hover' and 'avoid collisions and impact with hazards/third parties'. The middle level, *purpose-related functions*, lists the functions that have the ability to influence one or more of the values and

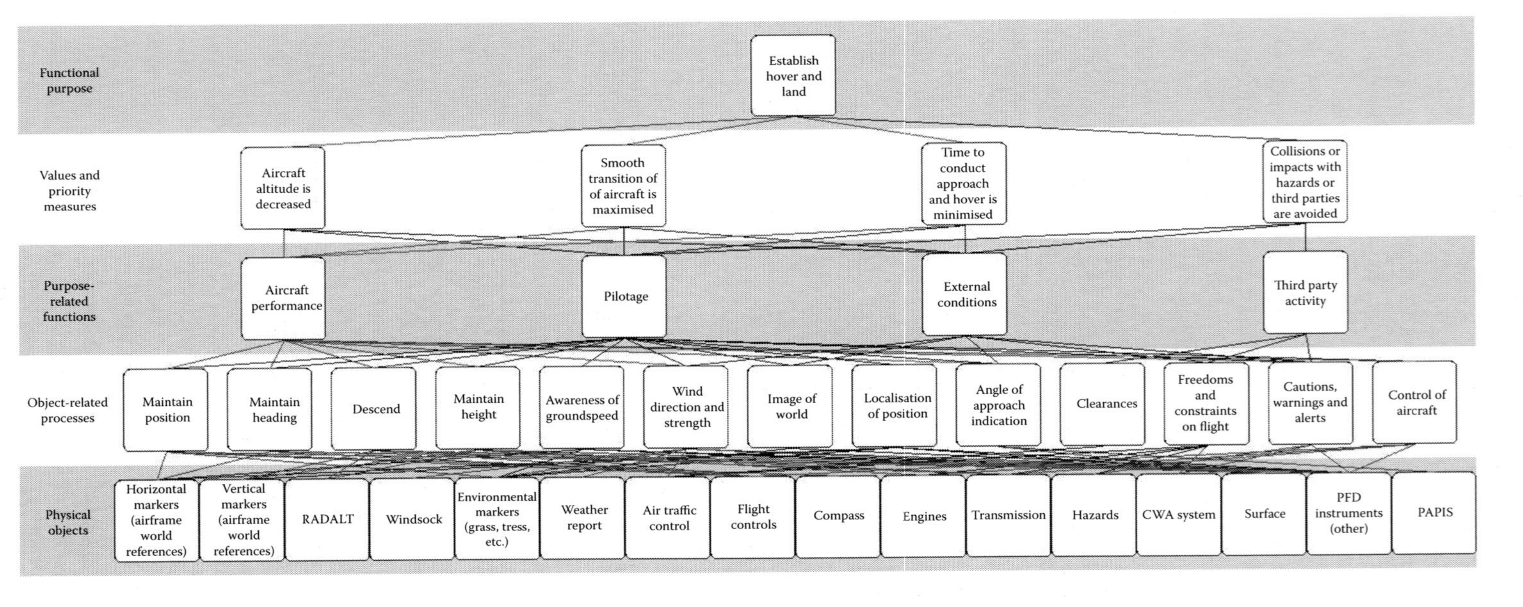

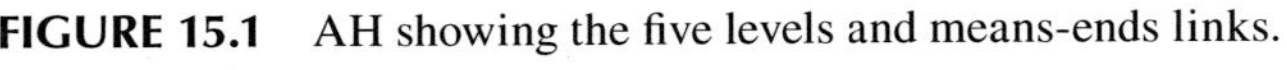
FIGURE 15.1 AH showing the five levels and means-ends links.

priority measures. They link the purpose-independent processes of the physical objects (described in the next level) with the more abstract measure of system performance (in the *values and priority measures*), thus joining the AH together. In this analysis the purpose related functions include; 'aircraft performance', 'external conditions', 'third party activity' and 'pilotage'. In the second level from the bottom, the *object-related processes* (O-RP) capture the affordances that are provided by the physical objects (described below) in order to perform the *purpose-related functions*. For example, the physical object of environmental markers (grass, trees, etc.) affords knowledge of maintaining position, awareness of groundspeed, wind direction and strength, image of world and localisation of position. The lowest level, *physical objects*, lists the physical components of the system. For this analysis, physical objects consisted of any piece of information, internal or external to the cockpit, which enables the pilot to achieve the functional purpose. Fundamental cockpit objects such as seats, harnesses and windows all provide important affordances to the cockpit; however the analysis needs to be kept as manageable as possible so the boundary was set to omit them. Example physical objects include; windsock, weather report, compass, surface, radio altimeter, flight controls and airframe world references.

15.2.2 HUD Development

Foyle et al. (1992) has argued that the challenge for Human Factors engineers is to design visual displays that represent the most useful visual cues that pilots most naturally use. The WDA focused on coming into an approach to maintain a hover in preparation for landing in a DVE. The tasks required to achieve the functional purpose of the system are represented by the O-RP and therefore these need to be represented on the HUD to ensure it adequately facilitates the task. The HUD concept consists of the physical objects necessary to achieve the functions defined in the AH and is presented in Figure 15.2. Table 15.1 depicts the mapping between the O-RP level of the AH and the symbology included on the HUD.

The HUD contained the following 2D flight instruments: conformal compass, heading readout, airspeed indicator, conformal horizon line, attitude indicator, vertical speed indicator, air speed indicator, wind direction and strength indicator and distance to go readout. To assist with the hover/landing task the HUD included the following elements:

1. Flight path vector (acts as a touch down indicator, this becomes visible to the pilot when losing altitude, it represents the point on the ground they will hit if the current velocity is maintained) (item 1 in Figure 15.2)
2. 3D augmented reality 'tree' (these provide a visual reference point for the pilot when landing) (item 2 in Figure 15.2)
3. Arrows against the trees provide the pilot with visual references for height and speed as the aircraft is descending (item 3 in Figure 15.2)
4. Runway grid and runway markings (item 4 in Figure 15.2)
5. Obstacles (3D augmented reality used to represent obstacles and hazards such as gas towers shown as item 5 in Figure 15.2)

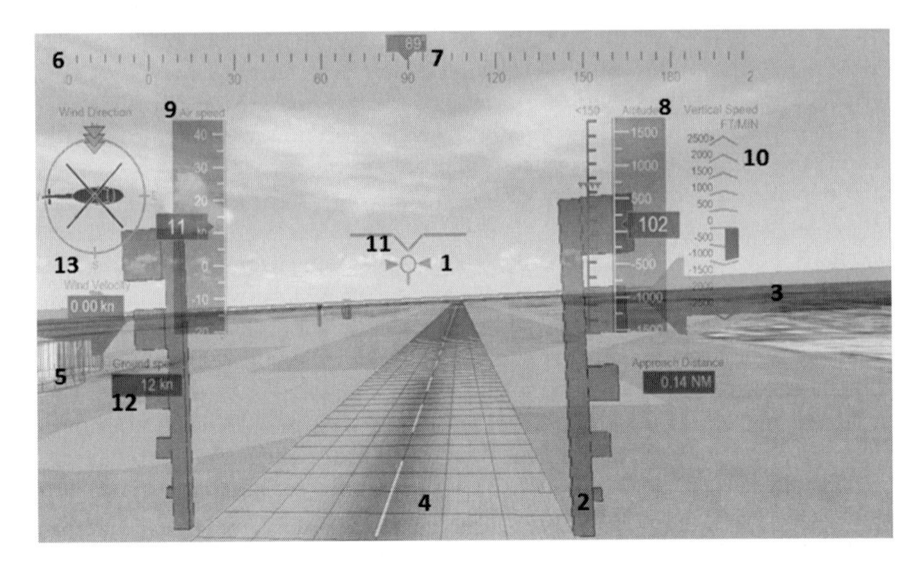

FIGURE 15.2 Landing symbology in the HUD (numbers relate to the description in subchapter 15.2.2 and Table 15.1).

15.2.3 Experimental Evaluation of the HUD

15.2.3.1 Design

The study employed a 2 × 2 within-subjects design. The independent variables were weather condition (clear sky or DVE/fog) and symbology used (with or without the HUD). The order of presentation of the conditions was counterbalanced between the participants.

15.2.3.2 Participants

Six male subjects aged 37–65 (mean = 51.00, SD = 10.29) were voluntarily recruited for the study via advertisement posters placed at local airfields and around the university campus. All subjects were qualified, instrument rated, rotary wing pilots with varying amounts of experience; flying hours ranged between 108 and 8300 hours (mean = 3804, SD = 3468). Pilots were recruited from a range of operational contexts including; search and rescue, private transport and the military. Ethical permission for this study was granted by the Research Ethics Committee at the University of Southampton.

15.2.3.3 Equipment and Materials

15.2.3.3.1 Flight Simulator

A fixed-based flight deck simulation facility at the University of Southampton was used in its rotary wing configuration. The simulator comprises a two-seater cockpit (including external cabin) with five multi-function display units. The external screen is projected onto three screens providing an 180° degree field of view. Participants were seated in the right-hand seat as which was configured for the

TABLE 15.1
HUD Symbology That Allowed the O-RP to Be Achieved

O-RP	HUD Symbology to Achieve O-RP Function
Maintain position	Conformal compass (item 6 in Figure 15.2)
	Heading readout (item 7 in Figure 15.2)
	Attitude indicator (item 8 in Figure 15.2)
	Airspeed indicator (item 9 in Figure 15.2)
	3D augmented reality trees (item 2 in Figure 15.2)
	Descent arrows (item 3 in Figure 15.2)
	Flight path vector (item 1 in Figure 15.2)
Maintain heading	Conformal compass (item 6 in Figure 15.2)
	Heading readout (item 7 in Figure 15.2)
Descend	Vertical speed indicator (item 10 in Figure 15.2)
	3D augmented reality trees (item 2 in Figure 15.2)
	Descent arrows (item 3 in Figure 15.2)
	Flight path vector (item 1 in Figure 15.2)
Maintain height	Altitude indicator (item 11 in Figure 15.2)
	Attitude indicator (item 8 in Figure 15.2)
	Descent arrows (item 3 in Figure 15.2)
Awareness of groundspeed	Groundspeed indicator (item 12 in Figure 15.2)
	Runway grid (item 4 in Figure 15.2)
Wind direction and strength	Wind direction and strength indicator (item 13 in Figure 15.2)
Image of world	Runway grid (item 4 in Figure 15.2)
	3D augmented reality trees (item 2 in Figure 15.2)
	Descent arrows (item 3 in Figure 15.2)
	Obstacles (item 5 in Figure 15.2)
Localisation of position	Runway grid (item 4 in Figure 15.2)
	3D augmented reality trees (item 2 in Figure 15.2)
	Descent arrows (item 3 in Figure 15.2)
Angle of approach indication	Runway grid (item 4 in Figure 15.2)
	3D augmented reality trees (item 2 in Figure 15.2)
	Descent arrows (item 3 in Figure 15.2)
	Attitude indicator (item 8 in Figure 15.2)
	Flight path vector (item 1 in Figure 15.2)
Clearances	*X*
Freedoms and constraints on flight	*X*
Cautions, warnings and alerts	*X*
Flight control	*X*

Note: *X* – Elements are not represented in the HUD because they are beyond the scope of this study.

rotary wing controls. The simulated environment runs on Prepar3D (previously Microsoft flight simulator software). The flight scenario was located over a runway at the Norfolk naval base, Virginia, USA, using the Bell 206 flight model. The Prepar3D software is highly customisable and allowed the required weather conditions to be simulated. In the clear sky condition the clear weather setting was selected and in the DVE the highest fog setting was used (approximately 300 m visibility).

15.2.3.3.2 Cockpit Displays

Head-down display: HDD was displayed to the pilots on the outer right multi-function display unit in the simulator. This was available to the pilots in all four conditions. The HDD was part of the Prepar3D software and consisted of analogue flight instruments for helicopters in a standard configuration, including: attitude indicator, airspeed indicator, a compass, heading indicator, altimeter, vertical speed indicator and torque indicator.

Head-up display: The HUD concept (Figure 15.2) was created using GL Studio. This is a software tool specifically developed for interface display design and provides the ability for its display instruments to be controlled from external applications (e.g. Prepar3D). A two-way data interface was developed to allow flight data to be transferred from Prepar3D and synchronised symbology to be transferred from GL studio. During the flight conditions with the HUD, the concept was overlaid onto the simulated environment using a ghost window application that is freely downloadable from the Internet.

15.2.3.3.3 Post-Task Questionnaires

Two questionnaires were administered after each experimental condition had been flown. The post-landing assessment (PLA) questionnaire was developed by an industry partner in the project funding this work. The questionnaire asks participants to rate their awareness of seven flight parameters (desired heading, desired rate of descent, desired groundspeed, power status, required landing point, drift and outside environment) from 1 (low) to 7 (high). These components represented different aspects of situation awareness under investigation. At the end of each flight a debrief session was held so that the researchers could probe the pilots about reasons for their ratings on the PLA.

The Bedford Workload Rating Scale (Roscoe and Ellis, 1990) is a uni-dimensional mental workload assessment technique developed to assess pilot workload. The technique involves a hierarchal decision tree to assess workload via an assessment of spare capacity whilst performing a task. Participants follow the decision tree to derive a workload rating for the task under analysis (Stanton et al., 2013). A scale of 1 (low workload – workload insignificant) to 10 (high workload – task abandoned) is used.

15.2.3.4 Procedure

The participants were briefed about the study and asked to complete the consent form and a questionnaire to gather demographic information. Participants were given an initial familiarisation session with the simulator which allowed them to get used to the flight model and flight controls. The participants were then familiarised with the HUD in a video talk provided by the software developer to explain each instrument and the conformal symbology. Participants were then given time to practice flying with the HUD. The familiarisation session lasted approximately 25 minutes and this aligned with previous studies that used both classroom and simulator training for a similar amount of time when introducing new cockpit symbology (e.g. Snow and Reising, 1999). The participants then flew each of the four experimental conditions (clear, clear + HUD, DVE or DVE + HUD). For each condition participants began 5 nm out to sea, at a height of 1500 feet and a speed of 50 knots. They were instructed to land on the runway which was visible in the clear sky conditions and a heading was provided in the DVE conditions. Participants were instructed to fly to the runway and land the aircraft. After each condition was flown the PLA and Bedford Workload questionnaires were administered and a debrief session probed pilots about the reasons for their ratings (each flight condition including questionnaires took approximately 15 minutes). When answering the questionnaires, the participants were instructed to detach from any feelings associated with the simulated environment (e.g. fidelity of the flight controls and flight model) and base their ratings purely on the HUD symbology and scenario under evaluation. At the end of the study a longer debrief session was held to allow pilots to comment on the symbology in the HUD.

15.3 RESULTS

15.3.1 Post-Landing Assessment

Figure 15.3 presents the average scores of the PLA ratings. Desired heading was rated higher in the two HUD conditions because pilots were able to use the conformal compass, heading readout and runway grid presented on the HUD to facilitate the O-RP of 'maintain heading'. Furthermore, three pilots stated that they used the runway grid during the HUD condition to maintain heading. Awareness of desired rate of descent came from the vertical speed indicator and trees with associated descent arrows. Awareness of rate of descent was highest in the two HUD conditions. This awareness parameter links to the O-RP of 'descend' and it is clear that the HUD was providing the pilot with enough information to achieve this function. 'Awareness of groundspeed' was an O-RP that related to the HUD symbology of the groundspeed indicator and runway grid (akin to the 'surface' in the physical objects

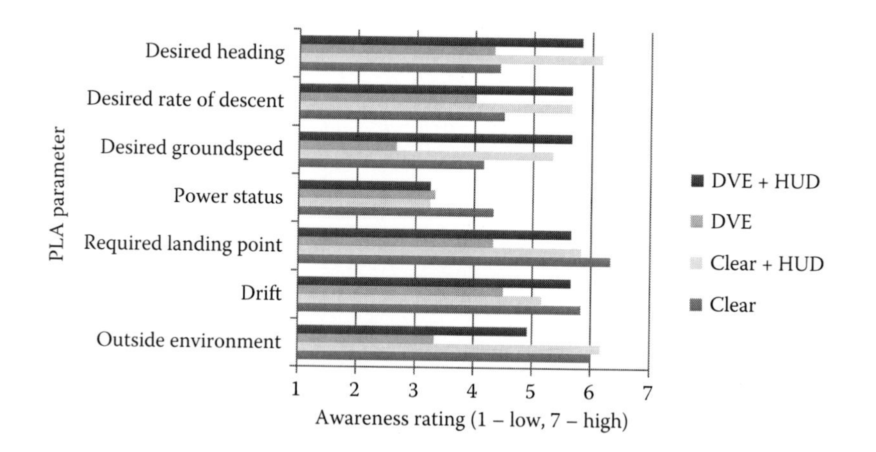

FIGURE 15.3 Mean scores of the PLA ratings.

of the AH). Awareness of groundspeed was highest in the DVE + HUD condition, followed by clear + HUD, clear and finally DVE condition. In the clear condition pilots stated that awareness increased as they got closer to the ground, whereas in the DVE condition awareness was low because they could not monitor the ground-rush, that is, speed of movement over the ground, this is why the runway grid proved so useful in the HUD conditions (particularly DVE + HUD). Awareness of power status was low across all conditions and this was a failing of the HUD. Power status was marginally higher in the clear condition when pilots had the spare capacity to scan the HDD for information they could not glean from the outside environment. Awareness of required landing point is associated with the O-RP of 'localisation of position' and 'angle of approach indication', the HUD symbology used to achieve these functions was the runway grid, 3D augmented reality trees and descent arrows. Awareness of required landing point was high in the clear condition, at the high end of moderate in the clear + HUD and DVE + HUD conditions and at the lower end of moderate in the DVE condition. As with power status, drift was not explicitly represented on the AH, the closest O-RP associated with drift was 'maintain position', the HUD symbology to achieve this function included; conformal compass, heading readout, attitude indicator, airspeed indicator and the trees with associated descent arrows (see Table 15.1). Specifically, the last of these, the trees with descent arrows, was the symbology the pilots were using to assess drift as it is only of concern as the aircraft gets closer to the ground. Awareness of drift generally received moderate to high awareness scores in all conditions. The outside environment was captured in the O-RP level of the AH as 'image of the world'. The HUD symbology to achieve this function was defined as the runway grid, 3D augmented reality trees, descent arrows and obstacles. In the DVE + HUD condition awareness of outside environment was generally rated at the higher end of moderate and high and was rated as high in both clear conditions.

15.3.2 Workload

Figure 15.4 clearly demonstrates that workload was highest in the DVE condition, averaging a rating of 6 which is defined on the Bedford Workload Scale as 'little

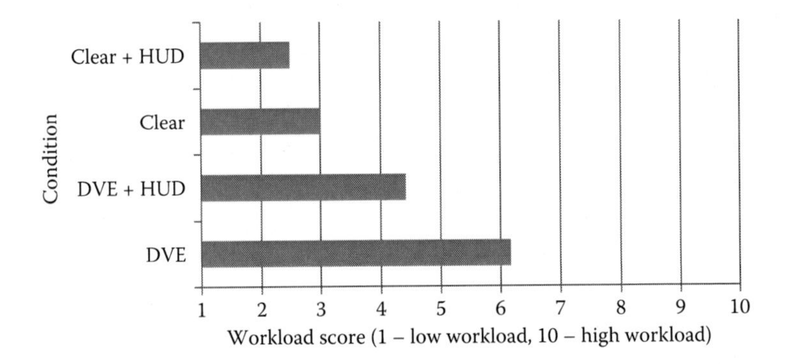

FIGURE 15.4 Mean scores of the Bedford Workload Scale.

spare capacity, level of effort allows little attention to additional tasks'. However this did range between pilots, for example pilot 1 rated workload at 3 ('enough spare capacity for all additional tasks') whereas pilot 3 rated it at 9 ('extremely high workload, no spare capacity. Serious doubts on ability to maintain level of effort'). Workload was lowest in clear + HUD condition, followed by clear (although they both averaged a rating of 2 'workload low') and then DVE + HUD (average rating of 4 'insufficient spare capacity for easy attention of additional tasks'). Pilots stated that they felt they needed more training with the HUD to decrease workload in the DVE + HUD condition.

15.4 DISCUSSION

15.4.1 Summary

The work presented in this chapter demonstrated how the WDA phase of CWA can be used to inform the design of a rotary wing HUD. Vicente (1999) argued that focusing on the work domain is the most important part of the information environment and will enable systems to be designed around the relevant information requirements. Specifically, the O-RP level of the AH was used to define the information requirements of HUD. This provides a structured approach to design that ensures that the interface captures the elements that are necessary to enable the functional purpose of the system to be achieved.

The HUD was evaluated in a flight simulator. The subjective ratings of situation awareness parameters and workload indicate that the HUD led to improvements in awareness and decreased workload. The results demonstrated that for three of the seven awareness parameters (rate of descent, heading and groundspeed) awareness was higher in the DVE + HUD condition compared to the clear condition. Compared to the DVE only condition, the HUD improved pilot's awareness in all seven parameters indicating that the HUD was providing sufficient external visual cues for flight. There was also a decrease in workload in the DVE + HUD condition, compared to the DVE only condition. This suggests that information on the HUD is being processed intuitively. However, workload in the DVE + HUD condition was still higher than in the clear conditions, with pilots stating that they needed more training time for workload to be significantly reduced as the DVE condition resulted in reduced capacity anyway.

It was also demonstrated that the HUD could be beneficial even in clear visual conditions. For example, awareness of heading, groundspeed, descent and outside environment was higher in clear + HUD condition compared to the clear only condition. Furthermore, although both the clear + HUD and clear conditions resulted in an average workload score of 2 ('workload low'), Figure 15.3 demonstrates that workload in the clear + HUD condition was marginally lower, compared to the clear only condition. Ververs and Wickens (1998) also found that presenting flight information in a HUD may be useful in clear visual conditions and attributed this to the ability of a HUD to allow different information forms to be integrated and the reduced visual scan it provided. This was confirmed by the pilots in this study who stated that the utility in the HUD came from having the information in front of them, rather than having to change their visual focus when assimilating information from the HDD.

15.4.2 Concept Development Based on WDA

For the concept development, the HUD had to be capable of representing the tasks detailed in the O-RP level of the AH as these allowed the functional purpose of the system to be achieved. The advantage of designing the HUD around the AH was that it provided a structured knowledge base and rationale for symbology development. It was clear from the analysis of the results that the symbology elements with the clearest coupling to the AH (e.g. landing site, outside environment, rate of descent, heading and groundspeed) assisted the most in enhanced awareness during the HUD conditions. However, the limitations of the AH, and thus its contribution to design, are also evident. For example, power was only represented via the HDD and not on the HUD, this was a design omission because power was not listed as an O-RP in the AH and therefore not mapped across to the symbology via a torque indicator. This was reflected in the PLA results as awareness of power was low across all conditions. In the debrief sessions all pilots stated that they would like to see power included on the HUD and it will therefore be included in future iterations of the HUD.

The other awareness parameter that could have been improved was drift. Although drift was somewhat represented via the trees and associated descent arrows, all pilots felt that this representation could be improved. As with power status, drift was not explicitly represented as an O-RP and the closest one associated with this was 'maintain position'. It is clear that the functions not represented via the O-RP level in the AH and therefore not adequately represented on the HUD scored the lowest awareness ratings. This demonstrates that the PLA questionnaire is sensitive to the HUD symbology and highlights areas for improvement and is backed up by the comments from the pilots. It also demonstrates the utility of using the WDA phase of CWA as a basis for design.

15.4.3 Evaluation

Effort was put into conducting a rigorous experimental procedure, although there are certainly areas for improvements and three valuable lessons have been learnt that can be implemented when future iterations of the HUD are tested. First, a primary limitation of the current study was the small sample size and therefore the decision not to conduct statistical analyses. Future studies will endeavour to recruit a larger sample size so that statistical analyses can be undertaken in order to increase the explanatory power of the results. However, it should be noted that recruiting professional helicopter pilots with time to participate is a difficult pursuit and the six pilots that did participate had much domain expertise and we are therefore confident in the meaning that can be derived from the results.

Second, some of the pilots (particularly those that did not have previous experience of using a HUD) stated that they needed longer training time to familiarise with the HUD. The pilots had approximately 20 minutes to familiarise with the flight controls and HUD. Additionally, a training video was also used to explain the features and functionality of the HUD and this is in line with previous studies that use both classroom and simulator training for a similar amount of time when introducing new cockpit symbology (e.g. Snow and Reising, 1999). Adequate training time is essential because new systems may be incorrectly used in the early stages of implementation and visual cues may not necessarily be intuitive or immediately apprehended (Foyle et al., 1992). From a practical point of view, however, it is unlikely that training time will be able to increase beyond the current allowance because of its effect on participant recruitment. This emphasises the importance of increasing the sample size so that statistical tests can determine whether factors such as previous experience with HUDs significantly impacts the results or not.

Third, due to software constraints the symbology on the HUD did not exactly match the symbology on the HDD. The pilots did not mention this as an issue but in future studies the HDD will be programmed with the HUD symbology in order to increase consistency across the four conditions.

15.4.4 Future Iterations of the HUD

The results demonstrated that the HUD in its current format improved situation awareness and reduced workload in DVEs. However, enhancements can be made to the HUD to increase its utility even further. For example, many studies demonstrate the effectiveness of 'Highway in the Sky' (HITS) symbology (Williams et al., 2001; Alexander et al., 2003; Thomas and Wickens, 2004). This is usually in the form of a graphical flight path representation. Whilst there are concerns that HITS symbology can lead to cognitive capture or attentional tunnelling (Thomas and Wickens, 2004), including a HITS type element into the HUD has the potential to enhance the utility of the HUD in DVEs and will therefore be considered in future research.

The HUD could also be enhanced by the use of a full terrain map. In this study a grid was displayed over the runway and the pilots consistently commented on how useful it was, helping them to achieve awareness of groundspeed, required landing point and drift. However, the utility of this could be enhanced with the provision of a whole terrain map; Snow and Reising (1999) found substantial increases in situation awareness with the addition of a synthetic terrain grid in a HUD and this was associated with a reduction in ground impacts during low level flying.

The representation of drift has already been raised as an area for improvement. One method to enhance awareness of drift is to provide a velocity vector which would show pilots whether they are moving forward or backward. However, a study by MacIsaac et al. (2005) found that a velocity vector did not adequately provide the pilots with lateral drift information because the angle of the vector only determines lateral velocity for large values and was hard to perceive for small values. Having an awareness of drift is essential for the pilots and was a concern raised by all of the pilots in the study and this will certainly be an area where re-design efforts are targeted.

It is also necessary to be mindful of the potential negative effect of HUDs as they have been shown to cause the detection of unexpected events to be degraded by attention capture when the pilot's attention shifts away from the outside environment and remains too focused on processing information presented by the HUD (Fadden et al., 1998; Thomas and Wickens; 2004; Jakus et al., 2014). An overly cluttered HUD can be detrimental to pilot situation awareness and the overlay of symbology can obscure objects and may disrupt effective scanning (Yeh et al., 2003; Harris,

2011). To optimise the benefits of HUDs, designers must preserve the most useful and unambiguous visual cues pilots naturally use so that information is processed intuitively by pilots (Foyle et al., 1992; Fadden et al., 1998; Harris, 2011).

15.5 CONCLUSION

Traditionally, pilots had two sources of flight information available to them; the cockpit instruments and out-of-the window visual references (Foyle et al., 1992). Technology today, primarily in the form of HUDs, allows these two sources of information to be amalgamated and integrated with the advantage of enabling the pilot to fly 'eyes-out' in a natural and intuitive manner. This chapter demonstrated how the WDA phase of the CWA method can be used to design a HUD solution to address the problem of landing in a DVE. It was demonstrated that this HUD enhanced situation awareness and had a positive impact on workload, although it has been acknowledged that there are areas where the HUD can be enhanced to facilitate performance even further. The provision of HUDs has the potential to increase helicopter operations by allowing them to fly in DVEs and, as demonstrated here, may also have utility in good visual conditions in order to enhance the safety of helicopter operations.

ACKNOWLEDGEMENTS

The work was supported by funding from the EU 7th Framework project ALICIA: All Condition Operations and Innovative Cockpit Architecture. We would like to thank the pilots who gave their time to participate in the study.

REFERENCES

Alexander, A. L., Wickens, C. D. and Hardy, T. J. 2003. Examining the effects of guidance symbology, display size, and field of view on flight performance and situation awareness. *Proceedings of the Human Factors and Ergonomics Society Annual Meeting*, 47(1), 154–158.

Allan, S. S., Beesley, J. A., Evans, J. E. and Gaddy, S. G. 2001. Analysis of delay causality at Newark International Airport. *4th USA/Europe Air Traffic Management R&D Seminar.* Santa Fe, New Mexico, USA, 1–11.

Baker, S. P., Shanahan, D. F., Haaland, W., Brady, J. E. and Li, G. 2011. Helicopter crashes related to oil and gas operations in the Gulf of Mexico. *Aviation, Space, and Environmental Medicine*, 82(9), 885–889.

British Helicopter Association, 2014. *The Future Role of Helicopters in Public Transport.* http://www.britishhelicopterassociation.org/?q=about-the-bha/helicopters. Accessed August 1, 2014.

Brook, R. and Dodge, M. 2014. Helicopter dreaming: The unrealised plans for city centre heliports in the post-war period. In: *Infrastructure and the Rebuilt City after the Second World War.* Working Paper series no. 22. Birmingham City University, 42–55. ISBN 978-1-904839-72-9.

Cornelissen, M., McClure, R., Salmon, P. and Stanton, N. A. 2014. Validating the strategies analysis diagram: Assessing the reliability and validity of a formative method. *Applied Ergonomics*, 45, 1484–1494.

Dudfield, H. 1991. Colour head-up displays: Help or hindrance? *Proceedings of the Human Factors and Ergonomics Society Annual Meeting*, 35(2), 146–150.

Fadden, S., Ververs, P. M. and Wickens, C. D. 1998. Costs and benefits of head-up display use: A meta-analytic approach. *Proceedings of the Human Factors and Ergonomics Society Annual Meeting*, 42(1), 16–20.

Foyle, D. C., Kaiser, M. K. and Johnson, W. W. 1992. Visual cues in low-level flight: Implications for pilotage, training, simulation, and enhanced/synthetic vision systems. *American Helicopter Society 48th Annual Forum*, 1, 253–260.

Harris, D. 2011. *Human Performance on the Flight Deck*. Aldershot: Ashgate.

Hart, S. G. 1988. Helicopter human factors. In E. L. Wiener and D. C. Nagel (Eds.). *Human Factors in Aviation*. San Diegom Ca.: Academic Press, 591–633.

Jakus, G., Dicke, C. and Sodnik, J. 2014. A user study of auditory, head-up and multi-modal displays in vehicles. *Applied Ergonomics*, 46, 1–9.

MacIsaac, M. A., Stiles, L. and Judge, J. H. 2005. Flight symbology to aid in approach and landing in degraded visual environments. *American Helicopter Society 61st Annual Forum*, Grapevine, Texas, 1–12.

McIlroy, R. C. and Stanton, N. A. 2011. Getting past first base: Going all the way with cognitive work analysis. *Applied Ergonomics*, 42, 358–370.

Pejovic, T., Williams, V. A., Noland, R. B. and Toumi, R. 2009. Factors affecting the frequency and severity of airport weather delays and the implications of climate change for future delays. *Transportation Research Record: Journal of the Transportation Research Board*, 2139, 97–106.

Roscoe, A. and Ellis, G. 1990. *A Subjective Rating Scale for Assessing Pilot Workload in Flight*. Farnborough: RAE.

Snow, M. P. and French, G. A. 2002. *Effects of Primary Flight Symbology on Workload and Situation Awareness in a Head-up Synthetic Vision Display*. Air Force Research Laboratory Wright-Patterson AFB, OH 45433.

Snow, M. P. and Reising, J. M. 1999. *Effect of Pathway-in-the-Sky and Synthetic Terrain Imagery on Situation Awareness in a Simulated Low-Level Ingress Scenario*. Air Force Research Laboratory Wright-Patterson AFB, OH 45433.

Stanton, N. A., Salmon, P. M., Rafferty, L. A., Walker, G. H., Baber, C. and Jenkins, D. P. 2013. Human factors methods. *A Practical Guide for Engineering and Design*. Second Edition. Ashgate: Aldershot.

Thomas, L. C. and Wickens, C. D. 2004. Eye-tracking and individual differences in off-normal event detection when flying with a synthetic vision system display. *Proceedings of the Human Factors and Ergonomics Society Annual Meeting*, 48(1), 223–227.

Ververs, P. M. and Wickens, C. D. 1998. Head-up displays: Effect of clutter, display intensity and display location on pilot performance. *The International Journal of Aviation Psychology*, 8(4), 377–403.

Vicente, K. J. 1999. *Cognitive Work Analysis: Toward Safe, Productive and Health Computer-Based Work*. Lawrence Erlbaum Associates: Mahwah, NJ.

Williams, D., Waller, M., Koelling, J., Burdette, D., Doyle, T. and Capron, W. 2001. *Concept of Operations for Commercial and Business Aircraft Synthetic Vision Systems (NASA Tech. Memo. No. TM-2001–211058)*. Langley, VA: NASA Langley Research Centre.

Yeh, M., Merlo, J. L., Wickens, C. D. and Brandenburg, D. L. 2003. Head up versus head down: The costs of imprecision, unreliability, and visual clutter on cue effectiveness for display signalling. *Human Factors*, 45(3), 390–407.

16 Designing Mission Communications Planning with Rich Pictures and Cognitive Work Analysis

Neville A. Stanton and Rich C. McIlroy

CONTENTS

16.1 ANALYSIS OF SOCIOTECHNICAL SYSTEMS

Despite the proliferation of ergonomics methods, with over 200 identified by Stanton et al. (2005), none appear able to do the actual work of design, although they may contribute to it. There are methods to analyse user and system requirements as they currently exist, or might exist in the future and there are methods that evaluate the performance of users and systems. In between the analysis and evaluation, the designer is left to explore solutions using their own experience and creativity (Jenkins et al., 2009). So whilst ergonomics methods can define the problem space, evaluate proposed design solutions and assess existing systems, they do not actually do the design work. The discipline of ergonomics serves to provide theories and metrics of human-system performance and highlights the need for a sociotechnical approach (Walker et al., 2008), often promoting the benefits of considering the user of the system (Stanton and Baber, 2003), but it also needs to demonstrate the effectiveness of the methods in a quantifiable manner (Stanton and Young, 1999).

In sociotechnical systems theory the unit of analysis is the sociotechnical system under investigation, rather than the technical or social system independently (Walker et al., 2008). Arguably the problem with most ergonomics methods is that they either address the social system or the technical system independently of each other. For example, cognitive task analysis methods tend to address aspects of human cognitive whereas interface analysis methods tend to focus on aspects of the technical system (Stanton et al., 2005). Whilst this is a slight over-exaggeration in order to make the point, it does suggest that these methods do not wholly embody the principle of equal consideration of both social and technical subsystems and their interaction. To this end, cognitive work analysis (CWA) offers a rather more balanced approach (Vicente, 1999; Jenkins et al., 2008a,b). CWA offers a useful approach to design, in defining the boundaries and constraints of sociotechnical systems, within which the design solution can be sought (Vicente, 1999; Jenkins et al., 2008a,b). This can help extend design thinking beyond incremental improvements toward new approaches for interaction design (Naikar et al., 2006).

CWA (Rasmussen et al., 1994; Vicente, 1999; Jenkins et al., 2009) represents a formative approach to the analysis of complex sociotechnical systems. Rather than analysing what a system currently does (*descriptive modelling*) or should do (*normative modelling*), the analysis offers a framework of methods that allow for the in-depth analysis of the properties of the work domain and the workers themselves, therefore defining a set of boundaries that shape activity within system. The approach leads the analysis into describing how the system *could* perform (*formative modelling*) given its constraints. The analysis comprises five phases; work domain analysis (WDA), control task analysis (ConTA), strategies analysis, social organisation and co-operation analysis and worker competencies analysis. Although each phase builds upon the last, not all of the phases must be used. The framework can be likened to a toolkit of methods, in that the analyst may apply any of the methods individually, or in combination, depending upon the nature and needs of the analysis. Within the context of this chapter, some aspects of CWA are explored in the design of a new communications planning system. First of all communications planning is introduced, with a Rich Picture to show how pilots think about planning. This is followed by analysis of the current system. The proposal for new system is then presented. A comparison of the current and proposed approaches was undertaken. Finally, conclusions for CWA, Rich Pictures and mission planning are drawn.

16.2 MISSION PLANNING

The mission planning system (MPS: Jenkins et al., 2008a) was developed in order to reduce the in-air workload of military pilots. The software is currently used by the United Kingdom to plan and assess helicopter single or multiple aircraft sortie missions. Mission plans are developed at MPS terminals prior to take-off and loaded onto a digital storage device called a data transfer cartridge (DTC). The DTC is used to transfer the completed plans to the helicopter's onboard data management system (DMS) which supports the mission subsequent to take-off. A more detailed explanation of the MPS and mission planning may be found in Jenkins et al. (2008a). Before commencing any mission, be it training or operations, all military pilots must be aware of a number of communication needs. These include the need to alert authorities on the ground to the presence of any aircraft passing through or travelling in close proximity to any controlled airspace; the need for effective and secure voice communications with other aircraft concurrently airborne and the need to effectively transfer data between concurrently airborne aircraft, whilst also maintaining security. Communications with authorities on the ground are dependent upon geographical orientation, whilst air-to-air communications are dependent upon mission requirements, command chain and the flight times and locality of other aircraft.

The main function of the MPS communications software is to allow helicopter pilots to load a collection of radio frequencies such that when airborne, pilots have easy access to all of the frequencies they will require, and that each of these frequencies is properly labelled with regard to where and with whom that frequency is associated. The MPS software contains a visual display of a map of the United Kingdom displaying the boundaries of all major controlled airspaces, including military danger zones and minor and major air fields and airports. By studying the proposed route, marked on the map by a solid black line, pilots must decide what frequencies they will need for their mission. These frequencies must then be looked up in one of the Royal Air Force (RAF) Flight Information Publications, for example the British Isles and North Atlantic en-route supplement (BINA). A detailed description of the process of planning communications in MPS can be found in McIlroy et al. (2012).

In its present guise, the communications planning software interface is perceived as difficult to use, which has implications for training and mission planning, as indicated by the following quotes from experienced pilots who also train others:

> The AH has an incredibly capable communications suite. However, it is regularly under-utilised by the front-line as the planning and setup process is overly complicated and error prone.
>
> Training aircrew to configure the wide variety of voice, data and frequency agile encrypted radios on the Helicopter is an overly intensive task for both staff and students – primarily due to the overly complex tools used to configure various radio, channels & Nets.

Some of the complexity is due to the inherent complexity of the system components (e.g. there are 4 radios, 60 call-signs, 10 presets, 4 boot-up channels – all doubled for a two-day communications plan) coupled with the constraints acting on communications planning (e.g. the standard operating procedures for communications settings, mission timings, changes in airspace authority and so on). Helender (2007) notes that computerisation of systems is making the coupling of systems even more complex, and that uncoupling of system elements may make interaction design more achievable. Although the current generation of communications planning starts with the presentation of a map with a route (as shown in Figure 16.1), communications planning resorts to a series of tables, fields and buttons (as shown in Figure 16.2). We suspect that this results in a disconnection between how pilots think about planning their communications and what the software requires of them to prepare a communications plan. This line of reasoning was explored in the research presented within the current chapter.

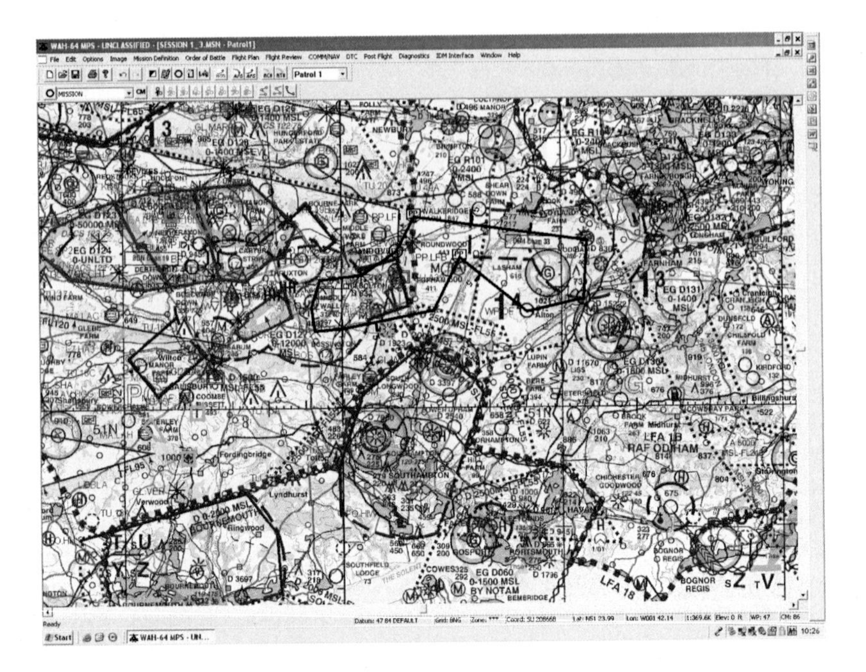

FIGURE 16.1 The map-based representation of the route for the communications planning.

COMM Day 1 - UNCLASSIFIED - [SESSION 1_3.MSN]

Call Sign Freq | Suffixes | Expanders | Auth/Reply | Preset Channels | Preset Networks

Call Sign Frequency Set

Number	ID	Call Sign	UHF	VHF	FM
16	XGAT	COPP1	226.125	0.000	0.000
17	XG G	COPP3	368.875	129.475	0.000
18	XG T	COPP4	370.975	119.225	0.000
19	XGTD	COPP5	234.875	123.300	0.000
20	XG D	COPP6	342.350	113.975	0.000
21	XG Z	COPP7	229.875	123.400	0.000
22	XG A	COPP8	273.700	125.425	0.000

Number: 1
ID: VPAT
Call Sign: SILV1
UHF (MHz): 240.975
VHF (MHz): 0.000
FM (MHz): 0.000

Preset 1 | Preset 2 | Preset 3 | Preset 4 | Preset 5
Preset 6 | Preset 7 | Preset 8 | Preset 9 | Preset 10

FIGURE 16.2 The call-sign frequency list for the communications planning.

16.3 RICH PICTURES

At the beginning of the project, two full day meetings were held with a subject matter expert (SME) in which the analysts were introduced to military communications in general and to the current software technology in particular. The analysts were supplied with a version of the MPS software to inform analyses. Across a further 6 days, the analysts worked with the SME to further their understanding of the mission planning process and the MPS software through SME walkthroughs of different communications planning tasks. During these meetings the SME produced a number of Rich Pictures (Checkland, 1981) of the communications planning task. A Rich Picture is graphical representation of a problem, concept, situation or work domain. It can include any kind of figure or text and has no prescribed rules or constraints. The Rich Picture has its origins in soft systems methodology (Checkland, 1981; Checkland and Scholes, 1990) and its primary purpose is to describe a system in such a way that is useful to both individuals external to and actors within that system; it serves to organise and structure the body of information provided by the expert. Rich Pictures are commonly used in information systems education (Horan, 2002), as they can often provide easily interpretable depictions of complex systems. Furthermore, the method is often adopted in the early phases of system analysis as it is regarded as a more easily understandable description of system relationships, connections and 'story-line' than more traditional representations, such as narrative text (Sutrisna and Barrett, 2007). The representations are inherently flexible, allowing for expert-lead integration of information from a variety of sources (Monk and Howard, 1998) thereby allowing the analyst to gain a broad view of the system in its entirety, from the perspective of the expert. As the method itself has no constraints (the expert is free to construct any representation they choose) the resulting diagram can only reflect the constraints of the system as interpreted by the expert; they capture the representations and concerns of the user, not the analyst. An example of the Rich Picture for the concept of air-to-ground communications is presented in Figure 16.3. The route is displayed by the solid line with waypoints and circles on the line. The dashed circles indicate airspace boundaries associated with an airfield (the cross on a circle in the centre of the airspaces). The dashed line indicates the planned diversion. The boxes represent collections of frequencies associated with each airfield.

This approach might be criticised because it was based on only one SME in the study. In this instance the pilot in question was the most experienced MPS person in the United Kingdom. Not only had he helped in the previous development of the MPS system but has also trained the system trainers in its use. As an experienced pilot and test pilot, he was well versed in the use and misuse of the MPS software and as such represented a super-expert. The use of only one SME is not necessarily being advocated; rather it was a constraint of the project. In any case, the Rich Picture drawn was a representation of how most pilots will typically brief a mission on a white board. For the purposes of the current study, the 'frequencies' box had been added to show that pilots conceptually attach frequencies to airfields when planning their communications. The Rich Picture presents the main constraints of the planning task, namely: the route, waypoints, airfield, airspace boundaries and frequencies.

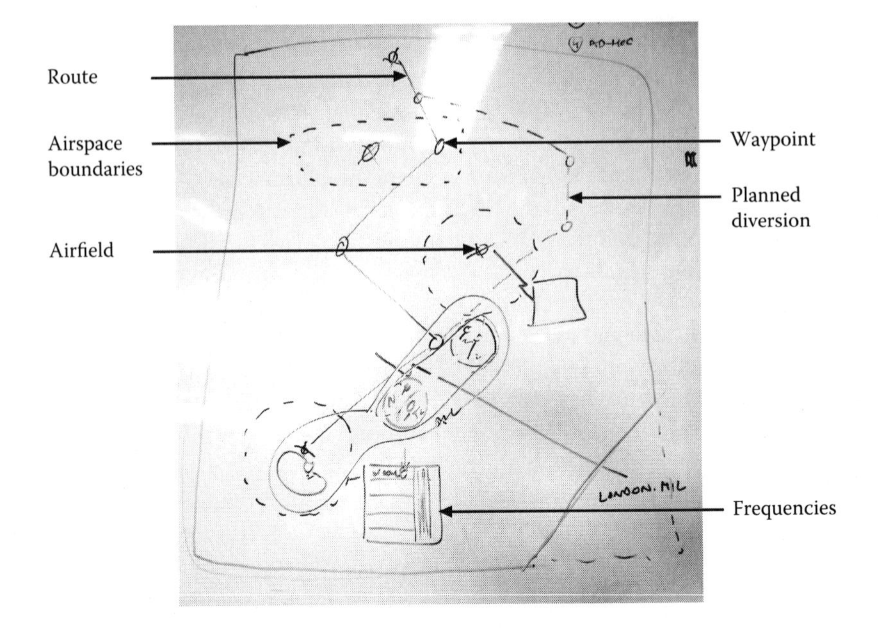

FIGURE 16.3 Rich Picture of air-to-ground communications.

In addition to the meetings with the SME, all based at the University of Southampton, a visit was made to the army flying school based at Middle Wallop. The data collected from all of the sessions were used to inform CWA. An abstraction hierarchy (AH), a contextual activity template (CAT) and social organisation and cooperation analyses (SOCA) were constructed, and subsequently refined and amended in further meetings with the SME.

16.4 COGNITIVE WORK ANALYSIS

The first phase of CWA is to develop WDA. In this phase the system is described in terms of the environment in which workers operate; the analysis is independent of activity, actors or goals. It identifies a fundamental set of constraints that shape the activity within the system, providing a foundation for the subsequent phases of CWA. The first stage of WDA is to construct an AH. The AH describes the system at a number of levels of abstraction. At the highest level the AH describes the reason for the system's existence; its functional purpose. At the lowest level the physical objects within the system are described. This is essentially an inventory of the physical resources in the system, both natural and man-made. The final version of the AH for the communications planning aspect of MPS is presented in Figure 16.4.

The functional purpose of the communications planning aspect of MPS is to 'support communications within the aircraft'. At the lowest level are the physical components of the system. In this case they are limited to the planning process; only those objects that are immediately involved in the communications planning

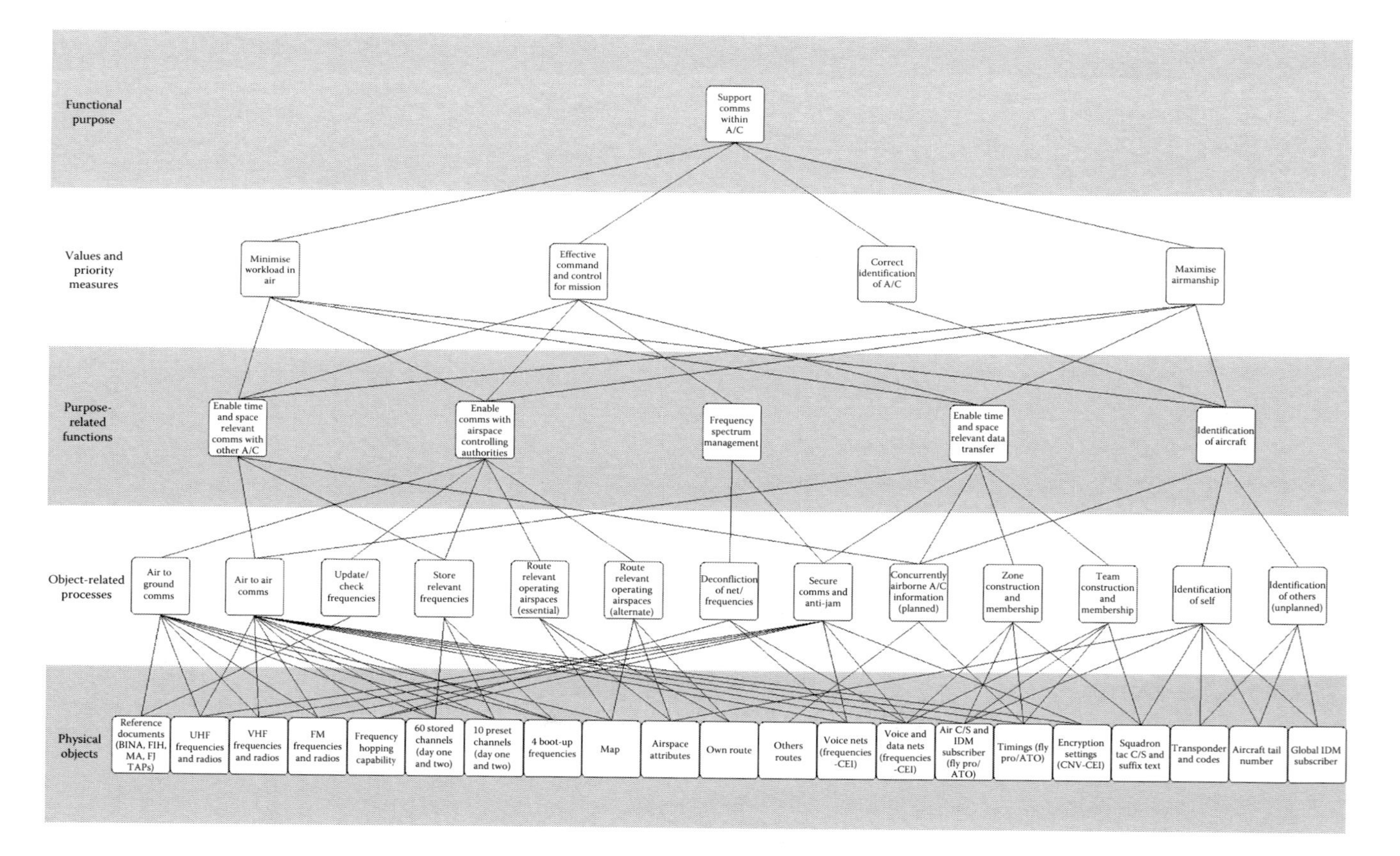

FIGURE 16.4 AH for communications planning.

process are included. Example physical objects include: reference documents, frequencies and frequency storage capabilities, maps, routes and airspace attributes, voice and data nets and the associated encryption settings and information on flight times, locality and data addresses of various aircraft. In the level above this, the object-related processes level, the affordances of the physical objects in the system are displayed. These affordances are directly tied to the physical objects and are independent of the overall system goals. For example, the physical object of the map affords information about the route-relevant operating airspaces, both essential and alternate. The second level from the top, *values and priority measures*, captures the criteria that can be used to judge whether the system is achieving its functional purposes. For this analysis, these include; minimise workload in air; effective command and control for mission; correct identification of aircraft and maximise airmanship. The middle level, *purpose-related functions*, joins the AH together. This level captures the general functions of the system necessary in order to achieve system purposes.

Each level of the AH can be linked together by means–ends relationships through the use of the how-what-why triad. Any node can be taken to answer the question of 'what' it does. In the level below, all of the connected nodes can be used to answer the question of 'how' this can be achieved. In the level above, the connected nodes are used to answer the question of 'why' it is needed.

The second phase of CWA is ConTA. Building on WDA, this phase considers recurring activities within the system. The analysis focuses on what is to be achieved independent of how the activity is to be conducted and by whom the activity is to be carried out. Naikar et al. (2006) developed the CAT for use in this phase of CWA. The template characterises system activity in terms of work functions and work situations. These situations may be temporal, spatial or a combination of the two. The activity that occurs is represented in terms of where it is able to be carried out and where it is likely to be carried out. An explanation is displayed in Figure 16.5.

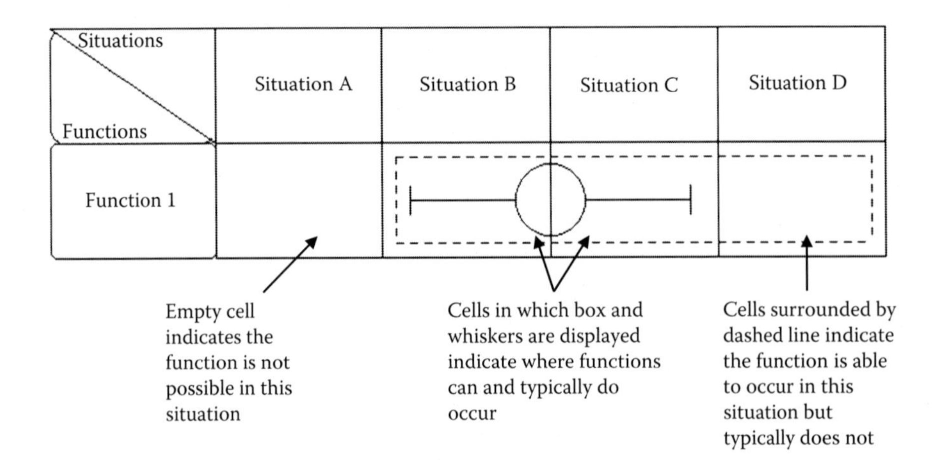

FIGURE 16.5 Explanatory figure of CAT.

The CAT developed for the communications aspect of MPS is presented in Figure 16.6. For this analysis the work functions, displayed on the horizontal axis, were imported from the *object-related processes* level of the AH. The work situations, defined on the horizontal axis, represent four stages of mission preparation and planning, namely preparatory support activities, template development, mission specific activities and individualisation of the communications plan. In this instance the work situations are temporally separated.

Here it can be seen that the majority of activity is carried out in the final two stages of mission planning; mission specific activities and individualisation of the communications plan. Although preparatory support activities have been included as a work situation, none of the activities are likely to occur here. Deconfliction of nets and frequencies is able to occur in this situation, though at present it is not a common, recurring activity.

The social organisation and cooperation analysis (SOCA) phase of CWA looks at the cooperation between actors within a system. It addresses the constraints imposed by organisational structures or specific actor roles and definitions. The aim of this section of the analysis is to analyse the constraints on the organisation of human and technical factors within a complex sociotechnical system, such that system performance is maximised. The analysis builds upon outputs obtained from previous sections of CWA through the use of shading; different actor roles are each assigned a colour and the various CWA outputs are shaded to indicate where each of the actor groups can conduct tasks. Allocation of function is described in this stage of CWA; different actors are allocated to different functions. The use of differential shading used in this stage of the analysis can be applied to any of the outputs obtained thus far, hence there are a number of different SOCA views.

In the MPS communications example three groups of actors were identified, namely pilots, template administrators and operations (Ops) staff. The pilots are the individuals that will be flying the aircraft. They are ultimately responsible for the mission plan. The template administrators' role is to assist the pilot in developing communications plans and data handling. The Ops staff are involved in planning missions and supplying the pilots with mission specific information, for example encryption settings and frequency hopping nets.

For the current analysis, the CAT was colour coded to indicate who carries out the work in the system and in which situation. The shade key for actors is presented in Figure 16.7. The shade-coded CAT (SOCA-CAT) is presented in Figure 16.8.

From Figure 16.8 a number of observations can be made. First it is clear that the majority of activity within the system is carried out by the pilot. Indeed the activity that occurs in the final situation, individualisation of the communications plan, is only carried out by the pilot. Furthermore, the pilot conducts all the activity in the mission specific activities situation, working with the Ops staff on 'air-to-air comms', 'secure comms and anti-jam', 'concurrently airborne aircraft (unplanned)' and 'identification of self'. This allocation of activity is in contrast to that seen in the first two situations, namely preparatory support activities and template development. The pilot is not involved in any of the activity in these stages, rather it is the responsibility of the template administrator for the majority of functions, with the Ops staff playing a role in only 'update/check frequencies' and 'deconfliction of

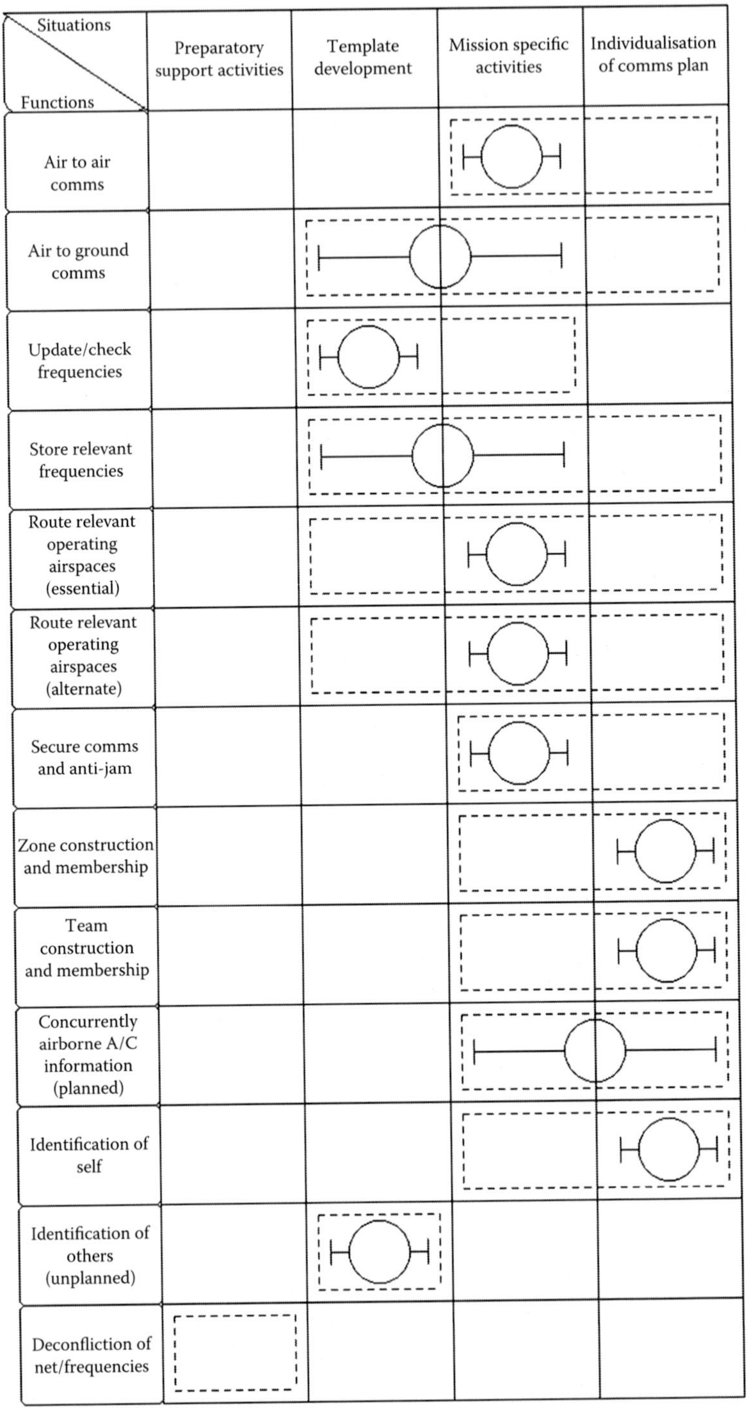

Situations / Functions	Preparatory support activities	Template development	Mission specific activities	Individualisation of comms plan
Air to air comms				
Air to ground comms				
Update/check frequencies				
Store relevant frequencies				
Route relevant operating airspaces (essential)				
Route relevant operating airspaces (alternate)				
Secure comms and anti-jam				
Zone construction and membership				
Team construction and membership				
Concurrently airborne A/C information (planned)				
Identification of self				
Identification of others (unplanned)				
Deconfliction of net/frequencies				

FIGURE 16.6 CAT for communications in MPS.

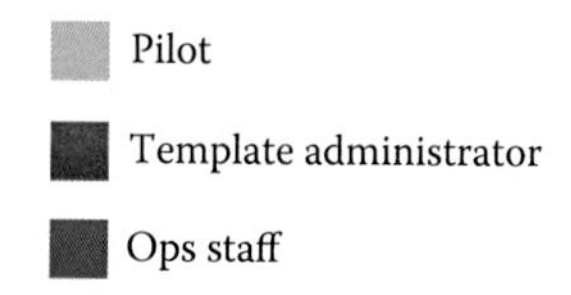

FIGURE 16.7 Colour key for actors.

net/frequencies'. The Ops staff and template administrators share the responsibility of 'update/check frequencies'. The diagram illustrates the nature of the division of work in the system; actors hand over to another as the communications plan is refined from the generic to the specific.

The AH developed in the first phase of the analysis allows for a new perspective on the system. By developing functional means–ends links across levels of abstraction it is possible to understand task flow within the system. As with a previous analysis of MPS (the scope of which did not include communications planning; Jenkins et al., 2008a,b) the AH highlighted the mismatch between the organisation of the task and the organisation of the MPS system. Individual tasks and functions described in the AH require multiple windows to be open in MPS, as currently the software is designed to support DTC data population, rather than to support communications planning. By re-designing the software package to support the communications planning task it will be possible to reduce this need for multiple windows to be open simultaneously. Additionally, much of the training pilots receive is in how to use the software rather than how plan communications. By redesigning training and redesigning the MPS interface it would be possible to focus on training pilots to plan communications instead.

The ConTA phase of CWA built upon the AH in describing specific recurring activities within the system. The CAT displayed where each of the object-related processes take place in terms of time and place. Although preparatory support activities were included as a possible work situation, no activity typically occurs here. Only deconfliction of nets/frequencies is able to occur in this situation. However in the current system state this activity is not typically undertaken. The majority of activity occurs in the final two situations; mission specific activities and individualisation of the communications plan.

The SOCA phase of the analysis identified three actors that are able to carry out the activities presented in the CAT. From this analysis it can be seen that the pilot carries out the majority of the activity in the system, and that all of the pilot's activity is carried out in the final two situations. The individualisation of the communications plan is carried out solely by the pilot. As expected, the MPS template administrator is responsible for all of the activities in the template development phase.

In summary, the CWA has highlighted some important issues with the current MPS. It is clear that the design of the interface does not optimally support the communications planning process; rather it is designed to populate the DTC. Although this goal is a valid one when considering data management and software programming (it is indeed important that all relevant information is fed into the DTC) this is not how the task of communications planning is set out. It is argued that a redesign of the software to support the planning tasks set out in the AH and CAT would reduce

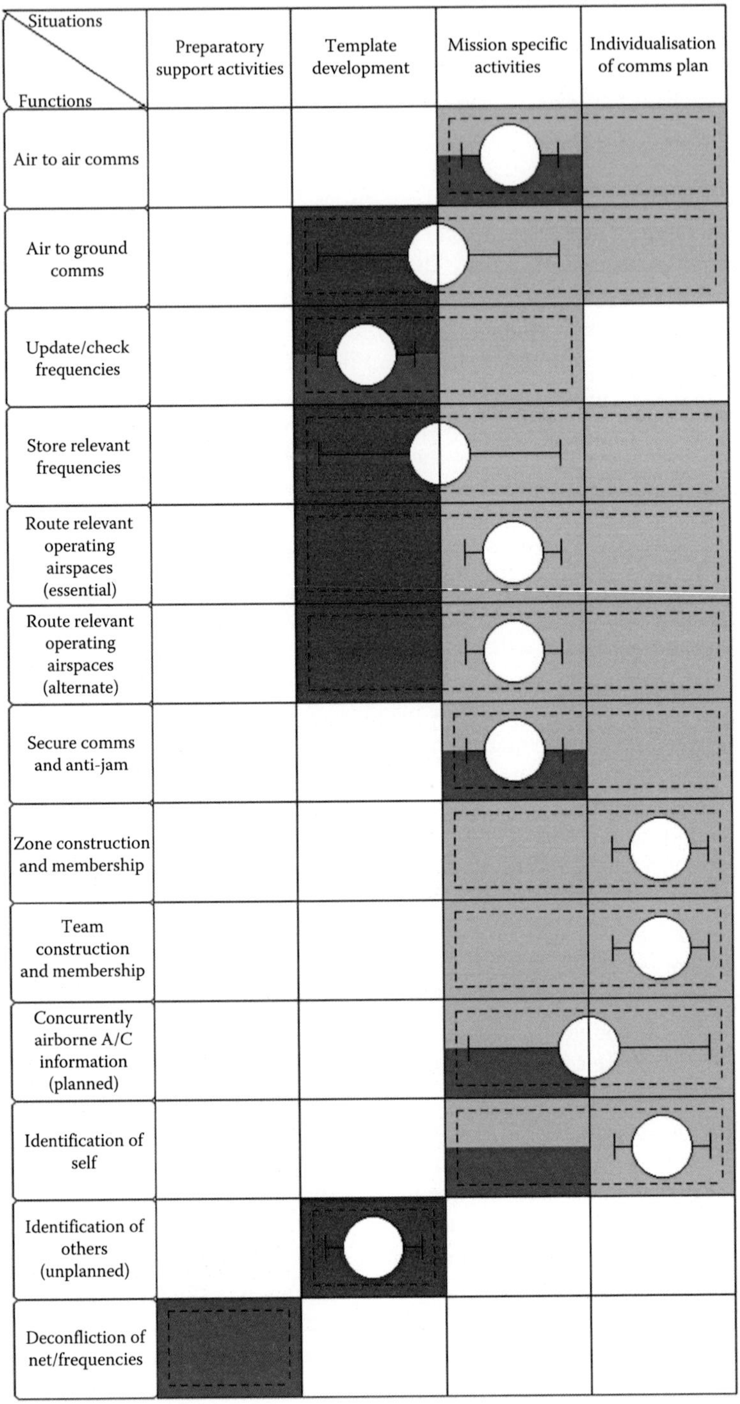

Situations / Functions	Preparatory support activities	Template development	Mission specific activities	Individualisation of comms plan
Air to air comms				
Air to ground comms				
Update/check frequencies				
Store relevant frequencies				
Route relevant operating airspaces (essential)				
Route relevant operating airspaces (alternate)				
Secure comms and anti-jam				
Zone construction and membership				
Team construction and membership				
Concurrently airborne A/C information (planned)				
Identification of self				
Identification of others (unplanned)				
Deconfliction of net/frequencies				

FIGURE 16.8 CAT shaded coded to show actors.

the need to have multiple windows open simultaneously. This would, in turn reduce the chance for omissions, as task-orientated groupings of information would prompt the user as to the information they should be considering. The SOCA phase of the analysis highlighted the uneven spread of activity across situations. In the current system the large majority of activity is carried out in the latter two situations. If more activity could be carried out in the preparatory support activities and template development situations, less time pressure would be put on the pilot immediately before take off. In addition, if more of the activity was carried out in the initial three situations, rather than in the individualisation of the communications plan situation, the activity could potentially be carried out by a number of different actors, hence dividing the workload.

16.5 DESIGN OF NEW INTERACTION

The primary purpose of this research was to analyse the communications aspect of the MPS software tool through the use of CWA in order to inform design recommendations for future iterations of the system. The AH, developed in the WDA phase of CWA, displays the organisation of the system from a functional perspective, that is, the physical objects can be grouped in such a way as to match their common object-related process, and the object-related processes can be grouped so that each group has a common purpose-related function. Through developing these functional means–ends links across levels of abstraction it is possible to understand the relationships between functions and affordances of artefacts within the system. From this it becomes clear that there is a mismatch between function organisation and the organisation of MPS. Individual functions require multiple windows to be open.

One of the most significant problems with MPS, as suggested by the results from this research, is the lack of correspondence between the user's mental model and the system model. A mental model is an internal, mental representation of the way things exist or work in the real world. A mismatch between the system model and user mental model results in a system that is unintuitive; the way the user thinks about the task is not reflected in the way the user must interact with the system. Rather than relying on a mental representation of the task to guide behaviour the user must use a learnt sequence of actions to progress through the system. For the air-to-ground communications planning stage of the task, results from a think-aloud study (reported in McIlroy et al., 2012) suggested that thought processes are map based. The expert explicitly stated that he was using the map to not only guide the decision-making process (when deciding which frequencies to include in the plan) but also when getting an overview of the plan so far and when reviewing the completed plan. The Rich Picture presented earlier underlines this mismatch. In this diagram the SME graphically represented his concept of the air-to-ground communications process. The route is presented in such a way as to draw attention to those airspaces that are passed through, and those that are close by. Furthermore frequencies are directly linked to an airspace (indicated by the boxes in Figure 16.3); they are not in a separate table only identified by ICAO (International Civil Aviation Organisation) codes (as is the case with the current version on the system, where the airspaces are shown in Figure 16.1 and the frequencies are in a separate window shown in Figure 16.2, which covers the map).

Another significant issue pertaining to the volume of information requiring management was relatively straightforward to address. In the current system the user must refer to a reference document (BINA) in order to get the frequencies associated with each airspace controlling authority. They must refer to a different reference document (RAF flight information handbook) to find out the name of an airport, airfield or special user airspace from the ICAO code given in the call sign frequency (CSF) or preset channels list, as these names are not contained within MPS. All of this information, although currently used from a paper-based format, is also held within an electronic database, the digital aeronautical flight information file (DAFIF). The information held in this file covers the entire globe, and all authorities contained within the file have both frequencies and coordinates associated with them. By taking advantage of this information it is possible to directly link frequency data with map data. In addition, the name of each airspace controlling authority can be displayed with the ICAO code. By linking communications data to the map, the problems of both task organisation and mental model mismatch are addressed, at least for air-to-ground communications. In designing a suitable interface, it was necessary to refer back to the AH to understand the task flow and task organisation. Connected to the 'air-to-ground comms' object-related processes node are a number of physical objects. These are displayed in Figure 16.9.

Re-organising the interface such that all of the physical objects shaded grey in Figure 16.9 are available to the pilot simultaneously removes the need to switch between windows when planning for air-to-ground communications. An example of this map-based interface is presented in Figure 16.10. Although the interface has been designed for the helicopter-specific MPS, it is contended that the communications planning principles and interaction methods described will remain the same, irrespective of platform or mission type.

Figure 16.10 displays a screenshot of the proposed interface; on it, a map with a route marked on it is displayed as it would be in MPS, with the exception that

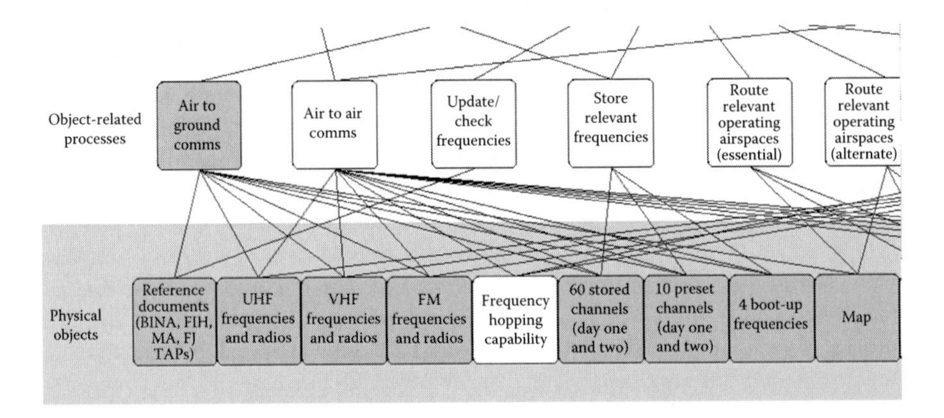

FIGURE 16.9 Section of the AH with 'air-to-ground comms' and associated physical objects shaded grey.

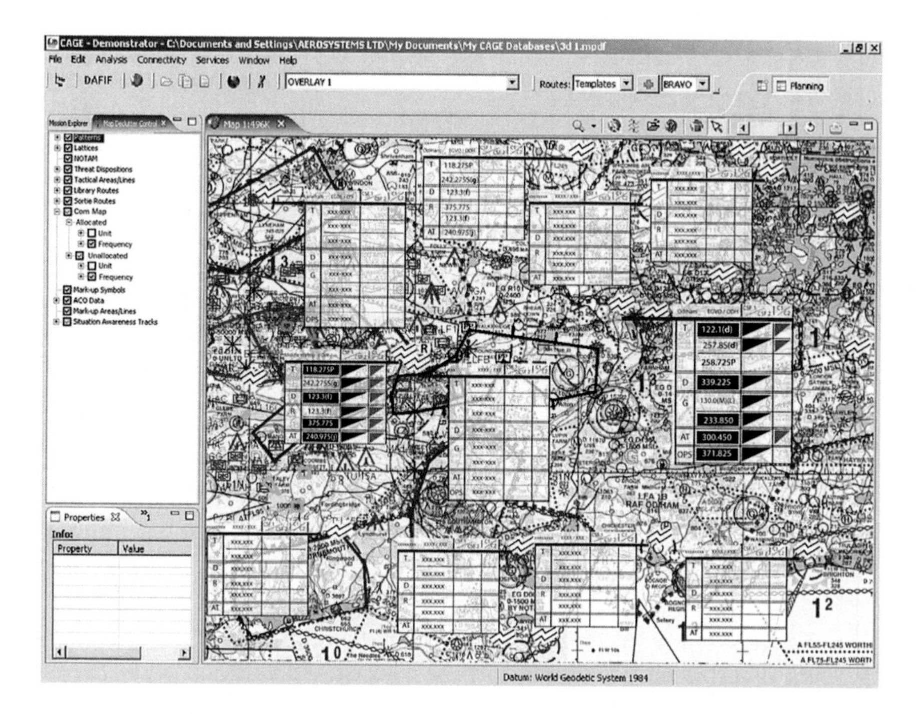

FIGURE 16.10 Example new communications planning interface.

there are communication symbols and frequency tables included. The symbols indicate that there is some communications information associated with that position on the map whilst the tables display that information. The symbols were designed and refined in an iterative process via repeated meetings with the SME. The symbol is displayed on a white background on the map so that even if the map is predominately a dark colour, the symbols will still be clear to the user. The symbols are displayed just off centre of the airspace such that runway information is not obscured. The distance of the icon from the centre of the airspace is based on distance on the screen, not map-based geographical distance.

On the left-hand side of Figure 16.10 a 'Map Declutter Control' toolbar is displayed. This allows the user to turn the communications symbols on or off. This should allow the user to match the task to his or her needs. For example if the user is currently planning mission routes rather than communications they can deselect the 'Comms' node. From this control area, the user can choose to display all the communications details for that section of the map that is in view. On the left of Figure 16.10, having only the 'Icons' box ticked will display the communications icons only. By selecting the 'Comms' box all the frequency information for the space on the map that is visible on the screen at that moment will be displayed.

For each airspace a label displaying all associated frequencies is presented alongside a number of options for selection. This presentation of frequencies is largely based on the depiction of frequencies in the Rich Picture for air-to-ground communications, an enlargement of which is displayed in Figure 16.11.

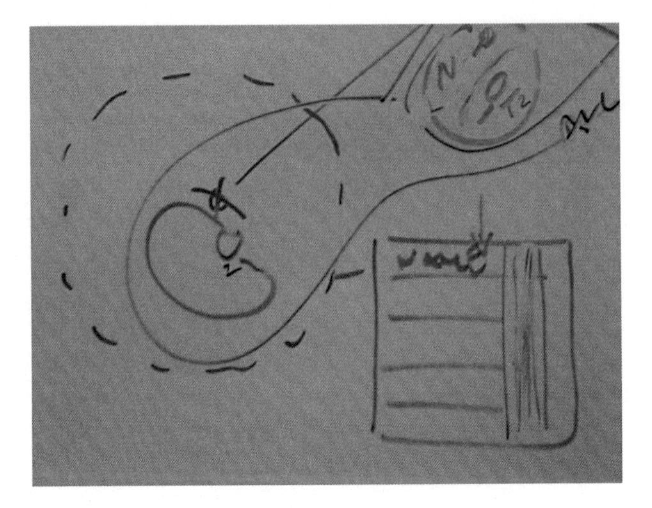

FIGURE 16.11 Enlargement of frequency boxes linked to airspace in the Rich Picture for air-to-ground communications.

An enlarged view of the label for Boscombe Down is displayed in Figure 16.12. The table is translucent so that the map is not completely obscured.

The frequencies are separated into channels, for example 'A' signifies the 'Approach' channel. The alternating translucent white and grey lines help distinguishes the channels. A single line separates the frequencies within a channel; a

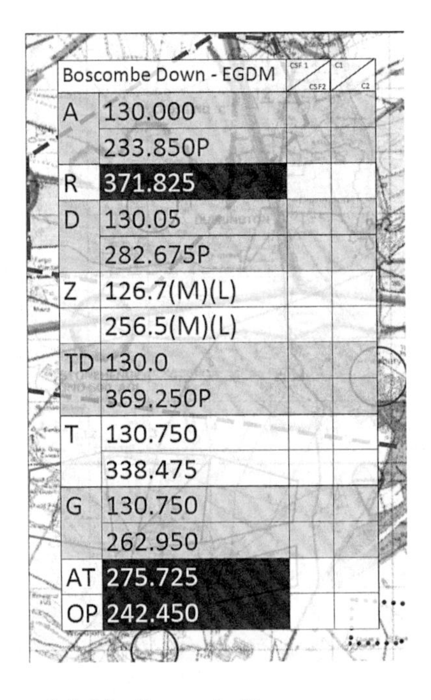

Boscombe Down - EGDM		CSF 1 / CSF2	C1 / C2
A	130.000		
	233.850P		
R	371.825		
D	130.05		
	282.675P		
Z	126.7(M)(L)		
	256.5(M)(L)		
TD	130.0		
	369.250P		
T	130.750		
	338.475		
G	130.750		
	262.950		
AT	275.725		
OP	242.450		

FIGURE 16.12 Frequency label for Boscombe Down.

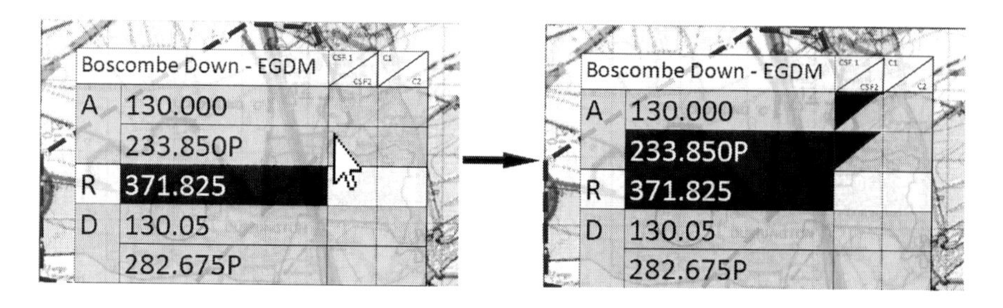

FIGURE 16.13 Selection of frequencies for CSF comms day one list, using the cursor to select the CSF1 column.

double line separates different channels. Where the frequencies are displayed in white text on a black background (for AT, OP and R) this indicates the primary band (usually either ultra high frequency [UHF] or very high frequency [VHF] for air-to-ground communications). When there is only one frequency for any given channel the selection of the primary band is made automatically. Manual selection of the primary band is described below.

The columns on the right of the label indicate where users make their selection for inclusion in the CSF list for 'comm day one' and 'comm day two' (CSF1/CSF2) and the preset channels list for 'comm day one' and 'comm day two' (C1/C2). To select a frequency for inclusion into one of the lists, the user must click in either the top left section (for comm day one – see Figure 16.13) or bottom right section (for comm day two) of the box. Upon selection of the UHF frequency (233.850) the VHF frequency (130.000) is automatically selected. Whichever frequency is clicked on will be the primary band. In this example, the UHF frequency was selected leading to automatic selection of the VHF frequency, hence the UHF is primary. If the VHF frequency were to be clicked on, this would show as the primary band with automatic selection of the UHF frequency.

Where there is more than one frequency within a band (e.g. two UHF frequencies) for a given channel, the top frequency will be automatically selected as a default (see Figure 16.14).

Here, the first frequency to be selected is 123.3. This is indicated as primary and the top UHF frequency (241.025) is selected automatically. The user then clicks in the box to select the lower UHF frequency (278.675), therefore deselecting 241.025. The user can then choose to change the primary band indication by clicking on the box in which the frequency is displayed.

To select a channel for inclusion in the preset channels list the required actions are almost identical, the only difference being rather than clicking in the boxes in

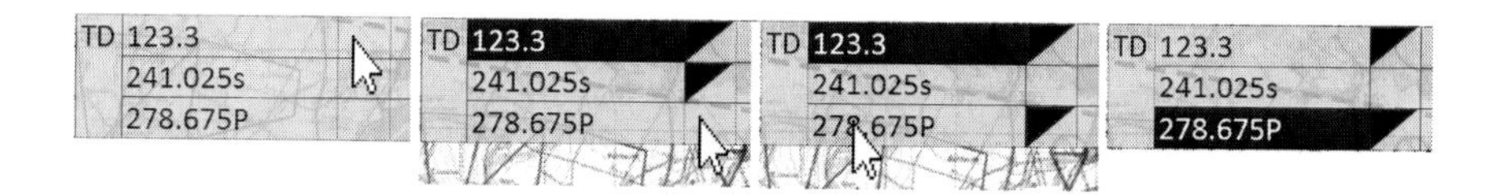

FIGURE 16.14 Manipulation of CSF frequency selections, using the cursor to select the CSF1 column and then the frequency.

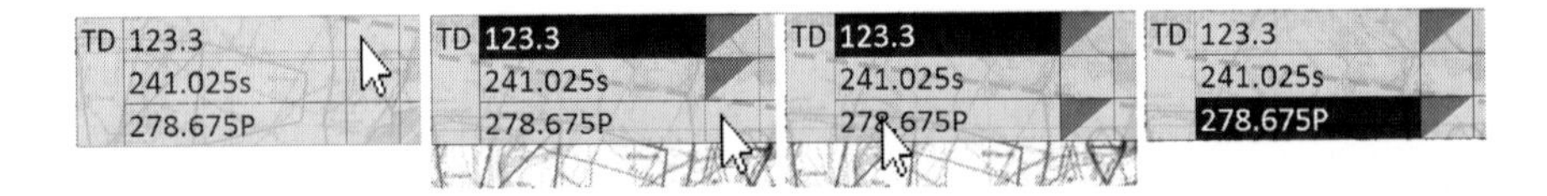

FIGURE 16.15 Manipulation of preset channels frequency selection, using the cursor to select the C1 column and then the frequency.

the CSF1/CSF2 column, the user must click in the boxes in the C1/C2 column (see Figure 16.15). The triangles that indicate selection are a slightly different shade of grey and of a different size so that the user is able to instantly distinguish between the two columns at a glance.

Further to selecting each individual frequency it is possible to select all the frequencies for a particular airspace with just one click. At the head of each column in the frequency labels are the titles; CSF1/CSF2 and C1/C2. These titles can themselves act as action keys. By clicking on one of these, all the frequencies for that column will be selected. An example for selection of CSF members for comm day two is presented in Figure 16.16.

When selecting all channels where there are more than two frequencies in a given band for a particular channel, for example the Ground (G) channel, the topmost frequency will be selected. This can be changed by the user by selecting the preferred frequency (middle and left of Figure 16.16). From the description presented, it has been shown that the map-based interface can support air-to-ground communications

Lyneham - EGDL/LYE		CSF1/CSF2	C1/C2
A	118.425P		
	278.7P		
	362.35		
Z	123.4		
	231.875		
D	118.425P		
	338.350Px		
TD	123.3		
	240.825P		
T	119.225P		
	122.15(f)		
	234.150		
G	122.15(f		
	129.475Px		
	369.2		
OP	337.975		
AT	233.125		

Lyneham - EGDL/LYE		CSF1/CSF2	C1/C2
A	118.425P		
	278.7P		
	362.35		
Z	123.4		
	231.875		
D	118.425P		
	338.350Px		
TD	123.3		
	240.825P		
T	119.225P		
	122.15(f)		
	234.150		
G	122.15(f		
	129.475Px		
	369.2		
OP	337.975		
AT	233.125		

Lyneham - EGDL/LYE		CSF1/CSF2	C1/C2
A	118.425P		
	278.7P		
	362.35		
Z	123.4		
	231.875		
D	118.425P		
	338.350Px		
TD	123.3		
	240.825P		
T	119.225P		
	122.15(f)		
	234.150		
G	122.15(f		
	129.475Px		
	369.2		
OP	337.975		
AT	233.125		

FIGURE 16.16 Selection of all frequencies for comms day two CSF list, using the cursor to select the CSF2 column.

planning and is a radical departure from the current approach. Before developing the software however, performance-based justification of the interfaces is necessary to test the implicit assumption that the new interface will be an improvement over the current approach.

16.6 VERIFICATION AND COMPARISON

To compare the proposed system with the current MPS software tool two simulations were created; a simple one comprising four contactable airspace authorities and a complex on containing eight contactable airspace authorities. The simulations, based on a series of possible screenshots, were constructed such that the amount of mouse travel, number of mouse clicks and number of keystrokes necessary to complete the plan could be recorded. Although the sequence of actions was fixed for each simulation, the method for advancing through the simulated plans was a realistic one. By moving the mouse cursor to the relevant area on the screen and only then clicking to progress it was possible to estimate the amount of mouse travel and number of mouse clicks necessary to complete the plan. The number of keystrokes was estimated from the amount of information requiring insertion by typing (Tables 16.1 and 16.2).

To aid in the comparison of the two systems a SOCA-CAT was constructed for communications planning in the proposed system and compared with the SOCA-CAT for communications planning in the current MPS. The two CAT diagrams are displayed side-by-side in Figure 16.17. From the diagrams it is clear that much more of the activity carried out within the system can be done in the early preparatory and

TABLE 16.1
Results for Simple Comms Plan

	Current System	Proposed System	% Decrease
Mouse travel (m)	19	9.17	52
Mouse clicks	206	86	58
Keystrokes	291	32	89
Time (mins)	22	5	77

TABLE 16.2
Results for Complex Comms Plan

	Current System	Proposed System	% Decrease
Mouse travel (m)	26.58	22	17
Mouse clicks	283	120	58
Keystrokes	534	116	78
Time (mins)	32	10	69

template development stages. In addition in the proposed system an engineer has been added to the list of relevant actors. Workload is shared across roles. A great deal more of the work is carried out by the Ops staff whilst the template administrator also sees an increase in activity loading. In the SOCA-CAT, by displaying a ball in each situation in which activity typically occurs (rather than just one ball and extended whiskers) it is possible use shading to indicate the relative level of activity that occurs here compared to the other situations in which that particular activity

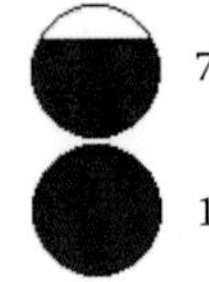

FIGURE 16.17 Modified SOCA-CAT-based comparison of systems to show activity weightings.

occurs. Figure 16.17 also displays the modified versions of the SOCA-CAT with estimated workload (gathered from discussions with SMEs).

To verify the validity of the proposed system a focus group with a number of SMEs was arranged. During the focus group, which lasted for approximately 6 hours, four helicopter pilots (responsible for training new recruits in MPS) and two software engineers were present along with the researchers. A presentation was given in which the research conducted and the resulting proposed system were explained at length. Comments, suggestions and criticisms were encouraged, and although a number of minor changes and additions were recommended, the proposed system was accepted as a vast improvement on the MPS software tool. All of the recommendations that came from the focus group were implemented, and all are contained within the above description of the proposed system.

A number of positive comments were received from the SMEs. In a correspondence giving feedback, two pilots wrote:

> The philosophy adopted brings sweeping changes to the present methodology. It aligns to the logical process that aircrew apply in their planning which is key to reducing errors. Progressively selecting those communications elements required for the mission – in the order of the mission sequence – and having those elements stored logically for easy reference and retrieval/manipulation is another fundamental time saver and error reducer.
>
> This feedback sheet doesn't itemize the considerable volume of beneficial features of this planning strategy. From the presentation, it is clear that the philosophy adopted appears to tick most of the boxes required by air crews to make their communications planning less cumbersome, prone to error and time consuming. It should free their time to other important aspects of mission planning, confident that their communications plan is robust and complete.

The proposed interfaces appear to be far more intuitive to the pilots evaluating the new system designs, probably because the information on frequencies are tied to the airfield locations on a map rather than being disembodied in a spreadsheet that covers the map. Prior to computerisation, pilots would develop their communications plan by scanning their route on a map and noting the airspace boundaries and airfields that they pass through or nearby. The new interfaces proposed are a return to the geographical approach to planning, and pilots seem to prefer to think about communications in terms of ground-based references. In addition, the new approach has the added advantage of linking the airfield frequencies to the map, which saves the pilot constantly referring to the BINA and prevents any transcription errors that might occur. The design capitalises on the affordances associated with maps and the appropriate use of computerised databases to present the frequencies associated with route-relevant airfields. This approach allows the pilot to focus on deciding upon the mission communications rather than attempting to populate spreadsheets with frequency numbers.

16.7 CONCLUSION

The purpose of this research was to conduct a CWA of the helicopter MPS, with the aim to guide design recommendations for future iterations of the system. Whilst CWA will not design the system, it does map out the problem space in an explicit manner in the CAT (Jenkins et al., 2008a,b) which encourages exploration. The outputs of CWA provide a detailed description of the system at a number of levels of abstraction. The work domain, contextual activity and social organisation were considered, allowing for a new perspective of the system. The AH developed in the WDA phase of the analysis gave an indication of the organisation of the communications planning functions. Based on functional mean-ends links within the hierarchy it was possible to describe the relations between functions within the system. From this description emerged recognition of the mismatch between function organisation and MPS organisation. Re-design of the software-based on activity flow would reduce the need for multiple windows to be open simultaneously and reduce the need for extensive training in MPS. The most apparent characteristic of MPS to emerge from this analysis was the requirement for last minute activity. Currently, only a small proportion of the activity carried out within the system happens in the first two situations, namely preparatory support activities and template development. The large majority occurs in mission specific activities and individualisation of communications plan. In SOCA-CAT it was shown that the majority of activity was conducted in the latter stages of the planning process, and that this activity is predominately carried out by the pilot. In summary, the CWA outputs not only highlighted the mismatch between task organisation and system organisation, but brought to attention the uneven spread of activity across situations and actors. This study has shown that the SOCA-CAT in particular to be useful for comparing of current and proposed approaches to system design (Niakar et al., 2006).

The Rich Pictures (Checkland, 1981; Checkland and Scholes, 1990) were constructed by an SME primarily to inform the CWA and to help the researchers gain a fuller understanding of the communications planning task independent of MPS. Not only did the pictures give a valuable insight into the way in which the expert conceives the task, and helps guide the CWA process, they also proved an invaluable tool in designing the interface for the proposed communications planning system. The air-to-ground Rich Picture was particularly useful in inspiring the design of the new system; it outlined the SME's representation of the task, showing that the arrangement of frequencies is not list-based but organised into groups of frequencies, each relating to an airspace controlling authority on the map of the ground. The representation of the aircraft's route as it travels through and around airspaces was similar to representation of the route on the map display in MPS; the major difference being that the selection of frequencies, that is, who will need to be contacted, is done *while* studying the route, not *after* studying the route. This has been capitalised on in the proposed system; frequencies are directly tied to the map and can be selected from that view. Mapping the structure of the system interface and interaction design to the user's conceptual models of communications planning has increased the usability of the mission planning software for the communications planning task.

A comparison of current communications planning in MPS and the proposed communications planning system revealed some significant differences in activity allocation across situations and actors. Where MPS-based planning requires users to perform more of the work later on in the planning process, the proposed system allows for the majority of activity to be carried out in the preparatory stages, namely preparatory support activities and template development. The proposed system also allocates more activity to the template administrator and the Ops staff, allowing the pilot to spend less time on the ground planning and more time flying the aircraft. This has benefits for both the individual helicopter pilot and communications planning in the military in general. At first sight, Rich Pictures and CWA might seem an odd pairing; there are certainly no other reports of the two approaches being used together before. Justification for the pairing is based on the assumption that eliciting the Rich Picture as well as constructing CWA is gaining two perspectives of constraints in systems. There is the added benefit that a Rich Picture can help put the SME in mind of system constraints from the outset as well as providing a visual representation for discussion. One needs the other however, as the analysis using CWA is more formal and allows the design team to explore the constraints in more detail. This additional level of fidelity was particularly useful in the SOCA-CAT representation, as it shows the way in which workload could be shifted from the pilot to other staff in the communications planning team. It is proposed that the general design principles derived in the study reported here could be used to improve communications planning across platforms in the armed services. The combined Rich Picture and CWA approach could be applied to other design projects, following the principles developed in the current chapter. In hindsight, it might have been more appropriate to involve more pilots in the design phase, in order to elicit more Rich Pictures (to analyse the similarity and differences) but time and resource constraints prevented this. Nevertheless, the general approach of exploring Rich Pictures coupled with analysing current and future systems constraints with CWA has produced novel design solutions. The development of interfaces together with formal and informal verification suggests that the approach has indeed worked. We will be applying the approach in different domains in our future research.

ACKNOWLEDGEMENTS

This work from the Defence Technology Centre for Human Factors Integration (DTC HFI) was partly funded by the Human Sciences Domain of the UK Ministry of Defence.

The authors would like to thank the pilots who took part in this study, acting as subject matter experts, both in showing us how communications planning works on the current system, sharing their understanding of the task with Rich Pictures, taking part in the evaluation and providing comments on the current and proposed system designs.

REFERENCES

Checkland, P. 1981. *Systems Thinking, Systems Practice*. Chichester: John Wiley & Sons.

Checkland, P. and Scholes, J. 1990. *Soft Systems Methodology in Action*. Chichester, England: John Wiley & Sons.

Helender, M. G. 2007. Using design equations to identify sources of complexity in human-machine interaction. *Theoretical Issues in Ergonomics Science*, 8(2), 123–146.

Horan, P. 2002. A new and flexible graphic organiser for IS learning: The rich picture. In *Proceedings of Informing Science Conference & IT Education Conference*. Cork, Ireland: Informing Science Institute, pp. 133–138.

Jenkins, D. P., Stanton, N. A., Salmon, P. M. and Walker, G. H. 2009. *Cognitive Work Analysis: Coping with Complexity*. Farnham, England: Ashgate Publishing Limited.

Jenkins, D. P., Stanton, N. A., Salmon, P. M., Walker, G. H. and Young, M. S. 2008a. Using cognitive work analysis to explore activity allocation within military domains. *Ergonomics*, 51, 798–815.

Jenkins, D. P., Stanton, N. A., Walker, G. H., Salmon, P. M. and Young, M. S. 2008b. Applying cognitive work analysis to the design of rapidly reconfigurable interfaces in complex networks. *Theoretical Issues in Ergonomics Science*, 9(4), 273–295.

McIlroy, R. C., Stanton, N. A. and Remington, R. E. 2012. Developing expertise in military communications planning: Do verbal reports change with experience? *Behaviour and Information Technology*, 31(6), 617–629.

Monk, A. and Howard, S. 1998. The rich picture: A tool for reasoning about work context. *Interactions*, 5(2), 21–30.

Naikar, N., Moylan, A. and Pearce, B. 2006. Analysing activity in complex systems with cognitive work analysis: Concepts, guidelines, and case study for control task analysis. *Theoretical Issues in Ergonomics Science*, 7(4), 371–394.

Rasmussen, J., Pejtersen, A. M. and Goodstein, L. P. 1994. *Cognitive Systems Engineering*. New York: Wiley.

Sutrisna, M. and Barrett, P. 2007. Applying rich picture diagrams to model case studies of construction projects. *Engineering Construction and Architectural Management*, 14(2), 164–179.

Stanton, N. A. and Baber C. 2003. On the cost-effectiveness of ergonomics. *Applied Ergonomics*, 34(5), 407–411.

Stanton, N. A., Salmon, P. M., Walker, G. H., Baber, C. and Jenkins D. 2005. *Human Factors Methods: A Practical Guide for Engineering and Design*. Aldershot, UK: Ashgate Publishing Ltd.

Stanton N. A. and Young M. S. 1999. What price ergonomics? *Nature*, 399, 197–198.

Vicente, K. J. 1999. *Cognitive Work Analysis: Toward Safe, Productive, and Healthy Computer-Based Work*. Mahwah, NJ: Lawrence Erlbaum Associates.

Walker, G. H., Stanton, N. A., Salmon, P. M. and Jenkins D. P. 2008. A review of socio-technical systems theory: A classic concept for new command and control paradigms. *Theoretical Issues in Ergonomics Science*, 9(6), 479–499.

Section V

Risk and Resilience

17 CWA vs SWIFT in a Nuclear Decommissioning Case Study

Guy H. Walker, Mhairi Cooper, Pauline Thompson and Daniel P. Jenkins

CONTENTS

17.1 INTRODUCTION

As many nuclear sites around the world make the transition from operations to decommissioning, the problem-space in which they operate is changing and so too is the appropriateness of methods aimed at managing risks therein. Stated simply, just because hazard identification methods have worked in the past no longer means they are guaranteed to work in the future. This chapter describes an opportunity which arose at a UK Magnox nuclear site to subject this broad question to a more specific test through the comparison of two methods: the Structured What If

(SWIFT) approach and cognitive work analysis (CWA). Underwood and Waterson (2013) make an interesting distinction that is relevant here. SWIFT might be termed a 'practitioner' method, one that was used extensively during the plant's operational phase. CWA, on the other hand, might be termed (incorrectly or otherwise) as an 'analyst' method of the sort more commonly found in the academic literature. The former is 'proven' in the present problem domain, but there are doubts about the extent to which it can cope with the uncertainty and novelty brought about by decommissioning. The latter is not proven at the present site – it has not been used before in this location – but it is designed to cope with uncertainty and novelty. This chapter explores how these trade-offs manifest themselves when both methods are applied in a real high-hazard setting.

17.1.1 Nuclear Decommissioning

17.1.1.1 Background and Context

The nuclear power station at which this study took place is a 360 MW Magnox site constructed in the early 1960s. Magnox reactors are fuelled with uranium fuel elements which are loaded into a graphite reactor core. The term Magnox refers to the non-oxidising magnesium alloy used to clad these elements. This design feature confers a number of technical advantages relating to containment of fissile material and the relative ease of material handling during reprocessing, and is a feature unique to British reactors of this era, albeit one that is now considered obsolete. The graphite in the reactor core, into which the Magnox encased fuel elements are placed, is known as the moderator. So-called 'fast neutrons' released in the fission process have to be slowed in order to sustain an on-going chain reaction, and graphite provides this function. Control rods are also raised and lowered into the core to control the reaction by absorbing excess neutrons. The heat energy released by the fission process is continually moved from the reactor core by a coolant which, in the case of Magnox reactors, is pressurised carbon dioxide gas. The coolant flows from the core to heat exchangers where water is converted into steam. This powers a number of conventional turbo alternators which supply electrical energy to the national grid system.

17.1.1.2 Cartridge Cooling Pond

Fuel rods have a finite lifespan and once they are 'spent', typically after a year, they are removed from the core, passed through a desplittering process to separate the fuel rods from the Magnox cladding, then loaded into crates and stored underwater in the cartridge cooling pond (CCP). The cooling process takes several months. Once cooled the fuel is removed from the CCP, placed in flasks and transported offsite for safe storage or reprocessing.

The CCP at the present site has a capacity of over one million gallons. It consists of a 1.2 m thick reinforced concrete slab with reinforced concrete internal and external walls and has a depth of 6.7 metres. The pond enclosure is divided into three sections; fuel handling bay (FHB) 1 and 2 and the CCP main enclosure. During the operational life of the station FHB 1 was used to accept and desplitter discharged fuel cartridges, while FHB 2 was available for emergencies andhas the capacity to store all fuel from the reactor should such an eventuality arise. The main enclosure

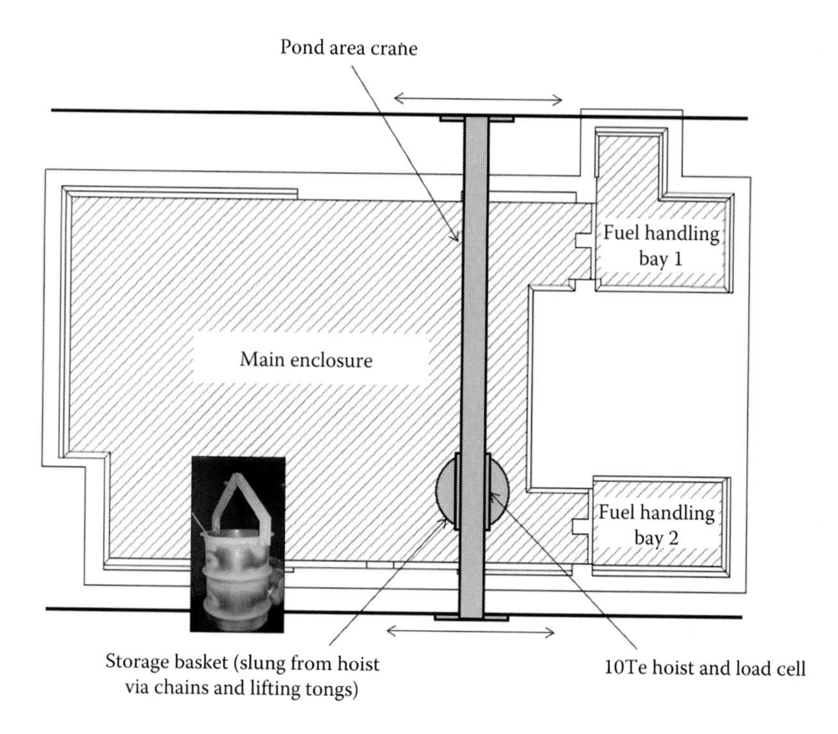

FIGURE 17.1 Layout of CCP showing two fuel handling bays (FHB 1 and 2) and the main enclosure.

was used to store spent desplittered fuel cartridges in skips. A skip crane runs the width of the pond and was used to transport the skips from one area to another (Figure 17.1).

17.1.1.3 Decommissioning

In the mid-1990s, following a successful period of base-load generation spanning 20 years, the power station ceased operation and preparations began for decommissioning. At the time of writing, the station is currently undertaking a programme to preserve the site in a 'safe state'. During the safe state no further projects will be undertaken and only care and maintenance regimes are required to assure the integrity of the civil structures on the station until such time that the structures can be safely removed and the site returned to its original condition. Under the programme of decommissioning a project has been established to decommission the CCP. This project involves retrieval, decontamination and safe disposal of pond furniture, followed by removal of sludge (desludging) which has built up on surfaces within the CCP, and then eventual draining of the CCP (dewatering). Before this stage is reached, however, two high-dose rate items (HDRIs) have to be removed. These HDRIs give a dose reading of 14Sieverts (Sv) which is the equivalent of a 5.25 second unshielded exposure time before the annual legal allowable dose uptake of 20mSv is exceeded. It should be pointed out that 20mSv is a maximum dose. In practice, workers would not be expected to exceed half of this value in a year. It is vital, therefore, that all required measures are taken to protect operators against such a scenario, and it is this specific risk that SWIFT and CWA are directed toward.

17.1.2 Hazard Identification

17.1.2.1 Practitioner Methods

The imperative to ensure operator safety when dealing with these HDRIs is compounded by the change in circumstances from 'operations' to 'decommissioning'. Many of the tasks required to decommission the CCP have, by definition, never been performed before on this site. Added to which, over the 40 years since the plant was commissioned it has naturally degraded and aged, creating as yet unforeseen conditions such as CCP sludge, unusual 'left-over' HDRIs, non-standard removal tasks and complex storage requirements. This increase in complexity also increases the chance of unexpected yet still highly credible hazards to personnel.

The first strategy for dealing with this is to use hazard identification methods that were routinely employed during the operational life of the facility. On this site the 'SWIFT' method was common. This method can be described as a practitioner tool and was originally developed for use in the chemical and petrochemical sectors, but has since become established as a useful technique throughout a number of high risk industries (BS EN 31010, 2010, p. 49). SWIFT is similar to the HAZOP technique in that it requires a multi-disciplinary team to work through the task in a systematic manner, asking questions of the form 'What if …' or 'How could …'. SWIFT relies on task decomposition and a systematic application of the 'what if' questions, which are applied to each and every identified task. In other words, the task is taken apart, assessed for hazards, then re-assembled on the tacit assumption that the reassembled 'whole' should not be more than the sum of its parts. SWIFT, therefore, has a large deterministic element well matched to stable problems involving well understood parts (Walker et al., 2010).

17.1.2.2 Analyst Methods

The presence of various age-related degradations in the system, combined with the need to perform novel tasks, makes it potentially more difficult to discern final states from initial conditions. This property is referred to as 'emergence' (Gleick, 1987; Halley and Winkler, 2008; Walker et al., 2010). Emergence creates an analytical problem in that 'calculation of system level emergent properties (such as unforeseen hazards and risks) from the component level rapidly becomes intractable' (Halley and Winkler, 2008, p. 12). In other words, it becomes increasingly difficult and effortful to deduce what 'could' happen via deterministic methods such as SWIFT which work at the component level. That being said, it is important to point out an important benefit inherent in SWIFT's team-based methodology. Personnel with close experience of the system in question are brought together and, in effect, prompted to construct narratives about how it works and the potential for different sorts of hazards to arise. Those narratives can reveal previously unthought of interactions and risks, however, equally importantly are the Decommissioning Safety Case Handbook (2011) recommendations that SWIFT is only used in cases where the nature of the proposed work is sufficiently broad to require input from a range of engineering disciplines or subject matter experts, but the hazard is intermediate or low and the design requirements are relatively straightforward (DSCH, 2011, p. B3). In other words, the ability of SWIFT to diagnose emergent risk issues is limited and recognised as such.

A further distinction can be drawn between practitioner methods like SWIFT, HAZOP, etc that are based around data on how the task 'should' be performed, and methods like CWA which focus on the constraints and purposes which define the 'space' within which all possibilities for how a task 'could' be performed are shown. To clarify, the possibilities for action that link these constraints and purposes together enable a much more systematic diagnosis of interactive complexity and the subsequent emergence of hazards to be made. Like all methods, this is dependent on the quality of the inputs. The point is that given a set of data of the same quality, CWA should be able to more systematically map the interactive complexity between system elements compared to SWIFT. Or at least, the chance to miss important linkages will be reduced. In simpler 'Rumsfeldian' terms, deterministic methods (like SWIFT) can only deal with 'known unknowns' (i.e. what the analysis team can think up for themselves) whereas formative methods (like CWA) can help to discover 'unknown unknowns'. Of course, CWA is noteworthy for another reason: it was originally developed by, and for use within, the nuclear sector. As we enter a new phase in the life-cycle of many nuclear facilities world-wide (i.e. decommissioning) and encounter new problems of complexity, emergence and potentially very high risk, perhaps now is a good time to investigate CWA's return to its original domain? Indeed, we pick up again on a theme recently presented by Underwood and Waterson (2013) that there is a growing gap between research (in which formative, systems methods are the dominant mode of thought) and practice (in which deterministic methods are much more common). The remainder of this chapter will explore these issues via two case studies. One deals with the application of SWIFT to the problem of removing a HDRI from a soon-to-be decommissioned CCP. The other case study applies CWA to the same problem. The practical and methodological outcomes of this exercise will then be discussed.

17.2 METHODOLOGY

17.2.1 Participants

Seven individuals took part in the SWIFT study, as detailed in Table 17.1. The work domain had been subject to several previous detailed studies including HAZOPs, SWIFTs and task analysis, and the analysis team was composed of individuals with experience of this previous work. The CWA analysis took place concurrently. A lead analyst who was part of the project, and who worked closely with the pond project engineer and safety case author, produced the results with the pond project engineer reviewing the outputs.

17.2.2 SWIFT Analysis Method

SWIFT provides a straightforward and logical approach to hazard identification. This makes it particularly attractive to project teams who are working within time

TABLE 17.1
SWIFT Participants

Role Title	Role Description	SWIFT	CWA
Chairman	This is a facilitator (independent to the project) who leads the team through the SWIFT process to ensure that it is undertaken in a thorough and systematic manner.	Yes	
Scribe	The role of the scribe is to accurately record the outcome of the SWIFT discussion and note down any actions arising.	Yes	
Safety Case Author	The safety case author prepares the safety case that identifies and addresses the specific project hazards.	Yes	
Engineering Manager	The engineering manager is responsible for all engineering work undertaken on site and attends the SWIFT to ensure all engineering issues are identified.	Yes	
Accredited Health Physicist	This role is responsible for managing the radiation dose to personnel and ensuring that all radiological hazards associated with the work are identified and addressed.	Yes	
Safety Case Officer	The safety case officer (SCO) has final sign off and therefore overall responsibility for completion of the safety case. The SCO attends the SWIFT to ensure they are confident that all hazards have been identified.	Yes	Yes
Pond Project Engineer	The pond project engineer is actively involved in the day to day project activities and is best placed to explain the intricacies of the project and can, therefore, assist in the identification of associated hazards. This individual also reviewed the CWA outputs.	Yes	Yes
Project Engineer	As above	Yes	
Principal Consultant	As above	Yes	Yes
Safety Case Verifier	This role is responsible for reviewing and verifying the safety case to ensure that the safety case itself meets the requirements of the *Decommissioning Safety Case Handbook.*	Yes	
	Number of method participants.	n = 7	n = 3

and cost constraints. There are seven generic steps for carrying out a SWIFT analysis in accordance with BS EN 31010 (2010) and these were adhered to for the purposes of the study.

1. Prior to the start of the study the facilitator (chairman) prepared a list of prompt words or phrases (see Table 17.2).
2. During the workshop the activity which will impact the system was agreed. This scoping step was important because all members of the team have to agree on the boundaries or 'nodes' of the system that were to be discussed. This aids focused discussion.
3. Managed by the facilitator (chairman) the SWIFT team discussed known risks and hazards using knowledge gained from previous experience

TABLE 17.2
SWIFT Checklist used to Ensure Consistent Coverage of 'What If' Questions

Item No.	SWIFT Checklist Item	Actual Prompt
1	Operating error/other human factors	'Is it possible to attach the lifting arrangements to the wrong hook?'
2	Measurement/ monitoring error	'What will the health physics controls be on the HDRI if the HDRI cannot be placed within basket?'
3	Equipment malfunction	'Is it possible that the tongs do not fully extend?'
4	Activity/dose rate	'What if the HDRI is not fuel?'
5	Utility failure (lighting, water supply, etc.)	Outside scope: back up lighting, etc. is provided
6	Maintenance	Outside scope of SWIFT as crane was subject to strict maintenance routines
7	Integrity failure or loss of containment	'What if pond water level is lowered?'
8	Loss of shielding	'What if pond water level is lowered?'
9	Spread of contamination	'Is it possible to contaminate the load cells or slings?'
10	Emergency operation	Outside scope of SWIFT as no emergency operations were required
11	External hazards	'How will personnel be restricted from entering the pond area?'
12	Environmental release	'What if pond water level is lowered?'
13	Design shortfall	'What if the HDRI will not fit in basket?'
14	Control of work	'How will personnel be restricted from entering the pond area?'
15	Training	Outside scope of the SWIFT as no training was required
16	Communication	Outside scope of the SWIFT as no special communication techniques were required
17	Access	'How will personnel be restricted from entering the pond area?'
18	Other (as suggested by the team)	No other checklist items were suggested by the SWIFT team

and incidents, bearing in mind any applicable regulatory or legislative constraints.

4. The facilitator (chairman) managed the discussion through each prompt word using questions beginning 'What if …?', 'What would happen if …?', 'Could something …?' with the team exploring potentially hazardous scenarios, causes, consequences and impacts.
5. The team identified any existing controls which mitigate against the potentially hazardous scenarios identified.
6. The team reviewed each control measure and decided whether they were adequate. Where existing control measures were deemed to be inadequate further action was agreed. These actions were ranked in order of priority for completion.

7. Further 'what if …?' led discussions took place until the team agreed that all potentially hazardous scenarios have been considered.

The SWIFT study considered all activities associated with the movement, identification and storage of the HDRI within the cooling pond. The process was divided into the following nodes for consideration:

- Node 1: Preparation for movement, movement under water and identification and placement of HDRI into designated receptacle within the pond.
- Node 2: Movement of basket containing HDRI in pond.

The following assumptions applied throughout the SWIFT study:

- No desludging or HDRI removal operations shall be carried out simultaneously in the same area of the CCP or FHBs.
- The existing personnel who are undertaking HDRI recovery operations will be used to conduct operations (along with any specialist assistance where required).
- Occupancy in the working area will be kept to minimum levels during the identification and isolation activities.

In summary, SWIFT is an extremely practical and expedient method but it cannot be used where unfamiliar tools and techniques are proposed, neither should it be used purely because it is simple and cost effective. Where novelty, uncertainty and complexity are recognised as being features of a problem, and where the need arises to go beyond SWIFT, it is recommended that more detailed methods of the same (deterministic) type, such as HAZOP and FMEA, are used. In this case, however, CWA was put forward as an alternative.

17.2.3 Application of CWA

SWIFT presents a structured method for taking a normative task description and subjecting it to interrogation via 'what if' questions. CWA is conceptually very different. Instead of looking at tasks and asking 'what if' it instead looks at the 'system' and ask what are the 'constraints'. Specifically, what are the

1. Constraints on people's behaviour imposed by the purposive context
2. Constraints on people's behaviour imposed by the physical situation
3. Constraints on activity that are imposed by specific situations
4. Constraints imposed by the available means by which activities can be performed
5. Constraints imposed by organisational structures, specific actor roles and definitions
6. Constraints imposed by human capabilities and limitations

CWA answers the 'what if' question with a 'space of possible outcomes', a multidimensional space bounded by six types of constraint. Multiple constraints require multiple dimensions of analysis and this is what CWA offers with five phases (as described at the beginning of this book).

17.2.4 The HDRI Lifting Task

The task under analysis is to safely lift the HDRIs out of the CCP and place them into storage baskets. The main items of plant to be used in the operation are as follows:

- Pond area crane
- 10Te hoist
- Load cell (for sensing lifting loads)
- Long lifting chains
- Storage basket
- Long lifting tongs
- Gamma probe (for radiation sensing)
- Camera and monitor

The planned tasks and their sequence are shown in Table 17.3.

17.3 FINDINGS

17.3.1 SWIFT Results

The SWIFT method described above generated 24 credible hazard scenarios for the HDRI lifting task, as shown in Table 17.4. The identified hazards were grouped into four generic hazard types

- Hazard 1 – Over-raising of HDRI out of pond caused by operator error.
- Hazard 2 – Drop of, or impact onto, HDRI caused by operator error, crane control system failure or mechanical failure.
- Hazard 3 – HDRI damaged by handling tool caused by compressor fault or operator error.
- Hazard 4 – Complete loss of shielding due to pond leakage or inadvertent use of pumps caused by operator error or civil structure failure.

The following actions (Table 17.5) were identified, all of which were to be addressed prior to the HDRI removal operation.

17.3.2 CWA Results

Phase 1 – Work Domain Analysis (WDA): The functional purpose of the system (the top level of the WDA) was defined by three 'reasons the HDRI task existed in the first place': these were to take the CCP out of use, deal with 'orphan' waste and discharge the pond water. Thirty five physical objects (the bottom level of the WDA) were defined, including skip crane, outer building structure, storage basket

TABLE 17.3
Procedure/Task List for HDRI Lift Operation

Step	Operation
1	Deploy all equipment to be used in the CCP.
2	Power up CCP crane and test each operation. Ensure the fully raised limit switch stops the 10Te hoist being over raised.
3	Using the 10Te hoist attach the load cell and storage basket long lifting chains.
4	Lift and position the storage basket on the floor of the CCP. Lift the basket clear of the pond floor and record the load cell reading. Detach the long lifting chains from the basket.
5	Detach the basket lifting chains from the crane hook, monitor, decontaminate and store.
6	Attach the long lifting tongs to the crane 10Te hoist.
7	Raise the crane hoist to fully raised and check that the tongs grab is beneath the water. Adjust as required.
8	Lower the hook until the tongs make contact with the bottom of the pond next to the storage basket to ensure that they are long enough.
9	Lower the gamma probe into the water above the HDRI and locate by gamma readings.
10	Traverse the crane slowly until the tongs are located above the HDRI and then lower the crane until the tongs are approximately 100 mm above the HDRI. Ensure the camera is working correctly and locate the HDRI visually.
11	Visually check on the monitors that the tongs are located directly above the HDRI in the north-south and east-west directions then gradually inch lower until the HDRI is within the jaws of the tongs.
12	Operate the tongs to close the jaws then visually check if HDRI has been captured.
13	On successful capture of the HDRI slowly raise the hoist, then slowly traverse the crane until the tongs are alongside the storage basket.
14	Once alongside the storage basket slowly raise the hoist until it is just clear of the basket then travel the hoist toward the centre of the basket.
15	Lower the hoist until the tongs make contact with the bottom of the basket then operate to open the jaws of the tongs and release the HDRI.
16	Raise the hoist and visually confirm the HDRI has been released.
17	Raise the tongs clear of the basket and traverse toward their park station, part the tongs and release them from the crane hook.
18	Clean, monitor and store all lifting equipment.

and so on. Structural means–ends links were established between these two outer layers of the abstraction hierarchy. All the nodes in the WDA were scrutinised for 'what' it does, then linked to all of the nodes in the level directly above to answer the question 'why' it is needed. The means–ends links knit together physical objects, object-related processes, purpose-related functions, values and priority measures and functional purposes to represent the 'affordances' embedded in the system. For example, underwater lighting in the cooling pond 'affords' visibility and allows operators to obtain visual information during the HDRI lift operation thus contributing to the functional purpose 'remove orphan waste'. Conversely, sludge 'affords' reduced pond water clarity, potentially impeding the functional purpose by making the HDRI more difficult to access and lift. The main purpose of the WDA was to feed into subsequent CWA phases. In particular, it was

TABLE 17.4
List of Identified Hazards from SWIFT Analysis

Node	Question	SWIFT Question/Hazard Scenario	Cause	Consequence(s)	Comments
1	1	How will personnel be restricted from entering the pond area?	Breach of security	Dose to personnel	
1	2	Is it possible that the crane could over raise?	Failure of raised limit switch	HDRI lifted out of pond	
1	3	Is it possible not to extend the tongs fully?	Human error using short tool	HDRI lifted out of pond	Monitoring will be carried out at maximum depth approximately 6 m
1	4	Is it possible to attach the lifting arrangements to the wrong hook?	Human error		Operational preference. To use the 10Te auxiliary hoist as it has a smaller hook, which improves visibility of the operation
1	5	Can the basket be dropped on the HDRI?	Dropped load	Damage to HDRI	No radiological consequences
1	6	Is it possible to contaminate the load cells or slings?	Load path line up	Contamination to equipment	The management and processing of contaminated pond equipment is covered under existing procedures
1	7	Is it possible to incorrectly line up load path?	Load path line up	Dropped equipment	No radiological consequences
1	8	Is it possible for the load path to drop?	Incorrectly connected load path	Potential damage to HDRI	Unlikely to severely damage HDRI. If HDRI was to be broken into several pieces then this would have an insignificant impact on the background dose in the pond area. Additionally any gases coming from the HDRI as a result of fracture would be dissolved into the pond water. It is noted that the weight of the tongs is approximately 15 kg
1	9	Is there a requirement for other cameras?			If possible existing techniques should be used as they are familiar and established with the operators to reduce human error. The use of additional cameras has also be used previously and is an established process

(Continued)

TABLE 17.4 (*Continued*)
List of Identified Hazards from SWIFT Analysis

Node	Question	SWIFT Question/Hazard Scenario	Cause	Consequence(s)	Comments
1	10	What if the tongs crush the HDRI		Damage to HDRI	Unlikely to severely damage HDRI. If HDRI was to be broken into several pieces then this would have an insignificant impact on the background dose in the pond area. Additionally any gases coming from the HDRI as a result of fracture would be dissolved into the pond water. It is noted that the weight of the tongs is approximately 15 kg
1	11	What if the HDRI is not fuel?			Presuming that HDRI is a fuel element bounds all other cases. SQEP operators would cease the process make safe the work and advise the engineering manager to determine the way ahead if HDRI is not solid or does not behave as expected
1	12	What if HDRI jammed in pond and cannot be manoeuvred?	Location of HDRI	Unable to move HDRI	If operations are not able to be completed for any reason then operations will cease, work made safe and engineering manager informed to determine the way ahead
1	13	What if pond furniture located adjacent to the HDRI was moved prior to removal of the HDRI?		Damage to or repositioning of HDRI	If the HDRI were to be jammed this would not cause any additional hazard, but would simply cause operational difficulties
1	14	What if HDRI is stuck on tongs?	Compressed Air Tool failure		
1	15	What if HDRI is raised too high?	See above (Item 1.2 and 1.3)		
1	16	What if pond water level is lowered?	Leak	HDRI is less than 2 m from surface of pond	
1	17	What if there is a dropped load whilst tongs are holding HDRI?	Load path failure	Potential damage to HDRI	
1	18	What if HDRI dropped onto 'knife edge'	Load path failure	Potential damage to HDRI	

(*Continued*)

TABLE 17.4 (*Continued*)
List of Identified Hazards from SWIFT Analysis

Node	Question	SWIFT Question/Hazard Scenario	Cause	Consequence(s)	Comments
1	19	What if the HDRI will not fit in basket?			
1	20	What will the HP controls be on the HDRI if HDRI cannot be initially placed within basket?			
1	21	If it is evident that the basket will not be big enough for HDRI is it ALARP to continue operations prior to larger basket being manufactured?			Doses are so low to operators within the pond building and time requirement for manipulation is such that it is considered acceptable to continue working on existing CCP projects (outside of FHB A) whilst any procurement lead time or fabrication work is underway to support the continued in-pond management of the 'High Dose Rate Item'
1	22	What if tongs are contaminated?			Expectation is that contamination levels will be slightly higher than during normal operations. This does not pose any additional hazard and will be controlled under normal monitoring and decontamination procedures
2	1	What if basket containing fuel is inadvertently moved during normal pond operations?		Potential damage to HDRI (see Node 1 Item 8)	The use of buoys above basket should be discounted to prevent possibility of inadvertent retrieval of HDRI
2	2	What if basket containing HDRI is inadvertently lifted out of the pond?	Human error	Potential dose to personnel	

Note: Node 1 refers to potential hazards occurring during the preparation phase, Node 2 refers to potential hazards occurring during the HDRI lifting phase.

TABLE 17.5
Hazard Mitigation Steps to Be Performed Prior to HDRI Removal Task Based on SWIFT Analysis

Action Number	SWIFT Node Question	Action	On	Due By
1	1.3	Consider marking tongs to provide visual indication at 2 m from grab end of tongs	Pond work supervisor	Prior to commencing operations
2	1.13	Perform tug test to determine whether or not pond furniture can be moved	Pond work supervisor	Prior to commencing operations
3	1.16	No other pond operations to be performed during this task. Check of pond level will be conducted prior to operation. This is normal operation	Pond work supervisor	Prior to commencing operations
4	1.19	Include a comment in the method statement to cover this eventuality. The HDRI will be placed beside the basket on an existing plate above the level of the sludge if it will not fit into the basket even after attempts at manipulating the HDRI into the basket	Pond work supervisor	Prior to commencing operations
5	2.1	Consider methods for isolating basket and marking the basket signage, etc., subsequent movement of basket containing HDRI should be covered in method statement, etc.	Pond work supervisor	Prior to commencing operations

important to show how different purposes and objects (layers of the system) related to different phases and actors involved in the task. Nevertheless, what the WDA also succeeded in doing was raising the analyst's perspective, helping them to consider how this individual task related to the higher level functional purpose of the 'total system' (Figure 17.2).

Phase 2 – Control Task Analysis (ConTA): The contextual action template (CAT) was used in this phase. It employed the steps of the established methodology for retrieval of HDRI as the 'situation' and the purpose-related functions identified through the abstraction hierarchy to provide the functions. The outputs are shown in Figure 17.3 and demonstrated the following:

- Shielding is needed, and provided, throughout the retrieval stages of the process.
- Cleaning, monitoring and storage of lifting equipment protects against external hazards and contamination in atmosphere, and the protection of personnel against injury and contamination.
- Water level indication typically occurs during lifting preparations and retrieval operations, effectively occurring during lifting of the HDRI, but

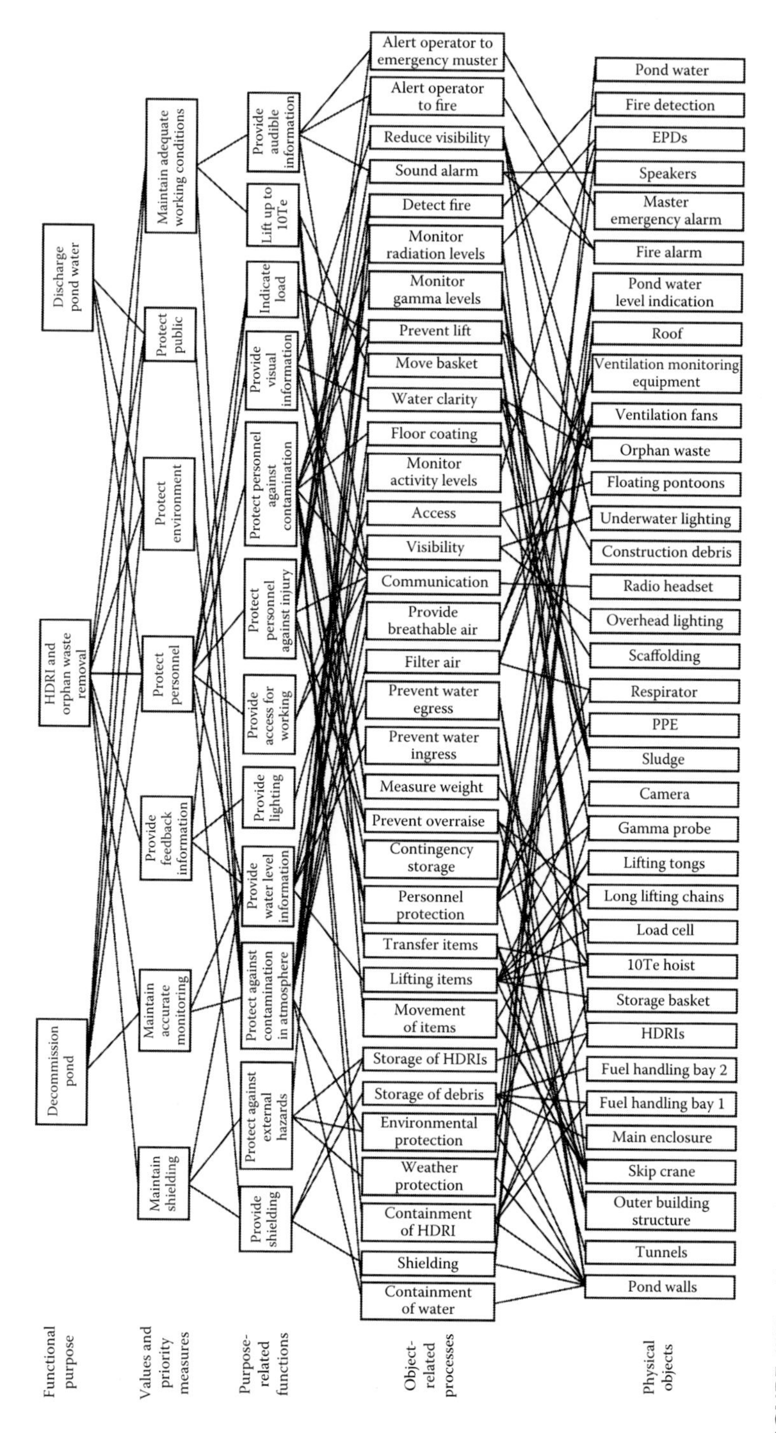

FIGURE 17.2 Abstraction hierarchy of the HDRI lifting task.

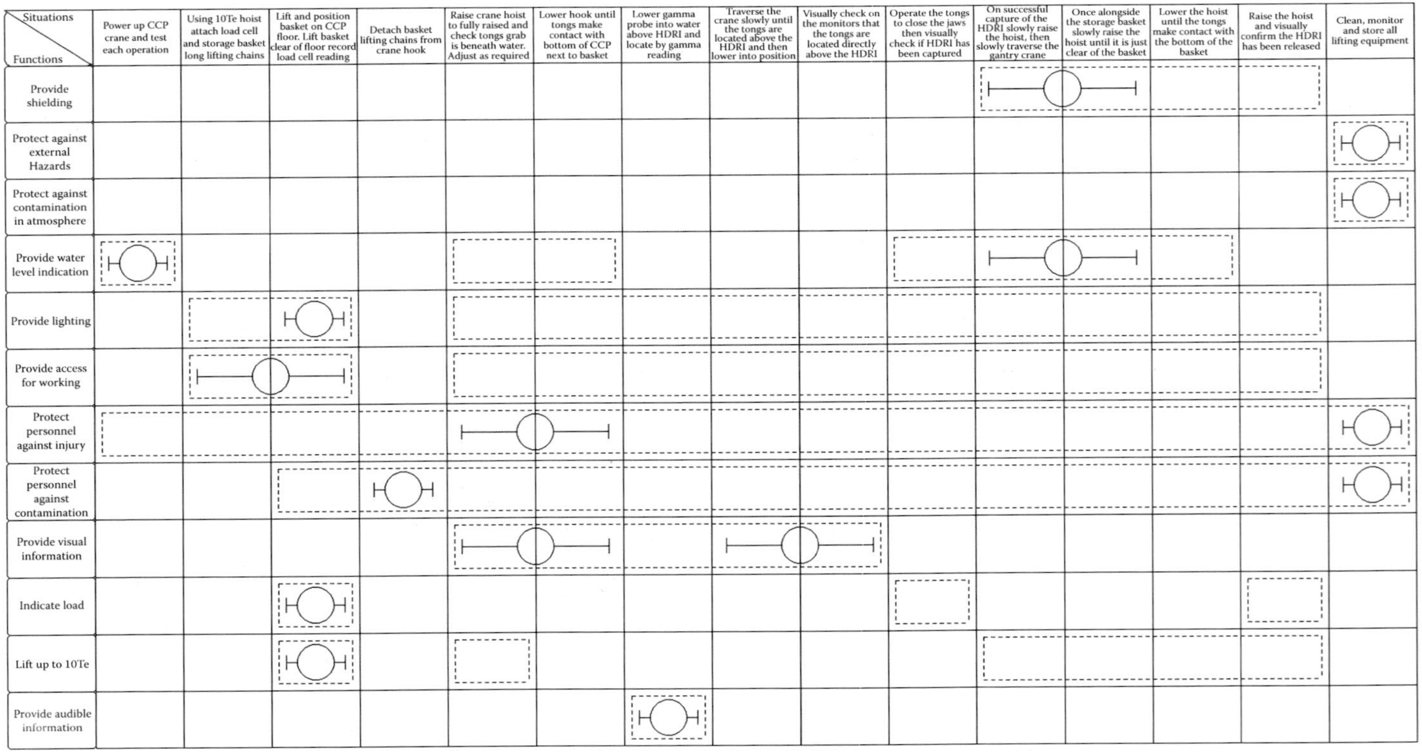

Situations / Functions	Power up CCP crane and test each operation	Using 10Te hoist attach load cell and storage basket long lifting chains	Lift and position basket on CCP floor. Lift basket clear of floor record load cell reading	Detach basket lifting chains from crane hook	Raise crane hoist to fully raised and check tongs grab is beneath water. Adjust as required	Lower hook until tongs make contact with bottom of CCP next to basket	Lower gamma probe into water above HDRI and locate by gamma reading	Traverse the crane slowly until the tongs are located above the HDRI and then lower into position	Visually check on the monitors that the tongs are located directly above the HDRI	Operate the tongs to close the jaws then visually check if HDRI has been captured	On successful capture of the HDRI slowly raise the hoist, then slowly traverse the gantry crane	Once alongside the storage basket slowly raise the hoist until it is just clear of the basket	Lower the hoist until the tongs make contact with the bottom of the basket	Raise the hoist and visually confirm the HDRI has been released	Clean, monitor and store all lifting equipment
Provide shielding															
Protect against external Hazards															
Protect against contamination in atmosphere															
Provide water level indication															
Provide lighting															
Provide access for working															
Protect personnel against injury															
Protect personnel against contamination															
Provide visual information															
Indicate load															
Lift up to 10Te															
Provide audible information															

FIGURE 17.3 Contextual activity template shows the purpose-related functions of the system on the y-axis (extracted from WDA) and the range of situational constraints these functions could occur in (x-axis). The dotted squares show the situations where the activity 'could' take place; the circle/bars show the situations where the function 'normally' takes place.

also at the initial testing stages of the process. This is particularly important because if the water level is not known prior to setting the lifting path, the incorrect length of lifting chains could be deployed resulting in potential risk to personnel from exposure to the HDRI.

- Provision of lighting and access for working typically occurs throughout crane movements, lifting and movement operations.
- Protection of personnel against injury and contamination typically occurs throughout most stages of the operations. Protection of personnel against injury occurs while the long lifting chain arrangement is tested. This is undertaken by raising the crane hook to the fully raised position while ensuring the tongs remain submerged underwater. This is important as over-raise of the HDRI would put operators at risk of injury through exposure to high-dose rates (just 5.25 seconds to reach a yearly maximum dose).
- Provision of visual information effectively occurs throughout hoist testing and HDRI retrieval to allow operators to ascertain that the long lifting chains are of suitable length to prevent over-raise and that the HDRI has been retrieved.
- The ability of the crane to lift up to 10Te and the provision of load indication are typically used during tong and hoist operation and during lifting and positioning of basket.
- The provision of audible information is typically used throughout the lifting and retrieval process. Operators, the health physicist and work supervisor communicate with one another to ensure operations are successful. Audible alarms are fitted to personal exposure rate monitors to warn operators of increasing dose rates.

The ConTA shown in Figure 17.3 demonstrates that the order in which activities are undertaken is important, moreover, there are a number of functions that influence all stages of the task. The contextual task analysis identified the importance of monitoring pond water level prior to undertaking any of the 'setting to work' steps detailed in the methodology. This was not identified during the course of the SWIFT. It was assumed that as pond water level is monitored on a daily basis this information would have been made available to the operators undertaking the HDRI retrieval task. However, it is vital information and should be considered as a documented step in the methodology as after many years of very little movement in the CCP, it may be easy to underestimate the importance of maintaining a suitable CCP water level.

Phase 3 – Strategies Analysis (StrA): StrA flowchart identified five different strategies which could be adopted in order to achieve the required end state of CCP decommissioning (Figure 17.4). Each strategy can be seen to follow the pattern of HDRI retrieval, with a period of short-term storage followed by long-term storage. The means by which these storage stages can be undertaken differ. A decision ladder was created for each strategy. It demonstrated that Strategy 1 required the most detailed planning and procedures. Strategies 2–5 relied on established procedures, thus simplifying the decisions required. Strategies 2–4 are potentially constrained

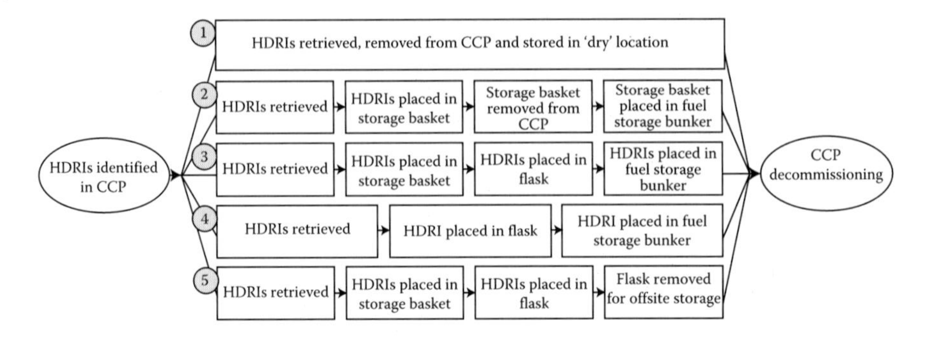

FIGURE 17.4 StrA flowchart.

by external factors such as availability of flasks for long-term storage and other project timescales. Given these it is likely that strategy 5 will be adopted as a means to allow progress toward CCP decommissioning.

This form of StrA is not covered by a SWIFT or similar hazard analysis techniques. It would be performed during an 'Optioneering' study which would assess various options to ensure appropriateness from a time and cost perspective. Using StrA is beneficial for understanding the larger objectives at an early stage and defining the constraints of each smaller component task. The StrA phase also helps to predict deviations from the planned process. If a simpler, more hazardous process is identified, control measures can be put in place to prevent their adoption during the operation, either as part of optimisation or in response to an unexpected event.

Phase 4 – Social, Organisational and Cooperation Analysis (SOCA): The SOCA is undertaken to show the way in which actors in the domain interact, communicate and cooperate. Both human and non-human actors were identified in the system as follows:

Human actors:

- Operators
- Accredited health physicists
- Work supervisors

Non-human actors:

- Physical pond structure
- Personnel protective equipment

The WDA, contextual activities template, decision ladder and strategies flowchart were annotated to demonstrate which stages of the system required actor interaction. This allowed the analyst to identify stages in the process where actors play key roles and, specifically, which actors are involved. The analysis showed the interacting nature of the operator, accredited health physicist and works supervisor roles. These individuals are required to work closely as a team. The SWIFT does not identify the stages in which key actors will be involved in the activity, rather, it is assumed that the existing team will be used and that they will perform in accordance with their normal role status. CWA enables these assumptions to be tested from first principles.

Phase 5 – Worker Competency Analysis (WCA): The final stage of the analysis looked at worker competencies in order to define the skills that each actor requires in order to complete the tasks required to reach a successful outcome.

Required skill-based behaviours can be summarised as

- Undertaking monitoring tasks
- Adhering to methodologies or method statements
- Understanding of importance of successful completion of the task

Required rule-based behaviours can be summarised as

- Using information as provided
- Adhering to process and procedures including methodology and document preparation processes
- Understanding working rules and adhering to them

Required knowledge-based behaviour can be summarised as

- Use previous experience to understand working environment and associated constraints, understand the way in which the operation may affect others and appreciate the importance of the operation as a means to achieve the overall goal of CCP decommissioning

WCA shows the actors are required to have knowledge of the operating procedures and working practices they must adhere to, combined with experience of conducting similar work within the confines of the CCP. The SWIFT recognised from the outset that the CCP Decommissioning Team would be used to undertake this activity, therefore, it was not necessary to identify team requirements as the team were classed as 'Suitably Qualified and Experienced'. The WCA went further by identifying the need for actors to have an understanding of the role their work plays in meeting the overall objective of the project.

17.4 DISCUSSION AND CONCLUSION

Both techniques identified the vital role that suitably qualified and experienced (SQEP) status would play in planning, organising and undertaking the retrieval operation. Both techniques also recognised that maintaining shielding was the most important factor during this operation as exposure to an unshielded HDRI for just 5.25 seconds would mean exceeding the legal annual allowable dose. The CWA method identified a range of more systemic factors to do with relating this individual task to wider objectives. The SWIFT method identified a range of physical hazards. CWA, however, was able to identify a further physical hazard not detected by SWIFT related to pond water level monitoring. As for the time and effort required to produce these outputs, the methods were broadly comparable (Table 17.6). SWIFT was undoubtedly easier and quicker, but required more people to be involved. CWA was more complex and lengthy, but required fewer people. In monetary terms the differences were not great.

We are not alone in noticing a gulf between the methods used by researchers and academia and those used by practitioners and industry. Practitioner methods rely on a common-sense deterministic logic, one that is easy to follow, well defined, explicit, all of which fits well with method end-users (who in this domain at least tend to be from engineering disciplines). They also have a long legacy of practical application, something that is very important in highly regulated industries. Analyst methods often have few of these attributes. Rather than logical and deterministic they appear complex and 'woolly', features that are at odds with end-user expectations and needs. Whilst they may have a long (sometimes longer) application track record in research domains they often appear 'unproven' in practical settings. Expressed in these terms the reason for a gulf between analyst and practitioner methods is understandable. But there is a problem. In some cases existing methods have difficulty adapting to new contexts, such as the case here, where a relatively stable operational phase gives way to a less predictable, more uncertain decommissioning phase. Answers to the challenges posed by different phases of a system life cycle are to be found in analyst methods but their benefits need to be demonstrated and the reason for their lack of uptake explored.

SWIFT is a normative process that considers problems, mitigations, actions and solutions, relying on the ability of experts to construct narratives about how the system works and the possibilities for hazards to arise. As with most analyses the quality of the output is dependent on the quality of the inputs. These two factors interact. Meeting quality objectives in the face of high uncertainty requires ever more detailed

TABLE 17.6
Approximate Time/Effort Comparison

	SWIFT	CWA
Personnel involved	7	3
Preparation time (e.g. compilation of data to form method inputs, recruitment of method participants, other preliminary/pre-analysis tasks, etc.)	4 hours[a]	4 hours
Data collection phase (e.g. gathering inputs for methods, running SWIFT workshop, etc.)	8 hours	8 hours
Analysis phase (e.g. process involved in producing results from data inputs		32 hours[b]
Total resource/effort requirements (personnel × [preparation + data collection] + analysis × basic consultancy rate)	£4200	£4760

[a] SWIFT preparation was undertaken by one analyst.

[b] In this case, one analyst performed the CWA based on data collected in tandem for the SWIFT exercise.

analyses, there is no stop rule, taken to its limit the effort required to undertake a fully exhaustive deterministic analysis on a complex (rather than simple) system rapidly becomes intractable. If the analysis stops too early then the possibility of failing to detect all credible hazard types increases, yet to be fully resilient all possible task deviations need to be considered. This is where certain analyst methods can be helpful. By focusing on the constraints in the system, CWA offers a mechanism for considering complex deviations to the standard operating procedures that may result in an unplanned emergency operation, but in a way that has an in-built stop rule. CWA is also a formative process that promotes abstract thinking and potentially more creative solutions. CWA requires more information, is more time consuming and is perhaps less well known. Despite this, it provides useful information and is flexible enough to be used alongside or within existing business practices. Taking the operation to retrieve the HDRIs as an example, Phase 1 of CWA could be used to establish a formative understanding of the work domain to be analysed. This would provide a picture with no assumptions or prejudice. Phase 2 could be used to determine the order that work should be performed and allow this to be changed and reordered, ensuring that focus is given to the correct situations. Phase 3 could be used alongside an optioneering study to demonstrate the possible strategies and the associated decision-making processes required to achieve the final outcome. Similarly, Phase 4 could be used in addition to appraisals or team design to ensure the required skills and awareness are recognised. Collectively, CWA provides a framework for exploring how users may respond in response to abnormal events, revealing critical areas of risk not easily available via existing approaches: this was certainly the case in the study reported here.

The reality, however, is that long-standing techniques are often used simply because it is 'just something we have to do', with little time being spent on considering the nature of the situation and what other techniques might be better suited. This is particularly true in highly regulated industries where oversight is based upon accepted and agreed methods that are clearly auditable. The potential risks inherent in this are significant. Using an inappropriate technique can result in a great deal of time and effort being wasted. A systems-based method such as CWA would be too resource intensive for many 'standard' risks, but so too could a deterministic method (like SWIFT) if applied to a problem with high degrees of complexity and emergence. The main risk, however, is failing to detect all credible hazards. The current study provides a powerful example of what is at stake. A 5.25 second exposure to the HDRI item would be all it takes for an individual to reach their annual yearly radiation uptake, a value they would almost never routinely receive. If the pond water level did fall unexpectedly then it would be extremely challenging for an organisation to subsequently discover that a more appropriate hazard identification method would have detected this in advance. The results, then, point strongly toward a contingent approach to method selection. Practically this involves a little more abstract thinking about the problem space and the degrees of complexity and emergence likely to be found therein. It may also require a larger repertoire of methods, going beyond the established norm, combined with greater awareness of their conceptual underpinnings (deterministic or systems-based, formative or normative). Ultimately, as the example used in this paper has shown, hazard identification methods are often complementary. In a highly regulated industry it is unlikely that a direct method replacement would gain universal acceptance (see Lundberg et al., 2012) thus, in the short term at least, the most practical option may be to introduce new methods alongside the accepted de-facto standards. Indeed, in this case the two worked in complimentary ways, with many of the analysis inputs required to 'drive' a SWIFT analysis also being useful in the production of a CWA.

Power station personnel in this study noted that CWA raised their awareness of the overall objectives of the project, and the roles that smaller projects play in achieving these objectives. This was seen as useful and novel. So is now a good time for formative methods like CWA to make a return to their domain of origin? Are the hazard identification methods used in the nuclear sector for 'operations' still appropriate for 'decommissioning'? The results of this study show that formative methods do have something to contribute but we will leave the final insight to those involved in the practical day-to-day realities of this site: 'I agree that CWA with its multiplicity could assist in these uncertain tasks, but the risk averse nature of our industry and relative lack of knowledge of CWA is not conclusive to its use on site. SWIFT is a clear, concise, well understood and a recognised methodology. CWA is less well known, rather abstract in its application, difficult to follow, not very certain in terms of its methodology and output. Perhaps it's these difficulties that need to be tackled in order to return CWA to use within the nuclear industry.' Or perhaps it is the further difficulties associated with how we all, practitioners and analysts alike, look upon the world of risk and the methods needed to tackle them.

REFERENCES

Ahlstrom, U. 2005. Work domain analysis for air traffic controller weather displays. *Journal of Safety Research*, 36, 159–169.

Bainbridge, L. 1993. Types of hierarchy imply types of model. *Ergonomics*, 36(11), 1399–1412.

Bisantz, A. M., Roth, E., Brickman, B., Gosbee, L. L., Hettinger, L. and McKinney, J. 2003. Integrating cognitive analyses in a large-scale system design process. *International Journal of Human-Computer Studies*, 58, 177–206.

BS EN 31010. 2010. *Risk Assessment Techniques*. London: British Standard.

Chalmers, D. J. 1990. *Thoughts on Emergence*. Available at: http://consc.net/notes/emergence.html

Clegg, C. W. 2000. Sociotechnical principles for system design. *Applied Ergonomics*, 31, 463–477.

Crone, D., Sanderson, P., Naikar, N. and Parker, S. 2007. Selecting sensitive measures of performance in complex multivariable environments. *Proceedings of the 2007 Simulation Technology Conference (SimTec 2007)*, Brisbane, Australia, 4–7.

Crone, D. J., Sanderson, P. M. and Naikar, N. 2003. Using Cognitive Work Analysis to develop a capability for the evaluation of future systems. *Proceedings of the 47th Annual Meeting Human Factors and Ergonomics Society*, Denver, CO, pp. 1938–1942.

DSCH. 2011. *S-259 Decommissioning Safety Case Handbook Issue 1 March 2011*. Internal Report.

Fidel, R. and Pejtersen, A. M. 2005. Cognitive work analysis. In K. E. Fisher, S. Erdelez, E. F. McKechnie (Eds.). *Theories of Information Behavior: A Researcher's Guide*. Medford, NJ: Information Today.

Gleick, J. 1987. *Chaos: Making a New Science*. London: Sphere.

Hajdukiewicz, J. R. 1998. *Development of a Structured Approach for Patient Monitoring in the Operating Room*. University of Toronto: Masters Book.

Halley, J. D. and Winkler, D. A. 2008. Classification of emergence and its relation to self-organization. *Complexity*, 13(5), 10–15.

Howie, D. E. and Vicente, K. J. 1998. Measures of operator performance in complex, dynamic microworlds: Advancing the state of the art. *Ergonomics*, 41(4), 485–500.

Jenkins, D. P., Stanton, N. A., Salmon, P. M. and Walker, G. H. 2009. *Cognitive Work Analysis: Coping With Complexity*. Farnham: Ashgate.

Lee, J. D. 2001. Emerging challenges in cognitive ergonomics: Managing swarms of self-organising agent-based automation. *Theoretical Issues in Ergonomics Science*, 2(3), 238–250.

Lindroos, O. 2009. Relationships between observed and perceived deviations from normative work procedures. *Ergonomics* 52(12): 1487–1500.

Lintern, G. and Naikar, N. 2000. The use of work domain analysis for the design of training teams. *Proceedings of the joint 14th Triennial Congress of the International Ergonomics Association/44th Annual Meeting of the Human Factors and Ergonomics Society (HFES/IEA 2000)*. San Diego, CA.

Lundberg, J., Rollenhagen, C., Hollnagel, E. and Rankin, A. 2012. Strategies for dealing with resistance to recommendations from accident investigations. *Accident Analysis and Prevention*, 45, 455–467.

Marmaras, N., Lioukas, S. and Laios, L. 1992. Identifying competences for the design of systems supporting complex decision-making tasks: A managerial planning application. *Ergonomics*, 35(10), 1221–1241.

McLeod, R. W., Walker, G. H. and Moray, N. 2005. Analysing and modelling train driver performance. *Applied Ergonomics*, 36, 671–680.

Naikar, N. 2006a. Beyond interface design: Further applications of cognitive work analysis. *International Journal of Industrial Ergonomics*, 36, 423–438.

Naikar, N. 2006b. An examination of the key concepts of the five phases of cognitive work analysis with examples from a familiar system. *Proceedings of the Human Factors and Ergonomics Society 50th Annual Meeting*. Santa Monica, CA. pp. 447–451.

Naikar, N., Pearce, B., Drumm, D. and Sanderson, P. M. 2003. Technique for designing teams for first-of-a-kind complex systems with cognitive work analysis: Case study. *Human Factors*, 45(2), 202–217.

Naikar, N. and Sanderson, P. M. 1999. Work domain analysis for training-system definition. *International Journal of Aviation Psychology*, 9, 271–290.

Naikar, N. and Sanderson, P. M. 2001. Evaluating design proposals for complex systems with work domain analysis. *Human Factors*, 43, 529–542.

Naikar, N. and Saunders, A. 2003. Crossing the boundaries of safe operation: A technical training approach to error management. *Cognition Technology and Work*, 5, 171–180.

Perrow, C. 1999. *Normal Accidents: Living with High-Risk Technologies*. New Jersey: Princeton University Press.

Rasmussen, J. 1983. Skills, rules, knowledge; signals, signs, and symbols, and other distinctions in human performance models. *IEEE Transactions on Systems, Man and Cybernetics*, 13, 257–266.

Reason, J. 1990. *Human Error*. Cambridge: Cambridge University Press.

Reason, J. 2008. *The Human Contribution: Unsafe Acts, Accidents and Heroic Recoveries*. Farnham: Ashgate

Sinclair, M. A. 2007. Ergonomics issues in future systems. *Ergonomics*, 50(12), 1957–1986.

Underwood, P. and Waterson, P. 2013. Systemic accident analysis: Examining the gap between research and practice. *Accident Analysis and Prevention*, 55, 154–164.

Vicente, K. J. 1999. *Cognitive Work Analysis: Toward Safe, Productive, and Healthy Computer-Based Work*. Mahwah, NJ: Lawrence Erlbaum Associates.

Walker, G. H., Stanton, N. A., Salmon, P. M., Jenkins, D. P. and Rafferty, L. 2010. Translating concepts of complexity to the field of ergonomics. *Ergonomics*, 53(10), 1175–1186.

Woods, D. D. 1988. Coping with complexity: The psychology of human behaviour in complex systems. In L. P. Goodstein, H. B. Andersen and S. E. Olsen (Eds.). *Tasks, Errors and Mental Models*. London: Taylor & Francis, pp. 128–148.

Woods, D. D. and Dekker, S. 2000. Anticipating the effects of technological change: A new era of dynamics for human factors. *Theoretical Issues in Ergonomics Science*, 1(3), 272–282.

Yu, X., Lau, E., Vicente, K. J. and Carter, M. W. 2002. Toward theory-driven, quantitative performance measurement in ergonomics science: The abstraction hierarchy as a framework for data analysis. *Theoretical Issues in Ergonomic Science*, 3(2), 124–142.

18 Deriving and Analysing Social Networks from SOCA-CAT Diagrams

Chris Baber, Neville A. Stanton and Robert Houghton

CONTENTS

18.1 INTRODUCTION

Social organisation cooperation analysis (SOCA)-contextual activity template (CAT) (Chapter 1) implies potential communications between individuals in a work setting. As functions are allocated to actors across situations, so the implication is that these actors might need to share information in pursuit of these functions or to agree which actor should perform a function (when there are multiple actors highlighted). We should begin this chapter by pointing out that the implied communications is neither an intended outcome of SOCA-CAT nor a motivation for constructing such a diagram. Indeed, if one wished to explore communications in a work setting, then it would make more sense to conduct observations with the explicit aim of recording who speaks to whom. However, the purpose of this chapter is to explore the potential for extracting social network analysis (SNA)-like analysis from SOCA-CAT. The reason why this might be useful is that it can provide a means of analysing the manner in which the actors in a work setting could interact and communicate.

CWA has used information flow maps (see Chapter 1) to address the manner in which information is employed in the performance of different strategies (Rasmussen et al., 1994; Vicente, 1999). Such an approach highlights the use of information across the system but does not directly indicate who communicates with whom. Previous research has used work domain analysis to explore collaboration and team work in, for example, the generation of maritime tactical pictures (Burns et al., 2009), military planning (Jenkins et al., 2008), teams designing microsystems technology (Durugbo, 2012) or teams in healthcare (Ashoori and Burns, 2013; Ashoori et al., 2014). While this can capture the nature of the work domain and the functions performed, it does not offer an analysis of the manner in which communication (i.e. in the form of person A communicating with person B, etc.) is undertaken. This chapter explores ways in which a description of communications amongst actors in a work setting could be quantified and analysed in order to support analysis of the communications networks which might occur in the work setting.

18.2 BACKGROUND AND CONTEXT

The potential relationship between CWA and SNA was raised by Pfautz and Pfautz (2008) in their exploration of social aspects of the work domain. Of particular interest is their discussion of what level of analysis to apply to the 'entity' which is communicating and to the 'relationship' between entities. They summarise the challenges facing the analysis of social and organisational aspects of the work domain as a series of questions (Table 18.1).

Table 18.1 highlights some of the wide range of issues that can arise when one considers the broader social and organisational environment in which work is performed. As Pfautz and Pfautz (2008) point out, many of these issues need to be dealt with using methods and approaches from outside the CWA domain. We are not disputing this or proposing that CWA can offer a one-size-fits-all solution to the exploration of social and organisational issues. In this chapter, our focus is primarily with the issues that Table 18.1 labels 'Communication or information flow relationships'. The question is, can such relationships be derived from SOCA-CAT diagrams, and how can this support analysis?

18.3 AIMS AND RESEARCH QUESTIONS

The primary question of interest is what additional benefit is gained from creating SNA diagrams on the basis of SOCA-CAT diagrams? If we assume that there is an association between the possibility of performing a function and the need to communicate with other actors, then it is possible to redraw SOCA-CAT as a social

TABLE 18.1
Questions about Social and Organisational Aspects of the Work Domain

Type of Relationship	Example Question
Communication or information flow relationships	Who communicates with whom in the work domain?
	How often do teamwork and/or taskwork communications occur?
	How is communication integrated or related to other events and activities?
	How does the content of the message influence its communication?
	What communication media are used?
Organisational or coordination/consensus	Who coordinates with whom in the work domain?
	Who participates in the work?
	Who depends on whom to support their work?
	Who is responsible for a task being completed?
	What is the structure of the group responsible for a task?
	What are the perceived organisational risks/consequences associated with the work?
	What specific information is communicated as part of trajectories through abstraction–decomposition space?
	How does a group's structure impact intragroup communication?
	Who distributes tasks?
	Who coordinates activities of a group?
	Who works to build consensus among entities?
	What structure has emerged to cope with the work?
Ecological relationships	What are the affective, familial, political, or cultural relationships likely to impact work?
	Which of these relationship types impair, enhance, or otherwise transform processes for communication, coordination, cooperation and/or consensus?
	What organisational structures are likely to influence these relationship types?
	Who share similar beliefs and attitudes about work and about the organisational structures needed to accomplish particular tasks?
	What personal affinities are present between workers in the domain?
	What are likely sources of heterogeneity in perspectives about work and organisational structure?
	What are the physical relationships that impact work performance?
	Which group members are distributed versus collocated?
	In what time zones are the collaborating entities?

Source: Adapted from Pfautz, S. L. and Pfautz, J. 2008. *Proceedings of the 52nd Annual Meeting of the Human Factors and Ergonomics Society Annual Meeting*. Santa Monica, CA: HFES, pp. 443–447.

network diagram. Translating between the terminology of Pfautz and Pfautz (2008) and that of SNA, we wish to extract entities as nodes and relationships as edges. In the first instance, we make the simple assumption that *all* actors associated with a given function will form a clique, that is, they will all communicate with each other for that function. We will discuss later some of the weaknesses of this assumption but the benefit for this analysis is that it would produce a fully-connected network that reflects all possible connections, given the content of the CAT.

18.4 INITIAL EXAMPLE

Assume that a team consists of five individuals and performs three functions {receive message, check data, send reply}. The team operates in four different situations {A, B, C, D}. While the SOCA-CAT provides an indication of who can do what in each situation (and thus is valuable to making sense of allocation of function and workload problems), it does not give a clear indication of what communications might be involved in this team.

18.4.1 Defining Metrics

SNA often takes the form of a link diagram, in which each actor is regarded as a node with edges (links) to other nodes. Such a diagram immediately suggests two ways in which the structure of the network can be defined. The first is in terms of the number of edges to and from a node. The second is in terms of the number of nodes between a given pair of nodes. Variations on these two measures define many of the metrics applied to SNA and represent different ways of defining the centrality of a node in the network. Thus, the first way of defining centrality (in terms of number of edges for a node) is degree centrality. This can be defined as a measure of all edges or can focus on the number in edges going in to a node (indegree) or edges coming out of a node (outdegree). This can be easily calculated by hand and will be used for the examples in this chapter. The second way of centrality relates to defining the shortest path between any pair of nodes. Readers interested in exploring this and related metrics are advised to consult Wasserman and Faust (1994).

Degree centrality is calculated simply as the total number of edges from a given node divided by the number of possible edges; each actor can communicate with N – 1 other actors.

In addition to considering metrics for individual nodes, it is possible to define global measures of network structure. For example, density is calculated by dividing the number of observed edges in the network by the total number possible, that is, N*(N – 1) = 12.

18.4.2 Communications × Situation

One could assume that actors in one situation (column) might need to communicate with each other. This could arise because the workflow, from one function to the next, involves information exchange. This would define a network in which actors are linked when they appear in the same column. From Figure 18.1, if one looks at the first situation ('A') all four actors are indicated. For the second situation ('B') only actors 1 and 2 are indicated. Assuming that each instance of actors being indicated in the same column implies a two-way link between the actors, one can produce the matrix shown in Table 18.2.

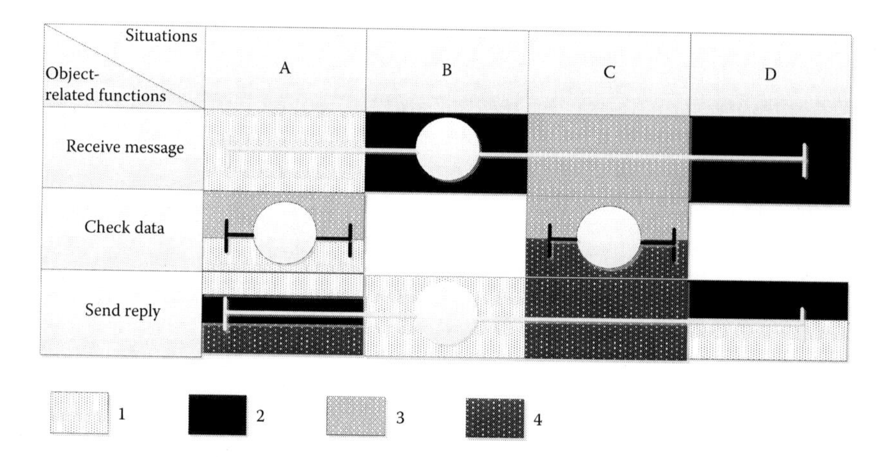

FIGURE 18.1 Example of SOCA-CAT for a four actor network.

Applying the equations for degree centrality and density to Table 18.2, we can produce the data shown in Table 18.3. This shows, for instance, that Situation A has a much higher density than the other situations as all actors are potentially involved in this network. This can be seen from Figure 18.1. Before discussing why the additional effort of performing these calculations might be useful, we consider an alternative way of reading Figure 18.1.

18.4.3 Communications × Function

One could assume that actors performing one function might need to communicate with each other – either to ensure cooperation during their performance or to reduce the possibility of duplication of effort.

Applying the equations for density and centrality to Table 18.4, we can produce Table 18.5. Here function 3 has maximal density and centrality for all actors.

18.4.4 Conclusions

The interpretation of the data in Tables 3 and 5 will depend on the questions one wishes to ask. If one considers the density metrics, it is possible to distinguish

TABLE 18.2
Links between Actors across Situations

Situation A					Situation B					Situation C					Situation D				
	1	*2*	*3*	*4*		*1*	*2*	*3*	*4*		*1*	*2*	*3*	*4*		*1*	*2*	*3*	*4*
1	–	0	1	1	*1*	–	1	0	0	*1*	–	0	0	0	*1*	–	1	0	0
2	0	–	1	1	*2*	1	–	0	0	*2*	0	–	0	0	*2*	1	–	0	0
3	1	1	–	1	*3*	0	0	–	0	*3*	0	0	–	1	*3*	0	0	–	0
4	1	1	1	–	*4*	0	0	0	–	*4*	0	0	1	–	*4*	0	0	0	–

TABLE 18.3
Centrality Statistics for Each Actor and Network Density for Each Situation in Table 18.2

Node	Situation A	Situation B	Situation C	Situation D
1	2/3 = 0.6	1/3 = 0.3	0	1/3 = 0.3
2	0	1/3 = 0.3	0	1/3 = 0.3
3	2/3 = 0.6	0	1/3 = 0.3	0
4	2/3 = 0.6	0	1/3 = 0.3	0
Density	10/12 = 0.83	2/12 = 0.17	2/12 = 0.17	2/12 = 0.17

between cliques (i.e. fully connected networks) in which all actors are connected to each other, from other network structures. Thus, function 3 constitutes a clique in this network (Table 18.5). This means that all actors could communicate with all other actors in this function. Given that function 3 is 'send reply', one might feel that this distribution of function and the implied high-level of teamwork might be inappropriate and so this might be a focus for redesign activity. Alternatively, as we have made no assumptions as to *how* the communication might be performed, function 3 might be supported by a shared display that any actor can view but which need not require the direct exchange of information.

The density metric can also be used to indicate how much spare capacity the team might have, for example, to handle unexpected communications or tasks. If all team members are engaged in communication, this might impact on workload. Thus, function 3 has a density of 1 and all actors are (potentially) engaged in this function, whereas functions 1 and 2 have lower density values, suggesting some capacity. Equally, situation A has high density, whereas situations B, C and D have much lower densities.

Of course, one could argue that the overhead of performing these calculations does not have much of a pay-off in terms of interpreting Figure 18.1. After all, these points about whether or not all actors are involved in a given situation or function can be readily seen in the figure (and this illustration of the allocation of functions is one of the benefits of the SOCA-CAT diagram). We would argue that

TABLE 18.4
Links between Actors across Functions

Function 1					Function 2					Function 3				
	1	2	3	4		1	2	3	4		1	2	3	4
1	–	1	1	0	1	–	0	1	1	1	–	1	1	1
2	1	–	1	0	2	0	–	0	0	2	1	–	1	1
3	1	1	–	0	3	1	0	–	1	3	1	1	–	1
4	0	0	0	–	4	1	0	1	–	4	1	1	1	–

TABLE 18.5
Centrality Statistics for Each Actor and Network Density for Each Situation in Table 18.4

Node	Function 1	Function 2	Function 3
1	2/3 = 0.6	2/3 = 0. 6	3/3 = 1
2	2/3 = 0.6	0	3/3 = 1
3	2/3 = 0.6	2/3 = 0. 6	3/3 = 1
4	0	2/3 = 0.6	3/3 = 1
Density	6/12 = 0.5	6/12 = 0.5	12/12 = 1

the advantage of the quantification of these links comes from the relative comparison of actors in functions or situations. For example, if one calculates average (mean) centrality for the four actors across situations or functions, one can produce Table 18.6.

From Table 18.6, one can appreciate some differences in the centrality of the actors. While actor 1 has highest centrality in both situations and functions, actor 3 has high centrality in the functions but not situations. Inspection of Figure 18.1 suggests that this could simply reflect the fact that actor 1 is involved in all three functions, while actor 3 is involved in two of them, and actor 1 is involved in all four situations, while actor 3 is involved in two of these. This seems obvious but consider actor 2, who is involved in two situations but has a lower average centrality than actors 3 and 4. The point at issue is whether the calculations are leading to a different way of reading Figure 18.1 and suggesting some differences in the roles and contributions that the actors could make to the network (over and above their performance of specific functions in specific situations).

18.5 CASE STUDY DESCRIPTION

Identifying members of criminal or covert networks is notoriously difficult. Not only do members of such networks seek to keep their membership secret but also any links between members can be carefully managed so as to protect the identity of certain individuals (Carley et al., 2003). Management could, for instance, be through

TABLE 18.6
Average Centrality of Actors

Actor	Situations	Functions
1	0.3	0.7
2	0.15	0.5
3	0.23	0.7
4	0.23	0.5

only linking to one other member of the network or through linking to the network through intermediaries. This can mean that it can be difficult for an outside observer to prove whether a given individual is a member of the network. Furthermore, such networks do not exist in vacuums but in the normal social interactions of individuals which, in turn, can obscure connections to these networks. As part of a project funded by UK Ministry of Defence Competition of Ideas, we explored ways in which covert networks could be modelled. In part this involved the use of genetic algorithms to 'crew' networks (Houghton et al., 2006). For example, if one has an idea of the type of roles and functions that a given mission requires, and the range of abilities required to perform these functions, then it is possible to map agents to roles on the basis of their abilities and whether they are available.

Rather than consider operational data, this section draws on an account of drug smuggling activity (Marks, 1997). The primary focus of this account was on the export and import rings which worked together to manage the shipment of cannabis (see Table 18.7).

The export ring has Daniels as the main hub. Daniels dealt with many other import rings, but had only one other rival export ring to compete with. Daniels put the main people together. He negotiated with farmers and landowners for the price of cannabis, and then bought off the military, police and narcotics control. He also put investors and financiers together to fund large shipments. His deals could be in the order of 20 tons of 'Thai Sticks' at a time. He arranged for all of the cannabis to be brought to a central warehouse where it was compacted and vacuum packed to preserve it. Daniels also had a system organised to prevent searching of his crates containing cannabis at the port. He had a blue crest stamped 'passed' onto a tag. All crates with this tag were not inspected by the customs men, in return for payment.

The import ring was the 'Coronado Company' in Table 18.7. This is one of many companies that Daniels set up. The four main players in the 'Coronado Company' were Villar (who handled procurement and negotiations), Weber (who was the pilot and mechanic of the equipment used to move the shipments), Otero (who was the beach-master and organised the landing and offloading of cannabis) and Acree (who was in charge of sales and distribution). They employed a captain of the ship (Chris) for bringing cannabis into the country, and an offload crew to get the cannabis off the ship to a safe house. Given the tonnage of cannabis being imported, their preference was to wholesale the shipment for distribution. The size of the shipments being imported required co-ordination to avoid detection and manage the offloading to the shore. The 'Coronado Company' had a communications house that had a high power antenna and a ship-to-shore radio, so that movement of the coast guard could be reported and the drop off could be arranged. There was a significant amount of equipment stored in a an equipment house, comprising 4×4 vehicles, boats, rafts, generators, conveyor belts and winches, all the equipment required to get the cannabis off the ship and to the safe stash house. The offload crew were housed in a beach house, in preparation for the unloading work. The 'Coronado Company' also acted as a mechanism for laundering the large amount of cash that came into the hands of the smugglers, to make the operation look legitimate.

TABLE 18.7
Actors and Functions

Actor	Function
Farmers	Grows marijuana using traditional farming techniques
Landowners	Own the land on which the farmer grows marijuana, may also grow marijuana as well
Military	Mostly army in Thailand, with key figures in pay of Daniel, may be used to protect transport of marijuana to dockside – may even do the transport work themselves for reward
Police	Key figures in pay of Daniels, may provide intelligence of impending operations and steer operations away from Daniels
Narcotics officers	Key figures in pay of Daniels, may provide intelligence of impending operations and steer operations away from Daniels – also do not search cargo with 'Passed inspection' labels on them in return for payment
Investors	People putting up money to finance shipment in return for cut of the profits
Financiers	People lending money to the investors
Daniels	Key link between the export ring and import ring. He specialises in putting all the people together and stays in the background of the operation, but takes large profits for himself
Chris (ship's captain)	Experienced sailor who is hired on an occasional basis for a large fee
Villar (procurement) Coronado Company	One of four key players in the import ring. He is a skilled negotiator and has multiple languages. Understands about the quality of marijuana
Weber (mechanic) Coronado Company	One of four key players in the import ring. Skilled mechanic, looks after all the equipment and keeps it in good working condition
Otero (beachmaster) Coronado Company	One of four key players in the import ring. In charge of offloading the marijuana from the ship and getting it to the stash house
Acree (sales) Coronado Company	One of four key players in the import ring. Has a wide sales and distribution network
Del Mar (distribution)	One of the distribution networks
Offload crew	The team that work under Otero
Truck drivers	Drivers of trucks
Guards	People used to guard the marijuana at various points on its journey

18.5.1 Presentation of CWA Outputs

Analysis of the activity of this set of actors (through the development of an abstraction hierarchy to describe the mission in more detail) resulted in a set of object-related functions which could be allocated to the different roles and then related to specific situations in SOCA-CAT (Figure 18.2).

18.5.2 Analysis

The SOCA-CAT diagram (Figure 18.2) shows the actors which *could* perform particular functions in the different situations (Table 18.8).

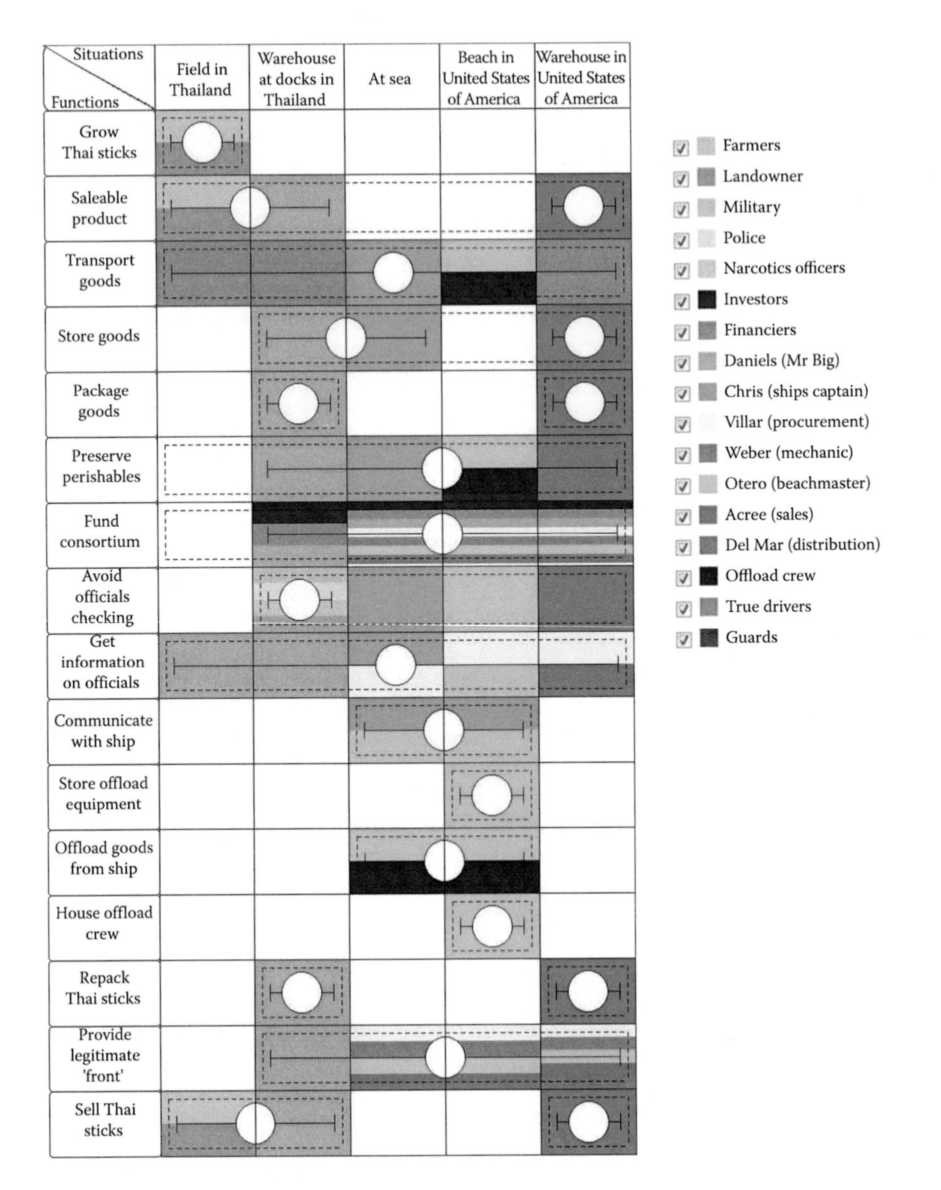

FIGURE 18.2 Contextual activity template.

18.5.3 Findings

From Table 18.9, Daniels has the highest centrality. This is not surprising as he is the ring-leader for this network. However, one might expect him to disguise his membership of the network (unless, of course, he was so confident that his ability to bribe officials would render him immune from prosecution and lead him to make no secret of his role in the network). If we remove Daniels from our calculations, then the

TABLE 18.8
Nodes and Edges Derived from Figure 18.2

	Fm	Lo	M	P	NO	I	F	D	C	V	W	O	A	DM	Oc	Td	Edges	Centrality
Farmers (Fm)	0	1	0	0	0	0	0	1	0	0	0	0	1	0	0	0	3	0.20
Landowners (Lo)	1	0	0	0	0	0	0	1	0	0	0	0	1	0	0	0	3	0.20
Militart (M)	0	0	0	1	1	0	0	1	1	0	0	1	1	0	0	0	6	0.40
Police (P)	0	0	1	0	1	0	0	1	1	0	0	1	1	0	0	0	6	0.40
Narcotics officers (NO)	0	0	1	1	0	0	0	1	1	0	0	1	1	0	0	0	6	0.40
Investors (I)	0	0	0	0	0	0	0	1	0	1	0	1	1	1	0	0	5	0.33
Financiers (F)	0	0	0	0	0	0	0	1	0	1	0	1	1	1	0	0	5	0.33
Daniels (D)	1	1	1	1	1	1	1	0	1	1	1	1	1	1	0	0	13	0.87
Chris (C)	0	0	1	1	1	0	0	1	0	0	1	0	1	0	0	1	7	0.47
Villar (V)	0	0	0	0	0	1	1	1	0	0	0	1	0	1	0	0	5	0.33
Weber (W)	0	0	0	0	0	0	0	0	0	0	0	0	0	0	0	0	0	0.00
Otero (O)	0	0	1	1	1	1	1	1	0	1	0	0	0	1	1	1	10	0.67
Acree (A)	1	1	1	1	1	1	1	1	1	0	1	0	0	1	0	0	11	0.73
Del Mar (DM)	0	0	0	0	0	1	1	1	0	1	1	1	1	0	0	0	7	0.47
Offload crew (Oc)	0	0	0	0	0	0	0	0	0	0	0	1	0	0	0	1	2	0.13
Truck drivers (Td)	0	0	0	0	0	0	0	0	1	0	0	1	0	0	1	0	3	0.20

TABLE 18.9
Ranked by Centrality Metrics

	Centrality
Daniels	0.87
Acree	0.73
Otero	0.67
Chris	0.47
Del Mar	0.47
Military	0.40
Police	0.40
Narcotics officers	0.40
Investors	0.33
Financers	0.33
Villar	0.33
Farmers	0.20
Landowners	0.20
Truck drivers	0.20
Offload crew	0.13
Weber	0.00

network itself remains intact, with Acree and Otero being the nodes with the highest centrality. By systematically removing the nodes with the highest centrality, we can plot the impact on the network, in terms of the number of connections (edges). This gives an impression of the resilience of the network in the face of loss of highly connected nodes (Figure 18.3).

Drawing the social network diagrams gives a different perspective on the structure of the network. Comparing the original network (i.e. with all nodes) with the last network which has connections across all actors shows how the node 'Chris' keeps

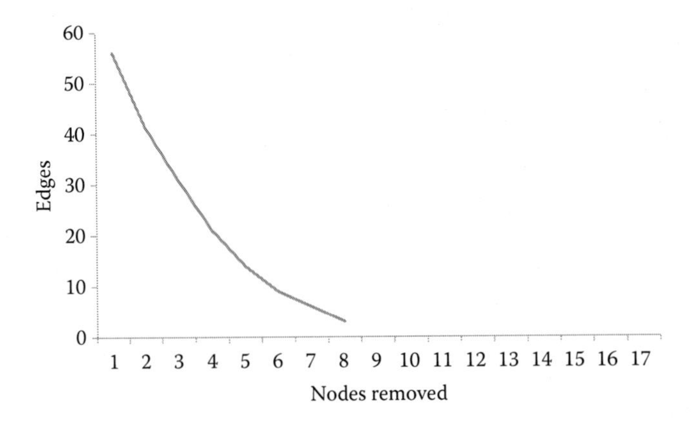

FIGURE 18.3 Plotting number of edges against nodes removed.

the network together. As soon as this node is removed, the network fragments into separate clusters (with the nodes 'Offload crew' and 'Truck drivers' separating from the rest of the network first).

18.6 DISCUSSION AND CONCLUSIONS

This chapter has illustrated how SOCA-CAT can be used to create SNA matrices and diagrams.

18.6.1 Practical Implications

The obvious question that this chapter raises is whether there is a benefit to be gained from recasting SOCA-CAT as a SNA description. The simple example presented at the start of the chapter might suggest that the effort involved in creating matrices and diagrams, and calculating centrality, density or other SNA metrics, might be relatively high for the insights offered. After all, reading the SOCA-CAT diagram in terms of communications and connections is not too onerous. Having said this, simply proposing that SOCA-CAT is a blueprint for communications raises some interesting questions on how functions are allocated. On the other hand, the case study example illustrates (we feel) how the application of simple SNA metrics can develop a line of enquiry into how a network might operate or how it might be resilient to threat or change.

18.6.2 Methodological Implications

We wanted to present a lightweight analysis of a work domain as a social network and to show how this could be derived directly from SOCA-CAT. Obviously if the aim of a project was to study in depth how communications were managed, then one ought to approach the work entirely from the perspective of building and analysing social networks. In this case, the use of SOCA-CAT (and CWA) might not be most appropriate.

18.6.3 Comparison with the Existing Literature

As we noted in the introduction to this chapter, developing CWA to explore team structures has been proposed by several authors. In terms of the link between CWA and SNA, there has been less work (cf. Euerby and Burns, 2012, 2013).

18.6.4 Recommendations

Recasting SOCA-CAT into a form of SNA, as we have done in this chapter, could offer a different way of the exploring of work domain. This might be particularly germane to discussions of workload, for example, where the allocation of function in SOCA-CAT can give an impression of how much task-work each actor has to perform, but creating the SNA matrices and analysing them could supplement this with a view of the amount of team-work that is added to the activity.

18.6.5 Future Research Areas and Study Limitations

Obvious limitations of the approach relate to the assumptions required to interpret SOCA-CAT in terms of links between actors. Just because more than one actor could perform a function, it does not follow that they have to engage in any communication (as opposed to simply carrying out the function). Thus, the basis for constructing the social network diagrams in the first place might need to be considered before embarking on such analysis. If one feels that it is appropriate to make the assumptions relating communication to allocation of function, then it is not obvious that all connections carry equal weight. In other words, some actors might have a higher probability of communicating than others, for example, because they are in the same room. To some extent, this could be influenced by (and could influence) the manner in which actors are assigned to 'situations' and a benefit of the approach could simply be to raise questions in the mind of the analyst as to how likely actors would be to speak to each other or otherwise share information.

In terms of future development, we offer the approach as an adjunct to SOCA-CAT and as an initial way of exploring work domains as social structures. If it appears that the communications and networking within a work domain might appear interesting, then we would suggest further study explicitly utilising SNA tools and metrics should be performed.

18.6.6 Conclusions

This chapter has illustrated how SOCA-CAT can be interpreted in terms of social network metrics. The intention is to show how a re-reading of SOCA-CAT is possible and how this provides insights into the manner in which a work domain might operate.

ACKNOWLEDGEMENTS

The case study reported in this chapter was supported by a grant from the UK MoD Competition of Ideas B1642.

REFERENCES

Ashoori, M. and Burns, C. M. 2013. Team cognitive work analysis: Structure and tasks. *Journal of Cognitive Engineering and Decision Making*, 7, 123–140.

Ashoori, M. Burns, C. M., Momtahan, K. and d'Entremont, B. 2014. Using team cognitive work analysis to reveal healthcare team interactions in a birthing unit, *Ergonomics*, 57, 973–986.

Burns, C., Torenvliet, G., Chalmers, B. and Scott, S. 2009. Work domain analysis for establishing collaborative work requirements. *Proceedings of the 53rd Annual Meeting of the Human Factors and Ergonomics Society*. Santa Monica, CA: HFES, pp. 314–318.

Carley, K. M., Dombroski, M., Tsvetovat, M., Remingan, J. and Kamnenva, N. 2003. Destabilizing dynamic covert networks, *8th International Command and Control Research and Technology Symposium*. Washington, DC: Department of Defense Command and Control Research Program.

Durugbo, C. 2012. Work domain analysis for enhancing collaborations: A study of the management of microsystems design. *Ergonomics*, 55(6), 603–620.

Euerby, A. and Burns, C. M. 2012. Designing for social engagement in online social networks using communities of practice theory and cognitive work analysis: A case study. *Journal of Cognitive Engineering and Decision Making*, 6, 194–213.

Euerby, A. and Burns, C. M. 2013. Improving social connection through a communities of practice inspired cognitive work analysis approach. *Human Factors*, 56, 361–383.

Houghton, R. J., Baber, C., McMaster, R., Stanton, N. A., Salmon, P., Stewart, R. and Walker, G. 2006. Command and control in emergency services operations: A social network analysis, *Ergonomics*, 49, 1204–1225.

Jenkins, D. P., Stanton, N. A., Salmon, P. M., Walker, G. H. and Young, M. S. 2008. Using cognitive work analysis to explore activity allocation within military domains. *Ergonomics*, 51, 798–815.

Marks, H. 1998. *Mr. Nice: An Autobiography.* London: Vintage.

Pfautz, S. L. and Pfautz, J. 2008. Understanding social and organizational aspects of the work domain using techniques and technologies from intelligence analysis. *Proceedings of the 52nd Annual Meeting of the Human Factors and Ergonomics Society Annual Meeting.* Santa Monica, CA: HFES, pp. 443–447.

Rasmussen, J., Petersen, A. M. and Goodstein, L. P. 1994. *Cognitive Systems Engineering.* New York, NY: Wiley.

Vicente, K. J. 1999. *Cognitive Work Analysis: Towards Safe, Productive, and Healthy Computer-Based Work.* Mahwah, NJ: Lawrence Erlbaum Associates.

Wasserman, S. and Faust, K. 1994. *Social Network Analysis: Methods and Applications.* Cambridge, UK: Cambridge University Press.

19 Using CWA to Understand and Enhance Infrastructure Resilience

Guy H. Walker, Lindsay Beevers and Ailsa Strathie

CONTENTS

19.1 INTRODUCTION

A nation's critical infrastructure needs to be able to withstand disturbances as they happen, and bounce back afterward. Most nations have the equivalent of a National Risk Register (e.g. Cabinet Office, 2013). In the United Kingdom it presents a range of civil emergencies with a greater than 1 in 20 chance of occurring in the next 5 years, and with the potential to yield impacts ranging from social disruption and economic harm through to widespread illness and fatalities. Flooding is a prime example (p. 10). The ability of daily life to continue in the face of disturbances like this does not depend on a single engineering solution, rather, on the ability of organisations, infrastructures and individuals to anticipate the changing shape of risk before failures and harm occur, then to respond in effective ways when it does. This chapter describes how the latest research on flood vulnerability was put in touch with CWA, specifically the first phase (work domain analysis/abstraction hierarchy [AH]), enabling this wider view to be captured explicitly. Several real towns were modelled and subject to a simulated 1 in 200 year flood event. The method shows how critical functions and processes at higher levels of system abstraction are progressively degraded as individual 'physical objects', and at low levels of abstraction, are knocked out. In addition, network metrics are extracted from the AH to enable each town to be characterised in terms of its vulnerability and positioned in a universal 'vulnerability space'. Solutions for improving resilience vary depending on what region of the space is occupied, and the method can be deployed to determine this for any town in any region of the world.

19.1.1 Defining Vulnerability

The concept of vulnerability is often used within natural hazard, disaster and environmental change research. Many authors have discussed and attempted to define this concept (e.g. Lewis, 1999; van der Veen and Logtmeijer, 2005; Adger, 2006). The IPCC definition (McCarthy, 2001) states that vulnerability is the degree to which a system is susceptible to, and unable to cope with, adverse climate change effects. A comprehensive discussion on vulnerability can be found in Balica et al. (2009) and Balica and Wright (2010), and its importance is difficult to overstate: according to some authors 'vulnerability is the root cause of disasters' (Lewis, 1999). This paper focuses on the issue of vulnerability to the natural hazard of flooding, and within this context a functional definition of vulnerability is required.

There have been numerous attempts to define flood vulnerability (e.g. Kaźmierczak and Cavan, 2011; Balica et al., 2013; Giupponi et al., 2013; Li et al., 2013). Common to them all is the division of vulnerability into certain component parts. There is some consensus about what these parts should be, with vulnerability incorporating concepts of susceptibility, exposure and resilience. Exposure can be considered the tangible and intangible goods and services, possessing value of some kind, which may be subject to flooding. Susceptibility is the extent to which such elements are exposed, which in turn influences the chance of being harmed at times of hazardous floods (Balica et al., 2009). Resilience relates to adaptive capacity and so-called 'bounce back' (Adger, 2006).

Balica et al. (2009, 2013) define flood vulnerability thus:

> the extent to which a system is susceptible to floods due to exposure, a perturbation, in conjunction with its ability (or inability) to cope, recover, or basically adapt (Balica et al., 2009, p. 2572).

According to this definition, vulnerability can be imagined as a three-dimensional space with exposure, susceptibility and resilience forming the main axes (Figure 19.1). A town (or any other system under analysis) will fall into a particular region of the three-dimensional space depending on the components which define the type of flood vulnerability. For example, a region may have high exposure combined with a population at risk (i.e. a city such as Rotterdam in the Netherlands or Ho Chi Minh City, Vietnam), leading to high susceptibility. A geographical area that has high exposure but is predominantly uninhabited, such as agricultural land prone to flooding (e.g. areas along the Mekong river used for rice production during the wet season), could be said to have low susceptibility and high resilience. The key issue is that the interventions required are likely to differ across vulnerability types. Using the three-dimensional vulnerability space, vulnerability 'solutions' can be scrutinised for their match to vulnerability 'type'. For example, high resilience in

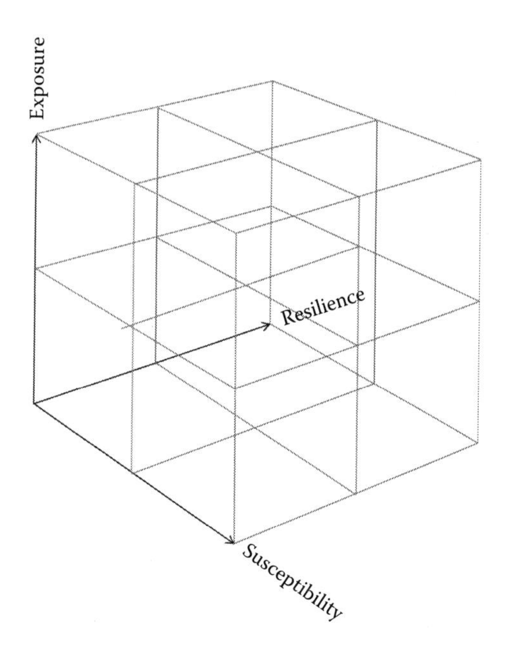

FIGURE 19.1 The three dimensions of vulnerability intersect to create a coordinate space, or cube, into which specific sites would fall.

a population can be encouraged through flood proofing of structures and businesses, education and robust flood warnings. Exposure can be reduced using hard engineering interventions such as flood defences or reservoir storage in the upper catchment. Susceptibility can be addressed via planning restrictions and other means that reduce the chance of damage and destruction due to flooding. The possible engineering solutions are, to some extent, plentiful and this reflects the typical focus of flood risk management. The conceptual challenge is to define what type of vulnerability a location possesses, and to respond with optimum (not necessarily engineering) solutions.

19.1.2 Traditional and Parametric Approaches to Modelling Flood Vulnerability

Floods are primarily the result of extreme weather events. The magnitude of extreme events has an inverse relationship with the frequency of their occurrence, so floods with high magnitude tend to occur less frequently than more moderate events. The relationship between the frequency of occurrence and its magnitude is traditionally established by performing a 'frequency analysis' of historical hydrological data using different probability distributions. Once the frequency, magnitude and shape of the hydrograph (a graph with flood depth on the y-axis and time on the x-axis) are established, computer models which discretise the topographical river and land form are used to estimate flood depth, flood elevation and velocity (Hartanto et al., 2011). The results from a computer model can then be used for loss estimation due to a particular design flood event. Loss estimation, however, does not cover the full remit of vulnerability. As a result, vulnerability estimation using more traditional approaches is difficult and open to interpretation (Balica et al., 2013).

To overcome these challenges a more recent innovation has been the development of 'parametric approaches' that attempt to quantify flood vulnerability. An example of this is the flood vulnerability index (FVI) method as developed by Balica et al. (2009). This is an indicator-based methodology that aims to identify hotspots related to flood events in data rich and/or data scarce areas around the world. The main concept consists of determining the spatial scale of the analysis (e.g. river basin through to urban area), then assessing the place on a battery of individual indicators in order to arrive at four high level characteristics (social, economic, environmental and physical). From this, actions to diminish focal spots of flood vulnerability can be identified and action plans to deal with floods and flooding put in place (for a full account see Balica et al., 2009). Methods such as these have an important role to play, and while they attempt to capture the softer and more difficult to define aspects of vulnerability, they still rely on an ostensibly deterministic logic. In other words, the focal spots of flood vulnerability are decomposed into elements that are counted or otherwise analysed separately. The FVI approach is taxonomic – which has many practical advantages – but in complex sociotechnical systems like the catchments and settlements that form the subject of flood risk analyses, there are opportunities to go further. CWA is one means to do so.

19.2 METHODOLOGY

19.2.1 Study Design

The concept of vulnerability will be explored in this chapter by applying a systems-based method to four towns. The purpose of the study was to derive measures of susceptibility, resilience and exposure and in doing so explore some wider research questions. Firstly, is it possible to discern specific vulnerability pathways, or areas where a settlement is more or less likely to be critically affected should a flood occur? Secondly, is it possible to derive a vulnerability 'profile' and associated flood risk counter measures? CWA, in particular the first phase (AH) is well placed to provide answers to these questions.

An AH enables any civil engineering system to be represented at any level of granularity (Stanton et al., 2006). As a method for exploring civil engineering systems it becomes possible to insert (or remove) concrete physical objects from the bottom of the hierarchy, and to analyse the effect in terms of higher-level intangible activities, effects, outcomes and end states. The approach is flexible. It becomes possible to analyse how changes at the bottom of the analysis, however small, might propagate up through the system to become magnified (or attenuated) at higher levels of abstraction. The means–ends links in the AH are an expression of all the 'affordances' in the system, or the 'possibilities for action'. Some of these will be readily apparent while others will be emergent. Using the AH representation these possibilities for action become visually manifest and can be explored.

19.2.1.1 Procedure

Step 1 – Four candidate towns in Scotland were selected: Dumbarton, Dumfries, Stranraer and Moffat. Scottish Environmental Protection Agency (SEPA) flood maps were accessed for each area, showing the flood extent for a 1 in 200 year fluvial (i.e. river-based) flood event.

Step 2 – Classes of 'physical object' that fell within the borders of the indicated fluvial flood extent were extracted. These included objects such as private housing, shops, infrastructure (e.g. roads, gas holders, bridges, etc.), industrial units and factories. These represented the 'physical objects' affected by the ingress of a flood at the bottom level of the AH.

Step 3 – The top level of the AH describes the system's 'functional purpose' or the system's fundamental 'reason(s) for being'. In this case the functional purpose of a town was defined as:

1. Meet housing/accommodation/shelter needs
2. Support economic activity
3. Provide safety and security
4. Protect cultural heritage
5. Support freedom of movement (people and goods)
6. Provide infrastructure needs (power, water, waste disposal, etc.)

Step 4 – Indicators used in the established parametric method (FVI; Balica et al., 2009) were used to populate the values and priorities layer of the AH and were linked to the achievement of functional purposes at the layer above. The rationale here is to use an established method for flood vulnerability assessment and embed it in the AH as an aid to practical application.

Step 5 – Object-related processes represent the next level of abstraction, showing the specific 'functions' to which physical objects (or groups of objects) contribute. Likewise, the middle layer in the AH analysis, purpose-related functions, raises the level of functional abstraction further still by defining the 'generic functions' that a town performs, as supported by object-related processes (in the layer below) and measurable by the FVIs (in the layer above).

Step 6 – Once the 'nodes' in the AH were inserted, the 'means-ends' links between levels were established. In this analysis the links represent the means–ends relations 'that are evident in a particular situation or set of situations' (Naikar, 2013, p. 105), specifically, those likely to be in place prior to and at the onset of a flood. These were established by asking three independent analysts to complete the means–ends linking task according to strict 'linking criteria'. Any node in AH can be taken to answer the question of 'what' it does. The node is then linked to all of the nodes in the level directly above to answer the question of 'why' it is needed. It is then linked to all of the nodes in the level directly below to answer the question 'how' this can be achieved. For each town, an inter-rater reliability analysis was performed using Cohen's Kappa to determine consistency between observers. The kappa values for Dumbarton ($\kappa = 0.64$, $p < 0.0001$), Dumfries ($\kappa = .753$, $p < 0.0001$), Moffat ($\kappa = 0.651$, $p < 0.0001$) and Stranraer ($\kappa = 0.658$, $p < 0.0001$) indicate that for all four towns, inter-rater agreement was 'substantial' (Landis and Koch, 1977).

Step 7 – The completed AHs were then subject to a numerical analysis based on graph theory. The constraints represented at each layer became nodes, and the means–ends relations became links. By these means the visual complexity of the raw AH diagrams was reduced into a tractable set of metrics which have underlying construct validity in relation to the vulnerability of the 'system' (i.e. the town in question). These metrics are described in full in the results. To anchor the results a baseline condition was created by 'fully connecting' an AH. This sets the upper limit for the various metrics that will be applied and allows comparisons to be made against an objective baseline.

19.3 FINDINGS

The completed AH for the town of Dumbarton is presented in Figure 19.2 as an example. In the three other towns subject to analysis the higher levels of abstraction remained the same but the physical objects that afforded them were changed. The following sections show how the AH can be explored using network analysis methods based on graph theory, and the novel findings connected to infrastructure resilience revealed.

19.3.1 Critical Flood Risk Nodes and Their Interactions

AH was used to identify which nodes, and at what level of abstraction, were more or less critical when subject to a flood induced disturbance. Criticality refers in this case not to a value judgement concerning the node's physical role in a town, but to its functional role within a network of system constraints and affordances. These might be 'common sense' nodes that arise from an intuitive understanding based on simple cause and effect logic (e.g. a gas holder sounds important and, for Dumbarton, is important), or they may also be 'emergent' properties arising from the complex systemic nature of the town (e.g. leisure facilities do not sound important but in Dumfries they were). The technique used to identify critical nodes within the AHs is to apply a graph theory metric called sociometric status, which is given by the formula:

$$\text{Sociometric status} = \frac{1}{g-1}\sum_{j=1}^{g}(\chi_{ji}+\chi_{ij})$$

where g is the total number of nodes in the network, i and j are individual nodes, and χ_{ij} are the number of links present between node i and j (Houghton et al., 2006). Network metrics of this kind are numerous (e.g. Monge & Contractor, 2003) and future work is aimed at exploring the construct validity of further variants. Sociometric status, however, served as a valid starting point in that (a) it provides an indication of the positional centrality of the node in the wider network, in particular, its ability to influence other nodes. Also (b) it is compatible with the bi-directional nature of the means–ends links. Table 19.1 summarises the results obtained by applying this network metric to the AH analyses.

Table 19.1 identifies the top three critical nodes for each of the four candidate towns at different levels of AH. Two key observations can be made. The first is a wider point about how the differences between towns become progressively 'damped

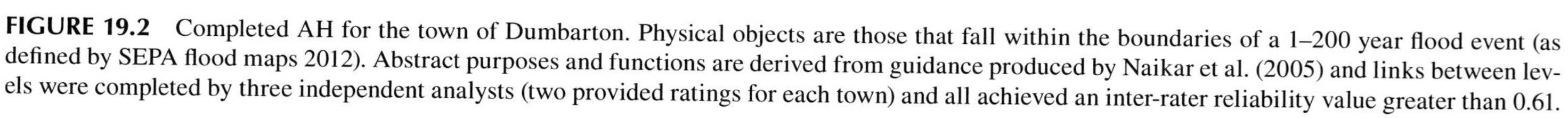

FIGURE 19.2 Completed AH for the town of Dumbarton. Physical objects are those that fall within the boundaries of a 1–200 year flood event (as defined by SEPA flood maps 2012). Abstract purposes and functions are derived from guidance produced by Naikar et al. (2005) and links between levels were completed by three independent analysts (two provided ratings for each town) and all achieved an inter-rater reliability value greater than 0.61.

TABLE 19.1
Top Three Nodes at Each Level of AH

	All Towns	
Functional Purpose	Housing/accommodation/shelter needs	81%
	Safety and security	66%
	Freedom of movement	62%
Values and Priority Measures	Insurance	79%
	Number of people working in emergency services	68%
	Dam storage	
	Child mortality	
	Human development index	
	Amount of investment	63%
Purpose-Related Functions	Business regulation	74%
	Employment	66%
	Industrial/economic output	63%

	Dumbarton		Dumfries		Stranraer		Moffat	
Object-Related Processes	Support distribution of goods	80%	Provide leisure facilities	100%	Provide education services	89%	Support distribution of goods	77%
	Provide housing services	67%	Provide housing services	91%	Provide housing services	87%	Provide education services	
	Provide social services	56%	Healthcare services		Produce goods and services		Provide housing services	
	Produce goods & services		Cultural heritage	76%	Store & distribute fuel and energy	81%	Store & distribute fuel and energy	66%
							Provide social services	60%

Source: The sociometric status values are expressed as a percentage of the maximum value (i.e. that derived from a fully connected AH).

out' the higher one progresses up the levels of abstraction. Specifically, the configuration of physical objects, and how they are affected by a flood event, cause individual vulnerabilities at the (lower) level of object-related processes, yet the ability of all towns to meet their highest level functional purposes are affected similarly. To clarify, this does not mean that all towns are identically affected in flood extent or damage but that the extent and damage degrade the housing/accommodation/shelter function to the greatest degree when the system as a whole is considered. This is certainly consistent with recent flood events in the United Kingdom and the response hierarchy of responding agencies (e.g. Hartwell-Naguib and Roberts, 2014). Table 19.2 ranks orders all of the higher-level systemic effects and also maps them onto the UK Environment Agency's response hierarchy of 'people-property-land'. The systemic insights discovered in this analysis map well on to the existing hierarchy, with the emphasis on people, then property, quite clearly manifest.

The order of criticality of these functional purposes can be correlated to the type of flood damage or impact felt. The top three critical functional purposes can be categorised as direct impacts of floods, while the lower three are representative of indirect impacts (Carrera et al., 2015). The observed correlation reflects the severity of the impact, whereby impacts to housing, accommodation, safety and security and freedom of movement are of great importance, and are often significantly impacted during and in the immediate aftermath of a flood event. Indeed, in terms of flood response this tends to be the first and immediate response of any government. This can be seen clearly in the case of the 2013/14 floods in the United Kingdom, where the Environment Agency delivered a hierarchical response to the population (people, property, then agricultural land; Hartwell-Naguib and Roberts, 2014 as shown in Table 19.2). In reflecting not only just this response hierarchy, but also the hierarchy of direct and indirect effects, the AH seems to be exhibiting good construct validity. Further analysis and development of the method to incorporate more complex

TABLE 19.2
Order of Criticality/Priority of Highest Level Functional Purposes

UK Environment Agency Response Hierarchy (People, Property and Agricultural Land)	Functional Purposes	Order of Criticality	Sociometric Status (% of Max)
People/Property	Housing/accommodation/shelter needs	1	81%
People	Safety and security	2	67%
People	Freedom of movement (people and goods)	3	62%
Property	Cultural heritage	4	57%
People	Infrastructure needs	5	47%
Land[a]	Economic activity	6	19%

[a] An approximate but related classification.

functional purposes may prove interesting. For example, it may be of interest to test whether tangible or intangible impacts are more critical to the network. Likewise, the current analysis shows only the immediate effect of the flood, but future work will examine the temporal aspects. Specifically, the ability of critical functions to restore themselves to pre-flood levels depending on the recovery levels occurring elsewhere in the wider system. For this other phases of CWA could prove very useful.

Whilst AH performs as a common model applicable across all towns at the level of functional purposes, at the level of object-related processes, more localised functional degradation is evident (Table 19.2). There is some uniformity, to the extent that individual processes appear consistently across all towns (e.g. the provision of housing services) but there are also some important differences (e.g. cultural heritage is degraded by a flood in Dumfries to the same extent that education is degraded in Moffat). Again, the list can be read as a form of flood-degradation priority/criticality list. This analysis identifies the services particularly vulnerable within a studied area, and highlights these in a way which may be overlooked by more traditional methods. This is particularly true if multiple geographical areas are assessed in a consistent manner, which is conducive to the approach, as comparisons between results will highlight competing vulnerabilities.

19.3.2 Vulnerability Profile

A systemic feature of the results so far is the extent to which changes at lower levels of the AH become progressively damped out as they propagate upward, in line with the critical functions degraded in real floods. A practical outcome of the work is that a vulnerability profile can be created, one that relates to a class of Scottish town similar to those modelled in this study. To do this, the values and priority measures are used. These are based on FVI metrics developed by Balica et al. (2009) and they belong to three vulnerability classes as shown in Table 19.3. Like all nodes in the AH the values and priority measures have an associated sociometric status value. It is again possible to derive a value based on the percentage difference in sociometric status values between a fully connected AH (the 'control' AH described above) and those relating to real-life locations. Because any differences between real-life locations have been damped out at this level of analysis, only one set of values need be displayed in Table 19.3.

A mean for each category of vulnerability can be derived (as shown in Table 19.3) that can then be used as a coordinate along separate axes described by resilience, susceptibility and exposure. The axes intersect to create a 'vulnerability cube' (see Figure 19.1 above) and the mean values provide coordinates to fix a point in this space (as shown in Figure 19.3).

If the coordinate space shown in Figure 19.3 is divided into eight zones (as shown) it can be seen that the modelled towns fall into a distinctive region. They measure low on exposure but also low on resilience and high on susceptibility (Zone 6). The eight zones create a form of taxonomy, which in turn represents the flood risk 'problem space' and where a given case study might fall depending on the structure and type of the associated AH, as shown in Table 19.4. This affords a different perspective and insight into the flood vulnerability of these Scottish towns and perhaps even

TABLE 19.3
Each FVI/Value and Priority Measure Has an Associated Sociometric Status Value (shown as a % of the maximum value) and Belongs to One of Three Vulnerability Categories

	Dumbarton	Stranraer	Moffat	Dumfries
Exposure				
Number of historic buildings, museums, etc.		15.79		
Topography (average slope of city)		47.37		
Land use/green areas inside urban area		31.58		
% Growth of population in urban areas in last 1 year		15.79		
Population density (people/km1)		36.84		
Disability		31.58		
Mean		29.82		
Resiliency				
Number of people working in emergency services		68.42		
Dams' storage capacity		68.42		
Number of insurance policies/provision		78.95		
Length of dikes/levees		26.32		
Number of shelters per km1		26.32		
Communication penetration rate		47.37		
Drainage system (km of canalisation)		57.89		
Awareness and preparedness		57.89		
Amount of investment/GDP		63.16		
% of asphalted roads		26.32		
Number of industries/economic activities in urban area		21.05		
Mean		49.28		
Susceptibility				
Child mortality		68.42		
Unemployment		47.37		
Inequality (Gini coefficient)		21.05		
Human development index		68.42		
Mean		51.32		

vulnerability more widely in the United Kingdom. What is interesting is that the towns studied do not have a significant risk associated with flood exposure, but that their high susceptibility and low resilience increase their vulnerability. Traditional approaches to flood risk management tend to focus on direct impacts and addressing exposure issues. What this research highlights is that susceptibility and resilience may be at least as important in the Scottish UK context. This certainly reflects current thinking in flood risk management which promotes a focus towards a more 'systematic' user/human orientated approach. Consequently this poses the challenge of identifying interventions that move these towns from Zone 6 to Zone 1 (Table 19.4).

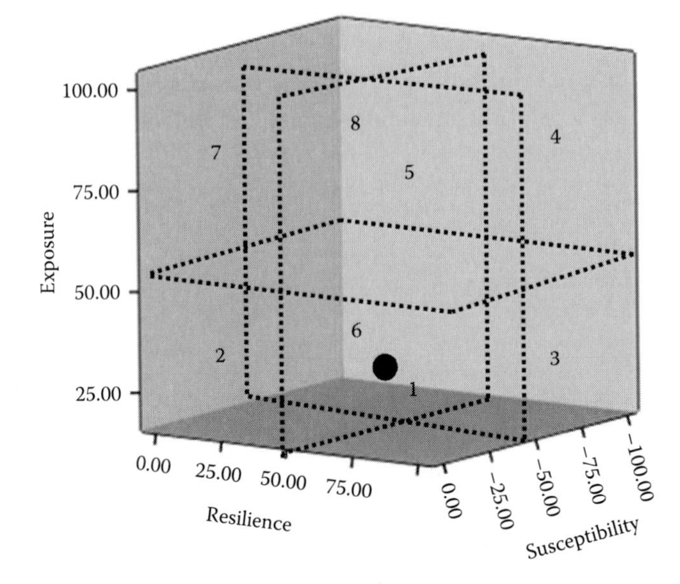

FIGURE 19.3 The coordinate space created by the intersecting resilience, susceptibility and exposure axes can be populated with the case study example(s). The coordinate space is divided into eight 'octants', associated with each is a distinctive flood response strategy.

Looking at recent UK events, and the subsequent responses, in a very general and simplistic way we can perhaps identify the impacts of current flood responses on the United Kingdom more widely. For example, the Association of British Insurers (ABI, 2010; UK Parliament, 2015) reported a significantly lower insurance payout (£500 M) as a result of the 2013–2014 flood events when compared to the 2007 events (approx. £3B). This is, in part, due to the impacted population being of a smaller number and in a rural rather than urban location. However, the public perception of the 2013–2014 events is that they were at least as bad as those experienced in 2007. Significant capital investment has been made in these intervening

TABLE 19.4
The Vulnerability Problem Space

Zone	Exposure	Resilience	Susceptibility	Description
Zone 1	Low	High	Low	Best combination
Zone 2	Low	Low	Low	Low resilience but low risk…
Zone 3	Low	High	High	High susceptibility but resilient…
Zone 4	High	High	High	High exposure but resilient…
Zone 5	High	High	Low	High exposure but resilient…
Zone 6	Low	Low	High	Low resilience and high susceptibility
Zone 7	High	Low	Low	Low resilience and high exposure
Zone 8	High	Low	High	Worst combination

years, focused primarily on reducing exposure. This may have had an impact on the numbers of those directly affected, although comparison is not possible due to differences in type, location and nature of the events experienced. At a high level this would fit the profile identified above of low exposure, but high susceptibility and low resilience. In other words, having tackled the former issues of exposure, the latter issues of susceptibility and resilience seem to require greater attention, and this is a feature which emerges from the present analysis.

19.4 DISCUSSION AND CONCLUSION

Flood vulnerability is a key issue, but there are still conceptual and methodological problems if the concept is to prove useful in informing policy and practice. This chapter offers a working definition of the term and relates it to three underlying concepts: susceptibility, exposure and resilience. The current state of the art is to use numerical measures and indices to show how catchments and settlements compare and where best to direct resources. In this chapter we present an argument to suggest that susceptibility, exposure and resilience are emergent outputs of a complex, non-linear sociotechnical system. To make further progress a more systemic method is required in order to understand the critical interactions and what they mean for a place's flood vulnerability. The approach adopted in this research is based on constraints and the idea that by modelling these constraints we create the 'behaviour space' of a place, and can begin to make sense not only of the behaviours which do occur, but also those behaviours which 'could occur' in major flooding episodes. By looking at the problem from a systems perspective the adaptability of the 'system' (i.e. the place) can be modelled and quantified to reveal a vulnerability profile, and a position within a typological flood risk model defined. Depending on the region a place falls within the typological model, also depends the strategy for improving vulnerability in that location.

This CWA application (and extension) has revealed some interesting discrepancies between what we currently do to mitigate flood risk and what we potentially should do. The United Kingdom has tended towards management measures that reduce exposure. The results of CWA (or AH at least) suggest that addressing urban resilience and susceptibility are at least as important, and this poses the question of how best to address them. What interventions are necessary to improve our current level of vulnerability, and what can we learn from other international examples (e.g. Bangladesh, Vietnam, etc.) to improve our resilience and lessen our susceptibility? Using CWA and the AH a methodological approach to testing interventions may now be possible.

REFERENCES

Adger, W. N. 2006. Vulnerability. *Global Environmental Change*, 16, 268–281.

Association of British Insurers (ABI). 2010. Massive rise in Britain's flood damage bill highlights the need for more help for flood vulnerable communities says the ABI. ABI, November, 24th. https://www.abi.org.uk/News/News-releases/2010/11/massive-rise-in-britains-flood-damage-bill-highlights-the-need-for-more-help-for-flood-vulnerable-communities-says-the-abi.aspx. Accessed 16th August 2013.

Balica, S. F., Beevers, L., Popescu, I. and Wright, N. G. 2013. Parametric and physically based modelling techniques for flood risk and vulnerability assessment: A comparison. *Journal of Environmental Modelling & Software*, 41(3), 81–92. doi: 10.1016/j.envsoft.2012.11.002.

Balica, S. F., Douben, N. and Wright, N. G. 2009. Flood vulnerability indices at varying spatial scales. *Water Science and Technology Journal, WST*, 60.10, 2571–2580.

Balica, S. F. and Wright, N. G. 2010. Reducing the complexity of flood vulnerability index. *Environmental Hazard Journal, EHJ*, 9(4), 321–339. ISSN 1747-7891.

Cabinet Office. 2013. *National Risk Register of Civil Emergencies: 2013 Edition*. London: Crown.

Carrera, L., Standardi, G., Bosello, F. and Mysiak, J. 2015. Assessing direct and indirect economic impacts of a flood event through the integration of spatial and computable general equilibrium modelling. *Environmental Modelling and Software*, 63, 109–122. doi:10.1016/j.envsoft.2014.09.016

Chalmers, D. J. 1990. Thoughts on Emergence. Available at: http://consc.net/notes/emergence.html

Giupponi, C., Giove, S. and Giannini, V. 2013. A dynamic assessment tool for exploring and communicating vulnerability to floods and climate change. *Environmental Modelling & Software*, 44, 136–147.

Halley, J. D. and Winkler, D. A. 2008. Classification of emergence and its relation to self-organisation. *Complexity*, 13(5), 10–15.

Hartanto, I. M., Beevers, L., Popescu, I. and Wright, N. G. 2011. Application of a coastal modelling code in fluvial environments. *Environmental Modelling and Software*, 26(12), 1685–1695.

Hartwell-Naguib, S. and Roberts, N. 2014. Winter Floods 2013/14 Standard Note: SN/SC/06809 Section Science and Environment. June 17th 2014.

Houghton, R. J., Baber, C., McMaster, R., Stanton, N. A., Salmon, P., Stewart, R. and Walker, G. H. 2006. Command and control in emergency services operations: A social network analysis. *Ergonomics*, 49(12-13), 1204–1225.

Jenkins, D. P., Stanton, N. A., Salmon, P. M. and Walker, G. H. 2009. *Cognitive Work Analysis: Coping with Complexity*. Farnham: Ashgate.

Kaźmierczak, A. and Cavan, G. 2011. Surface water flooding risk to urban communities: Analysis of vulnerability, hazard and exposure. *Landscape and Urban Planning*, 103(2), 85–197.

Landis, J. R. and Koch, G. G. 1977. The measurement of observer agreement for categorical data. *Biometrics*, 33, 159–174.

Lewis, J. 1999. *Development in Disaster-Prone Places: Studies of Vulnerability*. London, UK: Intermediate Technology Publications.

Li, C. H., Li, N., Wu, L. C. and Hu, A. J. 2013. A relative vulnerability estimation of flood disaster using data envelopment analysis in the Dongting lake region of Hunan. *Natural Hazards and Earth System Science*, 13, 1723–1734.

McCarthy, J. J. 2001. *Climate Change 2001: Impacts, Adaptation, and Vulnerability: Contribution of Working Group II to the Third Assessment Report of the Intergovernmental Panel on Climate Change*. Cambridge: Cambridge University Press.

Monge, P. R. and Contractor, N. S. 2003. *Theories of Communication Networks*. Oxford: Oxford University Press.

Naikar, N. 2013. *Work Domain Analysis: Concepts, Guidelines, and Cases*. Boca-Raton, FL: CRC Press.

Naikar, N., Hopcroft, R. and Moylan, A. 2005. *Work Domain Analysis: Theoretical Concepts and Methodology*. DSTO-TR-1165. Melborne: DSTO.

Rasmussen, J. 1974. *The Human Data Processor as a System Component, Bits and Pieces of a Model (Riso-M-1722)*. Roskilde, Denmark: Riso National Laboratory.

Rasmussen, J. 1986. *Information Processing and Human-Machine-Interaction – An Approach to Cognitive Engineering*. Amsterdam: North Holland.
Reber, A. S. 1995. *Dictionary of Psychology*. London: Penguin.
Simon, H. A. 1981. *The Sciences of the Artificial* (second edition). Cambridge, MA: MIT Press.
Stanton, N. A., Ashleigh, M. J., Roberts, A. D. and Xu, F. 2006. Levels of abstraction in human supervisory control teams. *Journal of Enterprise Information Management*, 19(6), 679–694.
UK Parliament. 2015. Current Parliamentary Material on Flooding. http://www.parliament.uk/topics/Flooding.htm. Accessed 22nd May 2015.
van der Veen, A. and Logtmeijer, C. J. J. 2005. Economic hotspots: Visualizing vulnerability to flooding. *Natural Hazards*, 36(1–2), 65–80.

Index

A

D

J

K

L

M

N

Q

R

S

T

X